D1064937

NONLINEAR ALMOST PERIODIC OSCILLATIONS

NONLINEAR ALMOST PERIODIC OSCILLATIONS

M. A. Krasnosel'skii, V. Sh. Burd, and
Yu. S. Kolesov

Translated from Russian by A. Libin

Translation edited by D. Louvish

A HALSTED PRESS BOOK

JOHN WILEY & SONS
New York · Toronto

ISRAEL PROGRAM FOR SCIENTIFIC TRANSLATIONS
Jerusalem · London

Sole distributors for the Western Hemisphere and Japan

HALSTED PRESS, a division of
JOHN WILEY & SONS, INC., NEW YORK

Library of Congress Cataloging in Publication Data

Krasnosel'skiĭ, Mark Aleksandrovich.
Nonlinear almost periodic oscillations.

"A Halsted Press book."
Translation of *Nelineĭnye pochti periodicheskie kolebaniiȃ.*
1. Oscillations. 2. Almost periodic functions. 3. Differential equations, Nonlinear. 4. Integral equations. I. Burd, Vladimir Shepselevich, joint author. II. Kolesov, I͡Uriĭ Serafimovich, joint author. III. Title.
QA871.K82813 515'.35 73-12277
ISBN 0-470-50700-4

Distributors for the U.K., Europe, Africa and the Middle East

JOHN WILEY & SONS, LTD., CHICHESTER

Distributed in the rest of the world by
KETER PUBLISHING HOUSE JERUSALEM LTD.
ISBN 0 7065 1348 7
IPST cat.no.22073

This book is a translation from Russian of
NELINEINYE POCHTI PERIODICHESKIE KOLEBANIYA
Izdatel'stvo "Nauka," Glavnaya Redaktsiya Fiziko-Matematicheskoi Literatury
Moscow 1970

Printed in Israel

Contents

PREFACE

This book is devoted to the method of integral equations as applied to the investigation of almost periodic solutions of nonlinear systems of ordinary differential equations.

The integral equations of the theory of almost periodic solutions differ in two essential respects from, say, those of the theory of periodic oscillations. Irrespective of whether they intrude explicitly or implicitly, these factors are the source of the special difficulties involved in the investigation of almost periodic oscillations.

First, even in the simplest cases the integral operators of nonlinear almost periodic oscillations do not possess the complete continuity property in spaces of almost periodic functions. This hampers the direct application of powerful and well-developed topological methods such as the Schauder principle, rotation of vector fields, and so on.

Second, the spectra of the integral operators of linear almost periodic oscillations do not have isolated points. Thus, linearization in bifurcation and branching problems for almost periodic oscillations leads to equations with operators which are not even normally solvable. One is thus prevented from using such direct methods of branching theory as the Lyapunov-Schmidt-Poincaré methods.

Thus the study of global (nonlocal) problems concerning nonlinear almost periodic oscillations and the relevant branching theory require special tools. The theory set forth here leans heavily on the recently developed methods of equations with monotone and concave operators.

The book comprises four chapters. The first is introductory in nature. The second is concerned with effective conditions for regularity of various types of linear differential operators with almost periodic coefficients, existence and constancy of sign of their Green's functions, and so on. The third chapter studies global theorems on nonlinear almost periodic oscillations (existence, estimated number of solutions, stability in the case of several almost periodic oscillations, regions of attraction, etc.). As an example, we consider in detail forced almost periodic oscillations in certain automatic control systems with several equilibrium states. In the fourth chapter attention is concentrated on the creation of almost periodic oscillations from an equilibrium state. The examples considered are almost periodic oscillations near the upper equilibrium state of a pendulum with vibrating point of suspension and oscillations of a pendulum of variable length.

The authors believe that the reader's work should be greatly facilitated by the appendix, which summarizes the entire content of the book, presenting a general description of the ideas and methods applied and developed in the book.

The methods expounded in the book are applicable without essential modifications to the investigation of almost periodic solutions of extensive classes of differential equations with lagging argument, differential equations in function spaces, integrodifferential equations, and so on. When all the terms of the equations are periodic (not merely almost periodic) with respect to time, the basic constructions of the book are substantially simplified and yield more powerful results. The authors have been unable to devote close attention to any of these questions.

No account is given of numerous important results from the qualitative theory of almost periodic oscillations (due to Amerio, Arnol'd, Bogolyubov, Demidovich, Zubov, Levinson, Malkin, Mitropol'skii, Moser, Halanay, Hale and others). The reason is either that the relevant problems and methods are beyond the scope of the book or that they have been considered in detail in other monographs and texts.

The reader is assumed to be familiar with the usual material from differential equations, linear algebra and elementary concepts of functional analysis. Concepts and theorems transcending these limits are given in full detail.

The manuscript was read in full or in part by V. V. Zhikov, P. P. Zabreiko, A. Yu. Levin, A. D. Myshkis, A. V. Pokrovskii and Ya. Z. Tsypkin. Their remarks and stimulating advice contributed greatly to the final version.

The authors

Chapter 1

LINEAR DIFFERENTIAL OPERATORS WITH ALMOST PERIODIC COEFFICIENTS

§ 1. ALMOST PERIODIC FUNCTIONS

In this section we shall present the basic properties of almost periodic functions. The systematic theory of scalar almost periodic functions is expounded in the monograph of Levitan /1/. Below we shall use almost periodic functions with values in euclidean n-space R^n. Sometimes almost periodic functions with values in a Banach or metric space will be also used; all basic definitions remain the same in more general spaces, as do most (but not all) of the properties of almost periodic functions. Further, instead of "almost periodic function" we shall use the abbreviation a p - f u n c t i o n .

1.1. Spaces of ap-functions

Let $\mathfrak{R}$ be a metric space with metric $\rho(x, y)$. Following Bohr, we call a continuous function $x(t)$ $(-\infty < t < \infty)$ with values in $\mathfrak{R}$ an a p - f u n c t i o n if for every $\varepsilon > 0$ there exists $l = l(\varepsilon) > 0$ such that every interval $[t_0, t_0 + l(\varepsilon)]$ contains at least one number τ for which

$$\rho[x(t), x(t+\tau)] < \varepsilon \qquad (-\infty < t < \infty). \tag{1.1}$$

The reader can easily prove that every ap-function is bounded and uniformly continuous on the entire real line and its range is compact in $\mathfrak{R}$.

Bohr's definition is not convenient for investigation of the solutions of differential equations. An equivalent definition, due to Bochner, is used more often.

We note by $C(\mathfrak{R})$ the space of continuous and bounded functions on $(-\infty, \infty)$ with values in $\mathfrak{R}$. The metric in $C(\mathfrak{R})$ is defined as

$$\rho_*(x, y) = \sup_{-\infty < t < \infty} \rho[x(t), y(t)] \qquad (x, y \in C(\mathfrak{R})). \tag{1.2}$$

The space $C(\mathfrak{R})$ is complete if and only if $\mathfrak{R}$ is complete; if $\mathfrak{R}$ is a linear space (for example, a finite- or infinite-dimensional Banach space), then $C(\mathfrak{R})$ is also a linear space.

A function

$$x_h(t) = x(t+h) \qquad (-\infty < t < \infty) \tag{1.3}$$

is called a *translate* of the function $x(t)$. If the function $x(t)$ belongs to $C(\mathfrak{R})$, then all its translates also belong to $C(\mathfrak{R})$.

Following Bochner, we call a function $x(t) \in C(\mathfrak{R})$ an *ap-function*, if its translates form a compact set in $C(\mathfrak{R})$. In other words, a continuous and bounded function with values in $\mathfrak{R}$ is almost periodic if from any sequence $h_k \in (-\infty, \infty)$ one can extract a subsequence such that

$$\lim_{i, j \to \infty} \sup_{-\infty < t < \infty} \rho\,[x(t + h_{k(i)}), x(t + h_{k(j)})] = 0. \tag{1.4}$$

As we have already said, almost periodicity in the sense of Bochner coincides with almost periodicity in the sense of Bohr.

The simplest examples of ap-functions are periodic functions. It follows immediately from Bochner's definition that a linear combination of almost periodic functions (when we speak of linear combinations, we imply that $\mathfrak{R}$ is a linear space) is an ap-function. It follows that trigonometric polynomials

$$x(t) = \sum_{i=1}^{n} a_i \cos \lambda_i t + b_i \sin \lambda_i t \tag{1.5}$$

are almost periodic. Note that trigonometric polynomials (1.5) with incommensurable λ_i are not periodic.

As a last example we consider a vector-function $x(t)$ with values in R^n:

$$x(t) = \{x_1(t), \ldots, x_n(t)\}; \tag{1.6}$$

it is almost periodic if and only if every component $x_i(t)$ is a scalar ap-function.

We denote by $B(\mathfrak{R})$ the space of all ap-functions with the metric (1.2).

If $\mathfrak{R}$ is a Banach space (for example, $\mathfrak{R} = R^n$), then $B(\mathfrak{R})$ (like $C(\mathfrak{R})$) is also a Banach space with the norm

$$\| x(t) \|_{B(\mathfrak{R})} = \| x(t) \|_{C(\mathfrak{R})} = \sup_{-\infty < t < \infty} \| x(t) \|_{\mathfrak{R}}. \tag{1.7}$$

The space $B(\mathfrak{R})$ contains the functions $a \cos \lambda t$, $b \sin \lambda t$ for all $a, b \in \mathfrak{R}$, $-\infty < \lambda < \infty$; if λ_1 and λ_2 are incommensurable, then

$$\| a \sin \lambda_1 t - a \sin \lambda_2 t \|_{C(\mathfrak{R})} = 2 \| a \|_{\mathfrak{R}},$$

whence it follows that $B(\mathfrak{R})$ is nonseparable. We note that the unit ball of $B(\mathfrak{R})$ is neither weakly complete nor weakly compact.

1.2. Approximation theorem

Formula (1.5) gives examples of ap-functions with values in $\mathfrak{R}$, if $a_i, b_i \in \mathfrak{R}$. It turns out that the set of functions (1.5) is dense in $B(\mathfrak{R})$, i.e., every ap-function can be approximated to within arbitrary accuracy by a trigonometric polynomial. This remarkable theorem was first proved (for the scalar case) by Bohr /1, 2/; a considerable simplification of the proof was proposed by De La Vallée-Poussin /1/. Various proofs, based on new ideas, were proposed by Bochner /1/, Wiener /1/, Weyl /1/, Bogolyubov /1/. All these proofs are presented in Levitan /2/. A proof of the theorem of approximation of ap-functions with values in Banach spaces by trigometric polynomials was proposed by Corduneanu /1/.

We shall point out a simple method that enables one to reduce the proof of the approximation theorem for an ap-function $x(t)$ with values in a Banach space $\mathfrak{R}$ to the scalar case, or, what is the same, to the finite-dimensional case.

Let $\varepsilon > 0$ be given. Since the range $\mathfrak{M}$ of the ap-function $x(t)$ is compact in $\mathfrak{R}$, we can construct a finite ε-net $y_1, \ldots, y_k$ in $\mathfrak{M}$. Set

$$y(t) = \frac{\sum_{i=1}^{k} \mu_i(t) y_i}{\sum_{i=1}^{k} \mu_i(t)} \qquad (-\infty < t < \infty), \tag{1.8}$$

where

$$\mu_i(t) = \begin{cases} 2\varepsilon - \| x(t) - y_i \|_{\mathfrak{R}}, & \text{if} \quad \| x(t) - y_i \|_{\mathfrak{R}} < 2\varepsilon, \\ 0, & \text{if} \quad \| x(t) - y_i \|_{\mathfrak{R}} \geqslant 2\varepsilon. \end{cases}$$

It is easily checked (using Bochner's definition) that every function $\mu_i(t)$ is almost periodic. Therefore $y(t)$ is also almost periodic. From (1.8) it follows that for any fixed t the value $y(t)$ belongs to the convex hull of the elements of the ε-net whose distances from $x(t)$ are at most 2ε. Consequently,

$$\| x(t) - y(t) \|_{\mathfrak{R}} \leqslant 2\varepsilon \qquad (-\infty < t < \infty).$$

Thus every ap-function with values in $\mathfrak{R}$ can be approximated to within arbitrary accuracy by an ap-function $y(t)$ with values in a finite-dimensional subspace (the linear span of the ε-net).

1.3. Differentiation and integration of ap-functions

In general, differentiation and integration do not preserve the property of almost periodicity. We leave it to the reader to construct examples. Below we shall use the following simple statement:

The derivative $x'(t)$ of an ap-function $x(t)$ is almost periodic if and only if it is uniformly continuous.

We understand the derivative here in the strong sense, i. e,

$$\lim_{\Delta t \to 0} \left\| x'(t) - \frac{x(t+\Delta t) - x(t)}{\Delta t} \right\|_{\mathfrak{R}} = 0.$$

The necessity of the above condition is obvious, while its sufficiency follows from the identity

$$\frac{x(t+\Delta t) - x(t)}{\Delta t} - x'(t) = \frac{1}{\Delta t} \int_0^{\Delta t} [x'(t+\theta) - x'(t)]\, d\theta. \tag{1.9}$$

Analogous arguments show that uniform continuity of the m-th derivative of an ap-function $x(t)$ implies almost periodicity of all the derivatives $x'(t)$, $x''(t)$, ... , $x^{(m)}(t)$.

An important characteristic of an ap-function is its mean value

$$M(x) = \lim_{T \to \infty} \frac{1}{2T} \int_{-T}^{T} x(s)\, ds. \tag{1.10}$$

The limit on the right always exists and is finite. The vanishing of the mean value is a necessary condition for almost periodicity of the integral

$$y(t) = \int_0^t x(s)\, ds \tag{1.11}$$

of an ap-function $x(t)$. However, this condition is not sufficient even for scalar ap-functions (construct an example!).

Bohl /1/ proved that if $x(t)$ is almost periodic and the function (1.11) is bounded, then it is also almost periodic.

1.4. The superposition operator

Let $f(t, x)$ be a continuous function of two variables $t \in (-\infty, \infty)$, $x \in E$ (where E is a Banach space) with values in a metric space $\mathfrak{R}$.

The function $f(t, x)$ is called *uniformly almost periodic*, if for every pair $\varepsilon > 0, r > 0$ there exists $l = l(\varepsilon, r) > 0$, such that any interval $[t_0, t_0 + l(\varepsilon, r)]$ contains at least one τ with

$$\rho[f(t, x), f(t + \tau, x)] < \varepsilon \quad (-\infty < t < \infty, \|x\| \leqslant r). \tag{1.12}$$

Later we shall encounter uniformly almost periodic functions in the case $E = \mathfrak{R} = R^n$. In this case uniform almost periodicity is equivalent to the condition that $f(t, x)$ be almost periodic in t for every fixed x and uniformly continuous in x in every ball $\|x\| \leqslant r$. In particular, a function $f(t, x)$ which is almost periodic in t is uniformly almost periodic if it satisfies a Lipschitz condition in x.

Every function $f(t, x)$ of two variables generates a superposition operator

$$\mathfrak{f}x(t) = f[t, x(t)]. \tag{1.13}$$

Lemma 1.1. If $f(t, x)$ is uniformly almost periodic, then the operator (1.13) acts in $B(R^n)$, is bounded and continuous.

The proof is left to the reader. We recall only that a nonlinear operator is said to be *bounded* if its range on every ball is bounded, and that a nonlinear operator may be bounded without being continuous and vice versa.

The assertion of Lemma 1.1 is well known; it is contained explicitly or implicitly in Zubov /1/, Corduneanu /1/, Amerio /1/, Perov /1/ and others.

The converse of Lemma 1 is also valid. Moreover, if the operator (1.13) acts in $B(R^n)$, then $f(t, x)$ is uniformly almost periodic.

§ 2. REGULAR AP-OPERATORS

2.1. Statement of the problem

Later several function spaces will be used systematically. We denote by $C(R^n)$ the space of functions continuous and bounded on $(-\infty, \infty)$ with the norm

$$\| x(t) \|_{C(R^n)} = \sup_{-\infty < t < \infty} | x(t) |,$$

where $|x|$ is the euclidean norm of x. By $C^m(R^n)$ we denote the set of functions $x(t) \in C(R^n)$ such that $x'(t), \ldots, x^{(m)}(t) \in C(R^n)$; this set becomes a Banach space if we introduce, for example, the following norm:

$$\| x(t) \|_{C^m(R^n)} = \\ = \| x(t) \|_{C(R^n)} + \| x'(t) \|_{C(R^n)} + \ldots + \| x^{(m)}(t) \|_{C(R^n)}.$$

By $B(R^n)$ we denote the subspace consisting of all ap-functions in $C(R^n)$. Similarly, $B^m(R^n)$, will denote the subspace consisting of all ap-functions in $C^m(R^n)$ such that the functions $x'(t), \ldots, x^{(m)}(t)$ are also almost periodic.

A matrix $A(t)$ with almost periodic elements $a_{ij}(t)$ $(i, j = 1, \ldots, n)$ will be called an a p - m a t r i x. The norm of an ap-matrix is defined by

$$\| A(t) \|_{C(R^n)} = \sup_{-\infty < t < \infty} | A(t) |,$$

where

$$| A | = \max_{x \in R^n, |x| = 1} | Ax |.$$

Let $A_i(t)$ $(i = 1, \ldots, m)$ be square matrices of order n with ap-elements. We consider the following differential expression

$$Lx = \frac{d^m x}{dt^m} + A_1(t) \frac{d^{m-1} x}{dt^{m-1}} + \ldots + A_m(t) x. \tag{2.1}$$

The differential expression (2.1) defines a differential operator L (as we shall say, an a p - o p e r a t o r), which can be considered in various function spaces. Usually we shall consider it in $C(R^n)$, assuming it to be defined on functions $x(t) \in C(R^n)$ such that $Lx(t) \in C(R^n)$. We shall show later that the domain of definition of L coincides with $C^m(R^n)$.

Our main interest will be to single out ap-operators with a number of special properties.

An ap-operator is said to be r e g u l a r if the equation

$$Lx = f(t) \tag{2.2}$$

has a unique solution $x(t) \in C^m(R^n)$ for any $f(t) \in C(R^n)$, and moreover $x(t) \in B^m(R^n)$ if $f(t) \in B(R^n)$.

An ap-operator L is said to be w e a k l y r e g u l a r if equation (2.2) has at least one solution $x(t) \in C(R^n)$ for any $f(t) \in B(R^n)$.

At first sight, the class of weakly regular operators seems broader than the class of regular operators. But this is not so. *As will be shown further on, weak regularity of an ap-operator implies (!) its regularity.* To fully appreciate this statement, one should bear in mind that the existence of bounded solutions of an equation (2.2) with bounded matrix-coefficients $A_i(t)$ and arbitrary bounded right-hand sides does not necessarily imply uniqueness of these solutions. An example is the scalar equation

$$\frac{dx}{dt} + a(t)\,x = f(t),$$

where

$$a(t) = \begin{cases} -1, & \text{if} \quad -\infty < t \leqslant -1, \\ t, & \text{if} \quad -1 \leqslant t \leqslant 1, \\ 1, & \text{if} \quad 1 \leqslant t < \infty. \end{cases}$$

In general, an ap-operator is not normally solvable, i.e., the range of an ap-operator on $C^m(R^n)$ (or on $B^m(R^n)$) is not necessarily a closed subspace of $C(R^n)$. An example is the operator $Lx = \frac{dx}{dt}$; its range consists of all bounded functions whose integrals are also bounded functions, and this set is not closed.

When studying equation (2.2) it is often convenient to write it as a system of first-order equations in the phase space R^{mn}.

Let

$$x = \{x_1, \ldots, x_n\}.$$

Set

$$u = \{u_1, \ldots, u_m\},$$

where

$$u_1 = \{x_1, \ldots, x_n\}, \quad u_2 = \{x'_1, \ldots, x'_n\}, \\ \ldots, u_m = \{x_1^{(m-1)}, \ldots, x_n^{(m-1)}\}.$$

Then equation (2.2) becomes

$$\tilde{L}u = \tilde{f}(t), \tag{2.3}$$

where

$$\tilde{L}u = \frac{du}{dt} + Q(t)\,u, \tag{2.4}$$

$$\tilde{f}(t) = \{0, \ldots, 0, f(t)\}, \tag{2.5}$$

with

$$Q(t) = \begin{Vmatrix} 0 & -I & 0 & \ldots & 0 \\ 0 & 0 & -I & \ldots & 0 \\ \cdot & \cdot & \cdot & \cdot & \cdot \\ 0 & 0 & 0 & \ldots & -I \\ A_m(t) & A_{m-1}(t) & A_{m-2}(t) & \ldots & A_1(t) \end{Vmatrix}. \tag{2.6}$$

A natural question arises: what is the connection between regularity of the operator (2.2) and regularity of the operator (2.4)? Below we shall prove the following proposition:

The ap-operator (2.2) is regular if and only if the ap-operator (2.4) is regular.

The basic results in the theory of ap-operators are due to Favard /1/; they are presented in Levitan /1/. The theorems presented below on regularity of ap-operators go back to Massera and Schäffer /1/; our proofs use their ideas.

Examples of ap-operators will be given below.

The regularity property of an ap-operator is stable under small perturbations of the matrices $A_i(t)$. In other words, the regular ap-operators of order m form an open subset (in the natural topology) of the set of all ap-operators of order m. The closed set of nonregular ap-operators is probably nowhere dense. This is easily proved for the case of first-order scalar operators.

Later we shall indicate some classes of nonregular operators of higher order which can be "regularized" by arbitrary small perturbations.

2.2. Esclangon's theorem

We introduce the notation

$$\left.\begin{aligned} \mu_0(T) = \max_{|t|\leqslant T} |\xi(t)|, \quad \mu_1(T) = \max_{|t|\leqslant T} |\xi'(t)|, \\ \mu_2(T) = \max_{|t|\leqslant T} |\xi''(t)|, \end{aligned}\right\} \tag{2.7}$$

where $\xi(t)$ is a scalar function defined and twice continuously differentiable on $(-\infty, \infty)$. The following lemma of Landau and Hadamard is well known (for a proof, see, e. g., Hardy, Littlewood and Pólya /1/, pp. 388 – 391 [of Russian translation]).

Lemma 2.1. If

$$T<\sqrt{\frac{\mu_0(T)}{\mu_2(T)}}, \tag{2.8}$$

then

$$\mu_1(T)\leqslant\frac{1}{T}\mu_0(T)+T\mu_2(T). \tag{2.9}$$

But if

$$T\geqslant\sqrt{\frac{\mu_0(T)}{\mu_2(T)}}, \tag{2.10}$$

then

$$\mu_1(T)\leqslant 2\sqrt{\mu_0(T)\mu_2(T)}. \tag{2.11}$$

These estimates almost immediately imply:

Lemma 2.2. Let T_k be a sequence of numbers such that $T_k\to\infty$ and

$$\lim_{k\to\infty}\frac{\mu_1(T_k)}{\mu_0(T_k)}=\infty. \tag{2.12}$$

Then

$$\lim_{k\to\infty}\frac{\mu_2(T_k)}{\mu_1(T_k)}=\infty. \tag{2.13}$$

Proof. If the numbers T_k satisfy inequalities (2.10) for large k, then (2.13) follows from (2.12) and (2.11). We shall therefore prove that (2.10) is valid for all sufficiently large k.

Suppose that this is false. Then there is a sequence of indices $k(i)$ such that

$$T_{k(i)}<\sqrt{\frac{\mu_0(T_{k(i)})}{\mu_2(T_{k(i)})}}.$$

By (2.9),

$$\mu_1(T_{k(i)})\leqslant\frac{1}{T_{k(i)}}\mu_0(T_{k(i)})+T_{k(i)}\mu_2(T_{k(i)}),$$

whence

$$\frac{\mu_1(T_{k(i)})}{\mu_0(T_{k(i)})} \leqslant \frac{1}{T_{k(i)}}\left[1 + T_{k(i)}^2 \frac{\mu_2(T_{k(i)})}{\mu_0(T_{k(i)})}\right] < \frac{2}{T_{k(i)}},$$

which contradicts (2.12). Q. E. D.

Theorem 2.1. Let the right-hand side $f(t)$ of equation (2.2) be bounded on the entire real line and suppose that equation (2.2) has a solution $x(t)$ bounded on the entire axis.

Then all the derivatives $x'(t), \ldots, x^{(m)}(t)$ are also bounded on the entire real line.

Proof. Consider first a scalar function $\xi(t)$ bounded and m times continuously differentiable on the entire real line. If the derivatives $\xi'(t), \ldots, \xi^{(m)}(t)$ are not all uniformly continuous, then by Lemma 2.2

$$\lim_{T\to\infty} \max_{|t|\leqslant T} |\xi^{(m)}(t)| = \infty. \tag{2.14}$$

Suppose that the statement of our theorem is false. Then some of the components of the solution $x(t) = \{x_1(t), \ldots, x_n(t)\}$ satisfy condition (2.14). To fix ideas, let these be $x_1(t), \ldots, x_l(t)$. It follows from Lemma 2.2 that there is a sequence $T_k \to \infty$ such that every component $x_i(t)$ $(i = 1, \ldots, l)$ satisfies the equalities

$$\lim_{k\to\infty} \frac{\max_{|t|\leqslant T_k} |x_i(t)|}{\max_{|t|\leqslant T_k} |x_i^{(m)}(t)|} = \lim_{k\to\infty} \frac{\max_{|t|\leqslant T_k} |x_i'(t)|}{\max_{|t|\leqslant T_k} |x_i^{(m)}(t)|} = \ldots$$
$$\ldots = \lim_{k\to\infty} \frac{\max_{|t|\leqslant T_k} |x_i^{(m-1)}(t)|}{\max_{|t|\leqslant T_k} |x_i^{(m)}(t)|} = 0$$

and a fortiori the equalities

$$\lim_{k\to\infty} \frac{\max_{|t|\leqslant T_k} |x_i(t)|}{\varphi(T_k)} = \lim_{k\to\infty} \frac{\max_{|t|\leqslant T_k} |x_i'(t)|}{\varphi(T_k)} = \ldots$$
$$\ldots = \lim_{k\to\infty} \frac{\max_{|t|\leqslant T_k} |x_i^{(m-1)}(t)|}{\varphi(T_k)} = 0, \tag{2.15}$$

where

$$\varphi(T) = \max_{|t|\leqslant T} |x^{(m)}(t)|.$$

The components $x_{l+1}(t), \ldots, x_n(t)$ also satisfy equalities (2.15), since they are bounded together with their respective derivatives, and $\varphi(T) \to \infty$ as $T \to \infty$.

Obviously, since

$$\frac{d^m x}{dt^m} = -\sum_{j=1}^{m} A_{m+1-j}(t) \frac{d^{j-1}x}{dt^{j-1}} + f(t)$$

and the elements of the matrices $A_j(t)$ and components of the function $f(t)$ are bounded, it follows that

$$|x^{(m)}(t)| \leqslant a\left[1 + \sum_{j=1}^{m} |x^{(j-1)}(t)|\right],$$

and therefore,

$$1 \leqslant a \frac{1 + \sum_{j=1}^{m} \max_{|t| \leqslant T} |x^{(j-1)}(t)|}{\varphi(T)}.$$

This inequality contradicts (2.15).

Q. E. D.

Theorem 2.1 is due to Esclangon /1/; the above proof is similar to that of Landau /1/.

Note that the proof of Theorem 2.1 did not use the fact that the matrices $A_i(t)$ are almost periodic, but only their boundedness.

2.3. Existence of bounded solutions when the right-hand sides are bounded

It follows immediately from Theorem 2.1 that the domain of definition of the operator L exhausts $C^m(R^n)$. Denote by $C_0^m(R^n)$ the set of functions $x(t) \in C^m(R^n)$ such that $Lx(t) \in B(R^n)$. Since L, as an operator from $C^m(R^n)$ to $C(R^n)$, is obviously continuous, $C_0^m(R^n)$ is a subspace of $C^m(R^n)$. Denote by E_0 the null subspace of L; the dimension of E_0 is at most mn (E_0 consists of the solutions of the homogeneous equation $Lx = 0$ bounded on $(-\infty, \infty)$). Since E_0 is finite-dimensional (see, e. g., Dunford and Schwartz /1/), $C_0^m(R^n)$ can be expressed as the direct sum

$$C_0^m(R^n) = E_0 \dotplus E_1$$

of E_0 and some infinite-dimensional subspace E_1. By definition, $LE_1 = LC_0^m(R^n)$; denote by L_1 the restriction of the operator L to E_1. L_1 has an inverse defined on $LC_0^m(R^n)$, which we denote by L_1^{-1}.

Assume that L is weakly regular. Then $LC_0^m(R^n) = B(R^n)$. By a well-known theorem of Banach (Dunford and Schwartz /1/), we see that L_1^{-1} is continuous from $B(R^n)$ to E_1. Hence follows

Lemma 2.3. Let the ap-operator L be weakly regular. Then there exists a constant γ such that for any $f(t) \in B(R^n)$ equation (2.2) has a solution $x(t) = L_1^{-1}f(t) \in C^m(R^n)$ with

$$\| x(t) \|_{C^m(R^n)} \leqslant \gamma \| f(t) \|_{C(R^n)}. \tag{2.16}$$

We now use Lemma 2.3 to prove the following important

Theorem 2.2. Let the ap-operator L be weakly regular. Then for any $f(t) \in C(R^n)$ there exists at least one solution of equation (2.2) which satisfies the estimate (2.16).

Proof. Let $f(t) \in C(R^n)$. Let us construct a sequence of ap-functions $f_k(t)$ converging uniformly on every finite interval to $f(t)$, such that

$$\| f_k(t) \|_{C(R^n)} \leqslant \| f(t) \|_{C(R^n)} \qquad (k = 1, 2, \ldots).$$

Let $x_k(t)$ denote the solutions $x_k(t) = L_1^{-1} f_k(t)$ of the equations

$$Lx = f_k(t),$$

satisfying inequality (2.16). Obviously,

$$\| x_k(t) \|_{C^m(R^n)} \leqslant \gamma \| f(t) \|_{C(R^n)}. \tag{2.17}$$

From these estimates and Arzela's theorem it follows that there exists a sequence of functions $x_{k(i)}(t)$ which converge to some function $x(t)$ together with their derivatives up to order $m-1$, uniformly on every finite interval. It follows from the equalities

$$\frac{d^m x_{k(i)}(t)}{dt^m} = f_{k(i)}(t) - A_1(t) \frac{d^{m-1} x_{k(i)}(t)}{dt^{m-1}} - \ldots - A_m(t) x_{k(i)}(t)$$

that the derivatives $x_{k(i)}^{(m)}(t)$ also converge to $x^{(m)}(t)$, uniformly on every finite interval. Hence $Lx = f(t)$. It follows from (2.17) that $x(t)$ satisfies the estimate (2.16). Q. E. D.

2.4. Uniqueness theorem

We continue our investigation of weakly regular ap-operators

$$Lx(t) = \frac{d^m x}{dt^m} + A_1(t) \frac{d^{m-1} x}{dt^{m-1}} + \dots + A_m(t) x. \qquad (2.18)$$

Set

$$L_h x(t) = \frac{d^m x}{dt^m} + A_1(t+h) \frac{d^{m-1} x}{dt^{m-1}} + \dots + A_m(t+h) x. \qquad (2.19)$$

All the ap-operators L_h are weakly regular, and they are continuous from $C^m(R^n)$ to $C(R^n)$. By Bochner's definition of ap-functions, the operators L_h $(-\infty < h < \infty)$ form a compact set in the space $\mathfrak{L}$ of continuous operators acting from $C^m(R^n)$ to $C(R^n)$; the closure of this set will be denoted by $H(L)$.

Let $h(j)$ be a sequence such that the matrices $A_i[t + h(j)]$ converge uniformly to certain matrices $A_i^*(t)$. Then the operators $L_{h(j)}$ converge in the norm of the space $\mathfrak{L}$ to the operator

$$L_* x(t) = \frac{d^m x}{dt^m} + A_1^*(t) \frac{d^{m-1} x}{dt^{m-1}} + \dots + A_m^*(t) x. \qquad (2.20)$$

The ap-operator (2.20) is also weakly regular. Indeed, suppose that $f(t) \in C(R^n)$. By Lemma 2.3, there exist solutions $x_j(t) = L_1^{-1} f[t - h(j)]$ of the equations

$$Lx = f[t - h(j)],$$

such that

$$\| x_j(t) \|_{C^m(R^n)} \leqslant \gamma \| f(t) \|_{C(R^n)} \qquad (j = 1, 2, \dots);$$

this means that the solutions

$$y_j(t) = x_j[t + h(j)]$$

of the equations

$$L_{h(j)} x = f(t)$$

satisfy inequalities

$$\| y_j(t) \|_{C^m(R^n)} \leqslant \gamma \| f(t) \|_{C(R^n)} \qquad (j = 1, 2, \dots).$$

Therefore (see the proof of Theorem 2.2) we can choose a subsequence of functions $y_{j(h)}(t)$ which converge on every finite interval, together with their derivatives up to order m, to a function $z(t)$. This function is a solution of the equation $L_*z(t)=f(t)$ and it is bounded together with its derivatives up to order m. This means that L_* is indeed weakly regular.

Lemma 2.4. Let the ap-operator L *be weakly regular and* $f(t)\in B(R^n)$.

Then every solution of the equation $Lx=f(t)$ ***which is bounded on*** $(-\infty,\infty)$ ***belongs to*** $B^m(R^n)$.

Proof. Let $x(t)\in C^m(R^n)$ and $Lx(t)=f(t)$. To prove the theorem we have to show that the family of translates $x(t+h)$ $(-\infty<h<\infty)$ is compact in $C^m(R^n)$.

Suppose that this is false. Let $h(i)$ $(i=1, 2, \ldots)$ be a sequence such that for $i\neq j$

$$\| x[t+h(i)] - x[t+h(j)] \|_{C^m(R^n)} \geqslant \varepsilon_0 > 0. \tag{2.21}$$

Without loss of generality we may assume that the operators $L_{h(i)}$ converge in the norm of $\mathfrak{L}$ to the operator (2.20) and that the functions $f[t+h(i)]$ converge in $B(R^n)$ to a function $f^*(t)$. From the identities

$$L_*x[t+h(i)] = [L_* - L_{h(i)}]x[t+h(i)] + f[t+h(i)]$$

it then follows that

$$\lim_{i\to\infty} \| L_*x[t+h(i)] - f^*(t) \|_{C(R^n)} = 0. \tag{2.22}$$

Let E_0^* denote the null subspace of L_* in the space $C^m(R^n)$. E_0^* is a finite-dimensional subspace; hence there exists a projection P_0^* on E_0^*. Set $P_1^*=I-P_0^*$ and $E_1^*=P_1^*C^m(R^n)$. Obviously,

$$L_*P_0^*x[t+h(i)] \equiv 0 \qquad (i=1, 2, \ldots).$$

Therefore, it follows from (2.22) that

$$\lim_{i\to\infty} \| L_*P_1^*x[t+h(i)] - f^*(t) \|_{C(R^n)} = 0. \tag{2.23}$$

The restriction of L_* to E_1^* has a continuous inverse L_*^{-1} on $B(R^n)$ (see the beginning of § 2.3). Hence, by (2.23),

$$\lim_{i\to\infty} \| P_1^*x[t+h(i)] - L_*^{-1}f^*(t) \|_{C^m(R^n)} = 0,$$

i. e., there exists i_0 such that for $i \geqslant i_0$

$$\| P_1^* x [t + h(i)] - L_*^{-1} f^*(t) \|_{C^m(R^n)} < \frac{\varepsilon_0}{3}. \tag{2.24}$$

Let $i, j \geqslant i_0$, $i \neq j$ Then, by (2.21) and (2.24),

$$\begin{aligned} \| P_0^* x [t + h(i)] - P_0^* x [t + h(j)] \|_{C^m(R^n)} &\geqslant \\ \geqslant \| x [t + h(i)] - x [t + h(j)] \|_{C^m(R^n)} &- \\ - \| P_1^* x [t + h(i)] - L_*^{-1} f^*(t) \|_{C^m(R^n)} &- \\ - \| P_1^* x [t + h(j)] - L_*^{-1} f^*(t) \|_{C^m(R^n)} &> \frac{\varepsilon_0}{3}. \end{aligned}$$

Thus the sequence $P_0^* x [t + h(i)]$ is not compact.

On the other hand, the sequence $P_0^* x [t + h(i)]$ belongs to the finite-dimensional subspace E_0^* and is bounded:

$$\| P_0^* x [t + h(i)] \|_{C^m(R^n)} \leqslant \| P_0^* \| \cdot \| x(t) \|_{C^m(R^n)} \qquad (i = 1, 2, \ldots),$$

so that it is compact. This is a contradiction. Q. E. D.

Lemma 2.5. Let the ap-operator L be weakly regular. Then the homogeneous equation $Lx = 0$ has no nontrivial solutions bounded on $(-\infty, \infty)$.

Proof. Let E_0 be the set of solutions of the equation $Lx = 0$ bounded on $(-\infty, \infty)$. Obviously, E_0 is a subspace of $C^m(R^n)$. The initial values (at $t = 0$) of these solutions form a subspace E^0 of the phase space R^{mn}. Let E^1 be the direct complement of E^0 in R^{mn}. Let E_1 denote the subspace of functions $x(t)$ in $C^m(R^n)$ such that $\{x(0), x'(0), \ldots, x^{(m-1)}(0)\} \in E^1$. It is easy to see that E_0 and E_1 have no nonzero common points and that their direct sum $E_0 \dotplus E_1$ is the entire space $C^m(R^n)$.

By Theorem 2.2, $LE_1 = C(R^n)$. By Banach's theorem, which we have already used several times, the restriction L_1 of L to E_1 has a continuous inverse L_1^{-1}, defined on $C(R^n)$.

Let $x(t)$ be a bounded solution of the equation $Lx = 0$. Set

$$y_\tau(t) = x(t) \int_{-\infty}^{t} \varkappa_\tau(s)\, ds \qquad (\tau \geqslant 2),$$

where

$$\varkappa_\tau(t) = \begin{cases} 0, & \text{if } -\infty < t \leqslant 1, \\ (t-1)^m [1 + (2-t)^m], & \text{if } 1 \leqslant t \leqslant 2, \\ 1, & \text{if } 2 \leqslant t \leqslant \tau, \\ [1 + (t-\tau)^m] (\tau + 1 - t)^m, & \text{if } \tau \leqslant t \leqslant \tau + 1, \\ 0, & \text{if } \tau + 1 \leqslant t < \infty. \end{cases}$$

It is easy to see that $y_\tau(t) \in C^m(R^n)$; since $y_\tau(t) = 0$ for $t \leqslant 1$, it follows that $y_\tau(t) \in E_1$.

Obviously,

$$Ly_\tau(t) \equiv [Lx(t)] \int_{-\infty}^{t} \varkappa_\tau(s)\, ds + f_\tau(t),$$

where $f_\tau(t)$ are functions in some ball of the space $C(R^n)$:

$$\| f_\tau(t) \|_{C(R^n)} \leqslant M \qquad (\tau \geqslant 2).$$

Since $Lx(t) \equiv 0$,

$$Ly_\tau(t) = f_\tau(t) \quad \text{and} \quad y_\tau(t) = L_1^{-1} f_\tau(t),$$

whence

$$\| y_\tau(t) \|_{C^m(R^n)} \leqslant \| L_1^{-1} \| \cdot M \qquad (\tau \geqslant 2).$$

In particular,

$$| y_\tau(\tau) | \leqslant \| L_1^{-1} \| \cdot M.$$

This inequality implies that

$$| x(\tau) | \leqslant \frac{M \| L_1^{-1} \|}{\tau - 2} \qquad (\tau > 2). \tag{2.25}$$

By Lemma 2.4, $x(t)$ is an almost periodic function. Hence it follows from (2.25) that $x(t) \equiv 0$. Q. E. D.

Lemmas 2.4 and 2.5 at once imply:

Theorem 2.3. Let the ap-operator L be weakly regular. Then L is regular.

2.5. Equivalence theorem

Theorem 2.4. The ap-operator (2.2) is regular if and only if the ap-operator (2.4) is regular.

Proof. Sufficiency is obvious. We need only to prove that regularity of the ap-operator L implies that of the ap-operator $\tilde{L}$.

Thus, let L be regular. By Theorem 2.3, it is sufficient to prove that the system

$$\left.\begin{aligned} \frac{du_1}{dt} &= u_2 + f_1(t), \\ \frac{du_2}{dt} &= u_3 + f_2(t), \\ &\dots\dots\dots \\ \frac{du_{m-1}}{dt} &= u_m + f_{m-1}(t), \\ \frac{du_m}{dt} &= -A_m(t)u_1 - \dots - A_1(t)u_m + f_m(t) \end{aligned}\right\} \tag{2.26}$$

has a solution $u(t) = \{u_1(t), \dots, u_m(t)\} \in C(R^{mn})$ for any function

$$f(t) = \{f_1(t), \dots, f_m(t)\} \in B(R^{mn}). \tag{2.27}$$

That such a solution is unique follows from Lemma 2.5.

We first show how to construct the solution of system (2.26) on the assumption that

$$f_1(t) \in B^{m-1}(R^n),\ f_2(t) \in B^{m-2}(R^n), \dots, f_m(t) \in B(R^n). \tag{2.28}$$

To this end, consider the equation

$$\begin{aligned} Lx = \{&f_1^{(m-1)}(t) + A_1(t) f_1^{(m-2)}(t) + \dots \\ &\dots + A_{m-2}(t) f_1'(t) + A_{m-1}(t) f_1(t)\} + \\ + \{&f_2^{(m-2)}(t) + A_1(t) f_2^{(m-3)}(t) + \dots \\ &\dots + A_{m-3}(t) f_2'(t) + A_{m-2}(t) f_2(t)\} + \dots \\ &\dots + \{f_{m-1}'(t) + A_1(t) f_{m-1}(t)\} + f_m(t). \end{aligned}$$

By (2.28) the right-hand side of this equation is in $B(R^n)$. Hence there exists a solution $x(t) \in B^m(R^n)$, which is unique by virtue of Lemma 2.5.

Set

$$\begin{aligned} u_1(t) &= x(t), \\ u_2(t) &= x'(t) - f_1(t), \\ u_3(t) &= x''(t) - f_1'(t) - f_2(t), \\ &\dots\dots\dots \\ u_{m-1}(t) &= x^{(m-2)}(t) - f_1^{(m-3)}(t) - \dots - f_{m-3}'(t) - f_{m-2}(t), \\ u_m(t) &= x^{(m-1)}(t) - f_1^{(m-2)}(t) - \dots - f_{m-2}'(t) - f_{m-1}(t). \end{aligned}$$

It is easily seen that these functions satisfy system (2.26).

The set $\mathfrak{F}$ of functions (2.27) satisfying condition (2.28) is dense in the space $B(R^{mn})$, because it contains all trigonometric polynomials. Hence, to complete the proof it will suffice to show that the solutions $u(t) \in C(R^{mn})$ of system (2.26) satisfy the inequality

$$\| u(t) \|_{C(R^{mn})} \leqslant \gamma \| f(t) \|_{B(R^{mn})} \qquad (f \in \mathfrak{F}), \tag{2.29}$$

where γ is a constant.

If this is not the case, there is a sequence $f_j(t) \in \mathfrak{F}$ such that

$$\lim_{j \to \infty} \| f_j(t) \|_{C(R^{mn})} = 0,$$

and the solutions $v_j(t)$ of the equations

$$\frac{dv}{dt} + Q(t) v = f_j(t) \qquad (j = 1, 2, \ldots)$$

are normalized:

$$\| v_j(t) \|_{C(R^{mn})} = 1 \qquad (j = 1, 2, \ldots).$$

Let us choose numbers t_j so that the inequalities

$$| v_j(t_j) | > 1/2 \qquad (j = 1, 2, \ldots) \tag{2.30}$$

hold and the ap-matrices $Q(t + t_j)$ converge uniformly to a matrix $Q^*(t)$. It follows from the equalities

$$\frac{dv_j(t + t_j)}{dt} + Q(t + t_j) v_j(t + t_j) = f_j(t + t_j) \tag{2.31}$$

that the sequence $v_j(t + t_j)$ is compact in the sense of uniform convergence on every finite interval. Thus, we may assume without loss of generality that $v_j(t + t_j)$ converges to a function $w(t)$ bounded on $(-\infty, \infty)$ uniformly on every finite interval. By (2.31), the sequence $v_j'(t + t_j)$ converges uniformly on every finite interval to $w'(t)$. Letting $j \to \infty$ in (2.31), we get

$$\frac{dw(t)}{dt} + Q^*(t) w(t) = 0. \tag{2.32}$$

The homogeneous equation corresponding to equation (2.32) is

$$L_* x \equiv \frac{d^m x}{dt^m} + A_1^*(t) \frac{d^{m-1} x}{dt^{m-1}} + \ldots + A_m^*(t) x = 0, \tag{2.33}$$

whose coefficients are the limits of the respective sequences $A_i(t + t_j)$. By Lemma 2.5, equation (2.33) has no nontrivial solutions in the space $C^m(R^n)$. Thus $w(t) \equiv 0$ and consequently

$$\lim_{j\to\infty} |v_j(t_j)| = w(0) = 0,$$

contradicting (2.30). Q. E. D.

Our proof of Theorem 2.4 implicitly used the following simple fact:

If the ap-operator L is regular, then all operators of $H(L)$ are regular.

§ 3. BEHAVIOR OF SOLUTIONS OF THE HOMOGENEOUS EQUATION

3.1. First dichotomy theorem

Consider the homogeneous equation

$$Lx = 0 \tag{3.1}$$

in R^n with a regular first-order ap-operator

$$Lx = \frac{dx}{dt} + A(t)\,x. \tag{3.2}$$

Let E_+ denote the set of initial values (at $t = 0$) $x \in R^n$ such that the corresponding solutions are bounded for $t \in (0, \infty)$, and E_- the set of initial values such that the corresponding solutions are bounded for $t \in (-\infty, 0)$. Since the operator (3.2) is regular, the supspaces E_+ and E_- have no nonzero common points.

Theorem 3.1. The direct sum of E_+ and E_- exhausts R^n.

Proof. We denote by $U(t)$ the fundamental matrix of solutions of equation (3.1) satisfying the condition $U(0) = I$, where I is the unit matrix. Then, as is well known, the solution $x(t)$ of the equation

$$Lx = f(t), \tag{3.3}$$

for initial value

$$x(0) = x_0, \tag{3.4}$$

is given by

$$x(t) = U(t)\left[x_0 + \int_0^t U^{-1}(s)\,f(s)\,ds\right]. \tag{3.5}$$

Suppose the statement of our theorem is false. Let $e_0 \equiv E_+ \dotplus E_-$. Set

$$f(t) = \xi(t)\, U(t)\, e_0,$$

where

$$\xi(t) = \begin{cases} 0, & \text{if} \quad -\infty < t \leqslant 0, \\ t(1-t), & \text{if} \quad 0 \leqslant t \leqslant 1, \\ 0, & \text{if} \quad 1 \leqslant t < \infty. \end{cases}$$

Then

$$x(t) = U(t)\left[x_0 + \int_0^t \xi(s)\, ds\, e_0\right]. \tag{3.6}$$

Since L is regular, there exists a unique $x_0 \in R^n$ such that the function (3.6) is bounded on $(-\infty, \infty)$. Obviously,

$$x(t) = U(t) x_0 \qquad (t \leqslant 0),$$

so that $x_0 \in E_-$. On the other hand,

$$x(t) = U(t)\left[x_0 + \frac{1}{6} e_0\right] \qquad (t \geqslant 1),$$

and hence $x_0 + \frac{1}{6} e_0 \in E_+$, i. e., $e_0 \in E_+ \dotplus E_-$.

This contradiction completes the proof.

If the set of all solutions of the homogeneous equation (3.1) is the direct sum of the subspace of solutions bounded as $t \to \infty$ and the subspace of solutions bounded as $t \to -\infty$, this direct-sum decomposition is known as a *dichotomy of solutions*.

3.2. Exponential dichotomy

Any decomposition of R^n into a direct sum $E_+ \dotplus E_-$ induces a corresponding decomposition $X_+ \dotplus X_-$ of the space X of solutions of the homogeneous equation (3.1) into the direct sum of subspaces of solutions whose initial values are in E_+ and E_-, respectively.

A dichotomy of solutions is said to be *exponential* if there exist positive constants $M_+, M_-, \gamma_+, \gamma_-$ such that for $x(t) \in X_+$

$$|x(t)| \leqslant M_+ e^{-\gamma_+(t-s)} |x(s)| \qquad (-\infty < s \leqslant t < \infty), \tag{3.7}$$

and for $x(t) \in X_-$

$$|x(t)| \geqslant M_- e^{\gamma_-(t-s)} |x(s)| \qquad (-\infty < s \leqslant t < \infty). \tag{3.8}$$

Theorem 3.2. Let the ap-operator (3.2) be regular. Then the dichotomy $X = X_+ \dotplus X_-$ is exponential.

Proof. Suppose first that $x(t) \in X_-$ and $x(0) \neq 0$. Set

$$\varkappa(t; \tau_1, \tau_2) = \begin{cases} 0, & \text{if } -\infty < t \leqslant \tau_1 - 1, \\ t - \tau_1 + 1, & \text{if } \tau_1 - 1 \leqslant t \leqslant \tau_1, \\ 1, & \text{if } \tau_1 \leqslant t \leqslant \tau_2, \\ 1 - t + \tau_2, & \text{if } \tau_2 \leqslant t \leqslant \tau_2 + 1, \\ 0, & \text{if } \tau_2 + 1 \leqslant t < \infty. \end{cases}$$

Then the function

$$y(t; \tau_1, \tau_2) = x(t) \int_0^\infty \varkappa(\sigma, \tau_1, \tau_2) |x(\sigma)|^{-1} d\sigma \tag{3.9}$$

is a solution, bounded on $(-\infty, \infty)$, of the equation

$$Ly = -\varkappa(t; \tau_1, \tau_2) |x(t)|^{-1} x(t).$$

It is obvious that

$$\| -\varkappa(t; \tau_1, \tau_2) |x(t)|^{-1} x(t) \|_{C(R^n)} \leqslant 1 \qquad (-\infty < \tau_1 \leqslant \tau_2 < \infty).$$

Thus, since L is regular, the functions (3.9) are uniformly bounded:

$$|y(t; \tau_1, \tau_2)| \leqslant b \qquad (-\infty < t < \infty; \ -\infty < \tau_1 \leqslant \tau_2 < \infty).$$

Letting $\tau_1 \to -\infty$, $\tau_2 \to \infty$ in these inequalities, we get

$$|x(t)| \int_t^\infty |x(\sigma)|^{-1} d\sigma \leqslant b \qquad (-\infty < t < \infty). \tag{3.10}$$

Since the matrix $A(t)$ is bounded, the solution $x(t)$ of the homogeneous equation (3.1) satisfies the inequality

$$|x(t)| \leqslant e^{a(t-s)} |x(s)| \qquad (-\infty < s \leqslant t < \infty), \tag{3.11}$$

where a is the norm of the matrix-function $A(t)$. From this inequality it follows that

$$|x(t)| \int_t^\infty |x(\sigma)|^{-1} d\sigma \geqslant \int_t^\infty e^{-a(\sigma-t)} d\sigma = \frac{1}{a} \quad (-\infty < t < \infty). \tag{3.12}$$

Consider the scalar function

$$\varphi(t) = \int_t^\infty |x(\sigma)|^{-1} d\sigma.$$

In this notation, inequality (3.10) becomes

$$\varphi'(t) \leqslant -\frac{1}{b}\varphi(t) \quad (-\infty < t < \infty),$$

and so

$$\varphi(t) \leqslant e^{-(t-s)/b}\varphi(s) \quad (-\infty < s \leqslant t < \infty). \tag{3.13}$$

Let $s \leqslant t$. Then, by (3.12) and (3.13),

$$|x(t)| \geqslant \frac{1}{a\varphi(t)} \geqslant e^{(t-s)/b} \frac{1}{a\varphi(s)}$$

and further, by (3.10),

$$|x(t)| \geqslant \frac{1}{ab} e^{(t-s)/b} |x(s)| \quad (-\infty < s \leqslant t < \infty).$$

This is precisely inequality (3.8)

In order to investigate the solutions belonging to X_+, let us consider the auxiliary equation

$$L_1 y = 0, \tag{3.14}$$

where

$$L_1 y = \frac{dy}{dt} - A(-t)y.$$

It is evident that regularity of either of the ap-operators L_1 and L implies that of the other. As we have already proved, the solutions of equation (3.14) which are bounded as $t \to -\infty$ satisfy inequality (3.8). Thus it follows that the functions $x(t) = y(-t)$ satisfy inequalities (3.7). It remains to note that every function $x(t) \in X_+$ is representable in the form $x(t) = y(-t)$, where $y(t)$ is a solution of equation (3.14), bounded as $t \to -\infty$. Q. E. D.

220733

The proof of Theorem 3.2 implies the following proposition: *The constants M_+, M_-, γ_+, γ_- in estimates (3.7) and (3.8) depend only on the norm b of the operator L^{-1} (as an operator from $C(R^n)$ into $C(R^n)$) and the norm a of the matrix-function $A(t)$.*

The above proof is analogous to the reasoning of Massera and Schaffer /1/ in their investigation of the dichotomy of solutions considered on a half-line. Similar constructions were carried out by Krein /1/ in his investigation of the stability of solutions of differential equations in Banach spaces.

3.3. Uniformity of dichotomy

Let E_1 and E_2 be subspaces of R^n. We define the angle between E_1 and E_2 to be

$$\alpha(E_1, E_2) = \min_{x_1 \in E_1,\ x_2 \in E_2,\ |x_1| = |x_2| = 1} |x_1 - x_2|. \tag{3.15}$$

We continue our investigation of the solutions of the homogeneous equation (3.1).

Set

$$E_+(t) = U(t)E_+, \quad E_-(t) = U(t)E_- \quad (-\infty < t < \infty). \tag{3.16}$$

Theorem 3.3

$$\alpha[E_+(t), \ E_-(t)] \geqslant \alpha_0 > 0 \quad (-\infty < t < \infty). \tag{3.17}$$

Proof. Let

$$x_+(t) \in X_+, \quad x_-(t) \in X_- \quad (x_+(0) \neq 0, \quad x_-(0) \neq 0).$$

It follows from (3.11) that, for all $m \geqslant 0$ and all $t \in (-\infty, \infty)$,

$$\left| \frac{x_+(t+m)}{|x_+(t)|} - \frac{x_-(t+m)}{|x_-(t)|} \right| \leqslant e^{am} \left| \frac{x_+(t)}{|x_+(t)|} - \frac{x_-(t)}{|x_-(t)|} \right|, \tag{3.18}$$

and by virtue of (3.7) and (3.8)

$$|x_+(t+m)| \leqslant M_+ e^{-\gamma_+ m} |x_+(t)| \quad (-\infty < t < \infty), \tag{3.19}$$

$$|x_-(t+m)| \geqslant M_- e^{\gamma_- m} |x_-(t)| \quad (-\infty < t < \infty). \tag{3.20}$$

Let us choose m so that

$$M_{-}e^{\gamma_{-}m} > M_{+}e^{-\gamma_{+}m}. \tag{3.21}$$

Hence it follows that from (3.18) – (3.21) that

$$\left| \frac{x_{+}(t)}{|x_{+}(t)|} - \frac{x_{-}(t)}{|x_{-}(t)|} \right| \geqslant e^{-am} \left\{ \frac{|x_{-}(t+m)|}{|x_{-}(t)|} - \frac{|x_{+}(t+m)|}{|x_{+}(t)|} \right\} \geqslant$$
$$\geqslant e^{-am}\left(M_{-}e^{\gamma_{-}m} - M_{+}e^{-\gamma_{+}m}\right) = \alpha_0 > 0 \qquad (-\infty < t < \infty).$$

This inequality implies (3.17). Q. E. D.

It is natural to call the property expressed by (3.17) *uniformity* of the dichotomy.

The decomposition of R^n into a direct sum $E_{+}(t) \dotplus E_{-}(t)$ means that every element $x \in R^n$ can be uniquely expressed in the form

$$x = x_{+} + x_{-} \qquad (x_{+} \in E_{+}(t),\ x_{-} \in E_{-}(t)).$$

This representation defines projection operators

$$P_{+}(t)x = x_{+}, \quad P_{-}(t)x = x_{-} \qquad (x \in R^n,\ -\infty < t < \infty) \tag{3.22}$$

onto $E_{+}(t)$ (in the direction of $E_{-}(t)$) and onto $E_{-}(t)$ (in the direction of $E_{+}(t)$), respectively. Henceforth we shall write P_{+} and P_{-} instead of $P_{+}(0)$ and $P_{-}(0)$, respectively. It is easy to see that

$$P_{+}(t) = U(t)P_{+}U^{-1}(t),\ P_{-}(t) = U(t)P_{-}U^{-1}(t) \qquad (-\infty < t < \infty). \tag{3.23}$$

Theorem 3.4. *Let the ap-operator (3.2) be regular. Then the norms of the operators $P_{+}(t)$ and $P_{-}(t)$ $(-\infty<t<\infty)$ are uniformly bounded.*

Proof. For every two nonzero elements $x, y \in R^n$, we have

$$\left| \frac{x}{|x|} + \frac{y}{|y|} \right| \cdot \max\{|x|, |y|\} \leqslant 2|x+y| \tag{3.24}$$

(the proof is left to the reader). Hence, by Theorem 3.3, it follows that for every such that

$$P_{+}(t)x \neq 0, \qquad P_{-}(t)x \neq 0, \tag{3.25}$$

we have

$$\max\{|P_{+}(t)x|, |P_{-}(t)x|\} \leqslant$$
$$\leqslant \frac{2}{\alpha_0}|P_{+}(t)x + P_{-}(t)x| = \frac{2}{\alpha_0}|x|.$$

The statement of the theorem is obvious if one of the operators P_+, P_- is zero. Otherwise, the elements satisfying conditions (3.25) are dense in R^n, and the last inequality implies the estimates

$$|P_+(t)x| \leqslant \frac{2}{\alpha_0}|x|, \quad |P_-(t)x| \leqslant \frac{2}{\alpha_0}|x| \quad (x \in R^n, \ -\infty < t < \infty). \tag{3.26}$$

Q. E. D.

Propositions analogous to the theorems of this subsection were proved by Massera and Schäffer /1, 2/ in a slightly different situation.

§ 4. THE GREEN'S FUNCTION

4.1. Description of the Green's function

Let the ap-operator

$$Lx = \frac{d^m x}{dt^m} + A_1(t)\frac{d^{m-1}x}{dt^{m-1}} + \ldots + A_m(t)x \tag{4.1}$$

be regular.

A matrix-function $G(t, s)$ $(-\infty < t, s < \infty)$ is said to be a Green's function of the operator (4.1) if it has the following properties.

1. For $t \neq s$,

$$\frac{\partial^m G(t, s)}{\partial t^m} + A_1(t)\frac{\partial^{m-1}G(t, s)}{\partial t^{m-1}} + \ldots + A_m(t)G(t, s) \equiv 0. \tag{4.2}$$

2. The matrix-functions

$$G(t, s), \ \frac{\partial G(t, s)}{\partial t}, \ \ldots, \ \frac{\partial^{m-2}G(t, s)}{\partial t^{m-2}} \tag{4.3}$$

are jointly continuous in the variables $t, s \in (-\infty, \infty)$, the derivative $\frac{\partial^{m-1}G(t, s)}{\partial t^{m-1}}$ is jointly continuous for $t \neq s$, and

$$\frac{\partial^{m-1}G(t+0, t)}{\partial t^{m-1}} - \frac{\partial^{m-1}G(t-0, t)}{\partial t^{m-1}} = I. \tag{4.4}$$

3.

$$|G(t, s)|, \left|\frac{\partial G(t, s)}{\partial t}\right|, \ldots, \left|\frac{\partial^{m-1} G(t, s)}{\partial t^{m-1}}\right| \leqslant Me^{-\gamma|t-s|} \tag{4.5}$$
$$(-\infty < t, s < \infty;\ t \neq s),$$

where $M, \gamma > 0$.

There is at most one matrix-function with properties 1—3 (because for fixed s the difference of any two such functions is a solution of the homogeneous equation $Lx = 0$, bounded on $(-\infty, \infty)$ and is therefore zero). It is easily seen that the integral operator

$$Tf(t) = \int_{-\infty}^{\infty} G(t, s) f(s)\, ds \qquad (f \in C(R^n)), \tag{4.6}$$

whose kernel is the Green's function, determines a solution, bounded on $(-\infty, \infty)$, of the inhomogeneous equation

$$Lx = f(t). \tag{4.7}$$

Hence the operator T coincides with L^{-1}, which acts from $C(R^n)$ to $C^m(R^n)$.

The principal aim of this section is to construct the Green's function.

4.2. Construction of the Green's function

Let

$$\tilde{L}u = \frac{du}{dt} + Q(t)\, u \tag{4.8}$$

be the ap-operator (2.4). By Theorem 2.4, this operator and L are either both regular or both nonregular.

We begin with construction of the Green's function of the operator (4.8).

By Theorem 3.1, the space R^{mn} is representable as a direct sum $E_+ \dotplus E_-$ of two subspaces, the first consisting of the initial values (at $t = 0$) of solutions of the homogeneous equation $\tilde{L}u = 0$ bounded as $t \to \infty$, the second of the initial values of solutions bounded as $t \to -\infty$; as usual, we denote by P_+ and P_- the projections onto E_+ (in the direction of E_-) and E_- (in the direction of E_+), respectively; $P_+ + P_- = I$. Let $U(t)$ be the fundamental matrix of solutions of the equation $\tilde{L}u = 0$ satisfying the condition $U(0) = I$.

Set

$$\tilde{G}(t, s) = \begin{cases} U(t) P_+ U^{-1}(s), & \text{if} \quad t \geqslant s, \\ -U(t) P_- U^{-1}(s), & \text{if} \quad t < s. \end{cases} \tag{4.9}$$

The function $\tilde{G}(t, s)$ obviously has properties 1 and 2. We shall prove that it satisfies estimates (4.5). In other words, we shall show that

$$|U(t) P_+ U^{-1}(s)| \leqslant M e^{-\gamma(t-s)} \quad (-\infty < s \leqslant t < \infty), \tag{4.10}$$

$$|U(t) P_- U^{-1}(s)| \leqslant M e^{\gamma(t-s)} \quad (-\infty < t \leqslant s < \infty), \tag{4.11}$$

where $M, \gamma > 0$.

Let $t \geqslant s$. Set

$$x_+(t) = U(t) P_+ U^{-1}(s) x.$$

Obviously, $x_+(0) \in E_+$, and by Theorem 3.2 (estimate (3.7))

$$|x_+(t)| \leqslant M_+ e^{-\gamma_+(t-s)} |x_+(s)| =$$
$$= M_+ e^{-\gamma_+(t-s)} |U(s) P_+ U^{-1}(s) x|.$$

This estimate and Theorem 3.4 imply (4.10).

Let $t \leqslant s$, and set

$$x_-(t) = U(t) P_- U^{-1}(s) x.$$

Since $x_-(0) \in E_-$, it follows from Theorem 3.2 (estimate (3.8)) that

$$|x_-(t)| \leqslant \frac{1}{M_-} e^{\gamma_-(t-s)} |x_-(s)| =$$
$$= \frac{1}{M_-} e^{\gamma_-(t-s)} |U(s) P_- U^{-1}(s) x|,$$

and the proof of (4.11) is completed by again referring to Theorem 3.4.

We have thus proved

Theorem 4.1. Formula (4.9) defines the Green's function of the operator (4.8).

From our proof it obviously follows that the constants M and γ may be determined in such a way that they depend only on the norms of the operator $\tilde{L}^{-1}$ and the matrix-function $Q(t)$.

Let us write the matrix-function (4.9) in the form

$$\tilde{G}(t, s) = \begin{Vmatrix} g_{11}(t, s) & g_{12}(t, s) & \dots & g_{1m}(t, s) \\ g_{21}(t, s) & g_{22}(t, s) & \dots & g_{2m}(t, s) \\ \cdot & \cdot & \cdot & \cdot \\ g_{m1}(t, s) & g_{m2}(t, s) & \dots & g_{mm}(t, s) \end{Vmatrix}, \tag{4.12}$$

where $g_{ij}(t, s)$ are square matrix-functions of order n. Fixing s, consider the vector-functions $\tilde{G}(t, s)u$, where $u = \{0, \dots, 0, x\}$; obviously

$$\tilde{G}(t, s)u = \{g_{1m}(t, s)x, \dots, g_{mm}(t, s)x\}.$$

When $t \neq s$ these functions are solutions of the homogeneous equation $\tilde{L}x = 0$. Therefore, for $t \neq s$,

$$g_{im}(t, s) = \frac{\partial^{i-1} g_{1m}(t, s)}{\partial t^{i-1}} \qquad (i = 2, \dots, m). \tag{4.13}$$

Equalities (4.13) for $i = 2, \dots, m - 1$ also hold when $t = s$. Set

$$G(t, s) = g_{1m}(t, s). \tag{4.14}$$

Since $\tilde{G}(t, s)$ is a Green's function, it satisfies conditions 1—3. It follows from (4.13) that (4.14) also satisfies these conditions. We have proved

Theorem 4.2. The matrix-function (4.14) is the Green's function of the regular ap-operator (4.1).

4.3. Dependence of the Green's function on a parameter

Let us consider two regular ap-operators

$$L_1 x = \frac{dx}{dt} + A_1(t)x, \tag{4.15}$$

$$L_2 x = \frac{dx}{dt} + A_2(t)x; \tag{4.16}$$

and denote their Green's functions by $G_1(t, s)$ and $G_2(t, s)$, respectively. The properties of Green's functions imply the existence of constants $M, \gamma > 0$ such that

$$|G_1(t, s)|, |G_2(t, s)| \leqslant Me^{-\gamma|t-s|} \quad (-\infty < t, s < \infty). \tag{4.17}$$

Lemma 4.1. There exist positive constants M_0 and γ_0, depending only on M and γ, such that

$$|G_1(t, s) - G_2(t, s)| \leqslant M_0 e^{-\gamma_0|t-s|} \| A_1(\tau) - A_2(\tau) \|_{C(R^n)}. \tag{4.18}$$

Proof. Let $f(t)$ be an arbitrary function in $C(R^n)$, and $x_1(t)$, $x_2(t)$ bounded solutions of the equations

$$L_1 x = f(t), \quad L_2 x = f(t).$$

Obviously, $L_1[x_1(t) - x_2(t)] = [A_2(t) - A_1(t)]x_2(t)$, and so

$$x_1(t) - x_2(t) = \int_{-\infty}^{\infty} G_1(t, s)[A_2(s) - A_1(s)] x_2(s)\, ds. \tag{4.19}$$

Consider the matrix-function

$$V(t, s) = G_1(t, s) - G_2(t, s) + \int_{-\infty}^{\infty} G_1(t, \sigma)[A_1(\sigma) - A_2(\sigma)] G_2(\sigma, s)\, d\sigma.$$

It is easy to see that $V(t, s)$ is jointly continuous in $-\infty < t, s < \infty$, and

$$|V(t, s)| \leqslant M_1 e^{-\gamma_1|t-s|} \quad (-\infty < t, s < \infty), \tag{4.20}$$

where M_1 and γ_1 are positive. By virtue of (4.20), we can define the function

$$v(t) = \int_{-\infty}^{\infty} V(t, s) f(s)\, ds.$$

Obviously,

$$v(t) = x_1(t) - x_2(t) + \int_{-\infty}^{\infty} G_1(t, \sigma)[A_1(\sigma) - A_2(\sigma)] x_2(\sigma)\, d\sigma.$$

This equality and (4.19) together imply that $v(t) \equiv 0$.

Since $f(t)$ was an arbitrary function in $C(R^n)$, it follows that $V(t, s) \equiv 0$. This proves the identity

$$G_1(t, s) - G_2(t, s) = - \int_{-\infty}^{\infty} G_1(t, \sigma)[A_1(\sigma) - A_2(\sigma)] G_2(\sigma, s)\, d\sigma. \tag{4.21}$$

This identity implies the inequality

$$|G_1(t, s) - G_2(t, s)| \leqslant$$
$$\leqslant M^2 \int_{-\infty}^{\infty} e^{-\gamma[|t-\sigma|+|\sigma-s|]} d\sigma \cdot \| A_1(\tau) - A_2(\tau) \|_{C(R^n)},$$

whence follows (4.18). Q. E. D.

We now proceed to an investigation of families of ap-operators

$$L_\mu x = \frac{dx}{dt} + A(t; \mu) x, \tag{4.22}$$

where μ ranges over some metric compact set M.

Theorem 4.3. Let the ap-matrices $A(t; \mu)$ be continuous in μ, uniformly in t. Let the ap-operators (4.22) be regular for all $\mu \in M$

Then the Green's function $G(t, s; \mu)$ of the operators (4.22) is continuous in μ and, moreover, the exist positive M_0 and γ_0 such that for any $\mu_1, \mu_2 \in M$

$$|G(t, s; \mu_1) - G(t, s; \mu_2)| \leqslant$$
$$\leqslant M_0 e^{-\gamma_0|t-s|} \| A(\tau; \mu_1) - A(\tau; \mu_2) \|_{C(R^n)}. \tag{4.23}$$

Proof. Since the matrices $A(t; \mu)$ depend continuously on $\mu \in M$, uniformly in $t \in (-\infty, \infty)$, and M is compact, the norms of the matrix-functions $A(t; \mu)$ and of the inverse operators L_μ^{-1} are uniformly bounded. Therefore

$$|G(t, s; \mu)| \leqslant Me^{-\gamma|t-s|} \qquad (\mu \in M; \ -\infty < t, \ s < \infty),$$

where the positive constants M and γ do not depend on μ. To prove (4.23) we have only to use Lemma 4.1. Q. E. D.

Define the projections (3.22) for the operators (4.22), denoting them by $P_+(t; \mu)$ and $P_-(t; \mu)$, set

$$E_+(t; \mu) = P_+(t; \mu) R^n, \qquad E_-(t; \mu) = P_-(t; \mu) R^n. \tag{4.24}$$

It follows from (4.9) that

$$P_+(t; \mu) = \lim_{h \to +0} G(t+h, t; \mu), \tag{4.25}$$

$$P_-(t; \mu) = -\lim_{h \to -0} G(t+h, t; \mu). \tag{4.26}$$

Hence, under the assumptions of Theorem 4.3, the operators $P_+(t;\mu)$ and $P_-(t;\mu)$ are continuous in μ (uniformly in t). It follows that *the dimensions of the subspaces* $E_+(t;\mu)$ *and* $E_-(t;\mu)$ *do not depend on* μ. These remarks complement the dichotomy theorems proved in § 3.

We shall continue our investigation of the dependence of the Green's function of ap-operators on a parameter in Chapter 4.

Let $A(t)$ be an ap-matrix. We denote by $H(A)$ the set of all translates of the matrix $A(t)$, i. e. the set of all matrices $A(t+h)$ $(-\infty < h < \infty)$ and all ap-matrices $A_*(t)$ which are limits (uniform in t) of sequences $A(t+h_k)$. $H(A)$ is compact in the uniform norm. It is easily checked that $H(A) = H(A_*)$ for any $A_*(t) \in H(A)$. It was shown in § 2.4, that regularity of the operator

$$Lx = \frac{dx}{dt} + A(t)\,x \tag{4.27}$$

implies regularity of all the operators

$$L_*x = \frac{dx}{dt} + A_*(t)\,x \qquad (A_*(t) \in H(A)). \tag{4.28}$$

Theorem 4.4. Let $A(t+h_k)$ *be a sequence of matrices converging to* $A_*(t)$. *Let the operator (4.27) be regular.*

Then

$$\lim_{k\to\infty} \sup_{-\infty < t,\, s < \infty} |G(t+h_k, s+h_k) - G_*(t,s)|\,e^{\gamma_0|t-s|} = 0, \tag{4.29}$$

where $G(t,s)$, $G_*(t,s)$ *are the Green's functions of the operators (4.27), (4.28) and* γ_0 *is a positive number.*

To prove the theorem it is sufficient to note that, for any h, $G(t+h, s+h)$ is the Green's function of the operator

$$L_h x = \frac{dx}{dt} + A(t+h)\,x,$$

and to refer to Theorem 4.3.

Statements analogous to those just proved can be formulated for ap-operators of order m. We remark only that the derivatives of a Green's function with respect to t, up to order $m-1$, satisfy the same estimates as the Green's function itself.

It follows immediately from Theorem 4.4 that the *matrix-functions* $P_+(t)$ *and* $P_-(t)$ *are almost periodic in* t.

4.4. Examples

As a first example, consider the operator

$$Lx = \frac{dx}{dt} + a(t)x, \tag{4.30}$$

where $a(t)$ is a scalar ap-function and $x(t)$ a scalar function. Conditions for this operator to be regular are given by a theorem of Massera /1/:

Theorem 4.5. The operator (4.30) is regular if and only if

$$M(a) = \lim_{T\to\infty} \frac{1}{2T} \int_{-T}^{T} a(t)\,dt \neq 0. \tag{4.31}$$

Proof. The proof that condition (4.31) is sufficient is quite simple.

Suppose first that $M(a) < 0$. Set

$$G(t, s) = \begin{cases} 0, & \text{if } t \geqslant s, \\ -e^{-\int_s^t a(\tau)\,d\tau}, & \text{if } t < s. \end{cases} \tag{4.32}$$

$G(t, s)$ is the Green's function of the operator (4.30), since for any $f(t) \in C(R^1)$ the expression

$$x(t) = \int_{-\infty}^{\infty} G(t, s) f(s)\,ds = -\int_t^{\infty} e^{-\int_s^t a(\tau)\,d\tau} f(s)\,ds \tag{4.33}$$

gives a solution, bounded on $(-\infty, \infty)$, of the equation

$$\frac{dx}{dt} + a(t)x = f(t). \tag{4.34}$$

Now let $M(a) > 0$. Then the Green's function of the operator (4.30) is given by

$$G(t, s) = \begin{cases} e^{-\int_s^t a(\tau)\,d\tau}, & \text{if } t \geqslant s, \\ 0, & \text{if } t < s. \end{cases} \tag{4.35}$$

We now prove the necessity of (4.30). Denote by $x_0(t)$ an ap-solution of the equation

$$\frac{dx}{dt} + a(t)x = 1.$$

We set $a_0 = \sup_{-\infty < t < \infty} |a(t)|$, and prove that

$$|x_0(t)| \geqslant \frac{1}{2a_0} \qquad (-\infty < t < \infty). \tag{4.36}$$

If this is false, there exists t_0 such that $|x_0(t_0)| < 1/(2a_0)$. Then for all $t > t_0$ such that $|x_0(t)| \leqslant 1/(2a_0)$ we have the inequality $\dot{x}_0(t) \geqslant 1/2$. Hence $x_0(t) \geqslant 1/(2a_0)$ for all $t \geqslant t_0 + 2/a_0$. It follows that $x_0(t)$ is not almost periodic.

By (4.36), $1/x_0(t)$ is an ap-function and

$$M\left[\frac{1}{x_0(t)}\right] \neq 0.$$

The equality

$$a(t) = \frac{1}{x_0(t)} - \frac{\dot{x}_0(t)}{x_0(t)}$$

implies

$$M(a) = M\left(\frac{1}{x_0}\right) + \lim_{T \to \infty} \frac{1}{2T} \int_{-T}^{T} \frac{\dot{x}_0(t)}{x_0(t)} dt = M\left(\frac{1}{x_0}\right).$$

Q. E. D.

Theorem 4.5 implies the following proposition:

The set of all nonregular ap-operators is nowhere dense in the space of ap-operators (4.30).

As our second example we take the operator

$$Lx = \frac{d^m x}{dt^m} + A_1 \frac{d^{m-1} x}{dt^{m-1}} + \ldots + A_m x \tag{4.37}$$

with constant coefficients. Obviously, the operator (4.37) is regular if and only if the characteristic polynomial

$$\det \|\lambda^m I + \lambda^{m-1} A_1 + \ldots + A_m\| = 0 \tag{4.38}$$

has no roots on the imaginary axis.

Investigation of operators of type (4.37) with variable coefficients involves considerable difficulties. Several relevant results will be presented in the following chapters.

To conclude, we offer a few general remarks on the ap-operator

$$Lx = \frac{dx}{dt} + A(t)x \tag{4.39}$$

where $x \in R^n$ and

$$A(t) = \left\| \begin{matrix} a_{11}(t) & \ldots & a_{1n}(t) \\ \cdot & \cdot\ \cdot\ \cdot & \cdot \\ a_{n1}(t) & \ldots & a_{nn}(t) \end{matrix} \right\|. \tag{4.40}$$

The matrix (4.40) is said to be triangular if $a_{ij} \equiv 0$ for $i > j$. Theorem 4.5 implies the proposition:

An operator (4.39) with a triangular matrix (4.40) is regular if and only if none of the mean values $M(a_{ii})$ $(i = 1, \ldots, n)$ *vanish.*

The operators (4.39) with periodic matrices $A(t)$ are particularly important. They are regular if and only if the homogeneous equation $Lx = 0$ has no nontrivial solutions bounded on $(-\infty, \infty)$.

Recently, Mukhamadiev has generalized the above proposition. His theorem is as follows: An ap-operator (4.39) is regular if and only if none of the homogeneous equations

$$\frac{dx}{dt} + A_*(t)\,x = 0 \qquad (A_*(t) \in H(A)) \tag{4.41}$$

has a nontrivial solution bounded on $(-\infty, \infty)$.

Mukhamadiev's proof is based on the estimate

$$|x(t)| \leqslant \delta \max\{|x(T_1)|,\ |x(T_2)|\} \qquad (T_1 \leqslant t \leqslant T_2) \tag{4.42}$$

valid for all solutions $x(t)$ of equations (4.41), where δ is a constant, the same for all these equations, independent of T_1 and T_2.

Mukhamadiev's theorem is substantially stronger than the well-known first theorem of Favard (see Levitan /1/) on the solutions of the inhomogeneous equation

$$\frac{dx}{dt} + A(t)\,x = f(t) \tag{4.43}$$

where the right-hand side is almost periodic.

4.5. The Green's function of a periodic boundary-value problem

Let the coefficients $A_i(t)$ of the regular operator (4.1) be ω-periodic matrix-functions. Consider the equation

$$Lx = f(t) \tag{4.44}$$

where $f(t)$ is ω-periodic. If $x(t)$ is a bounded solution of equation (4.44), then $x(t+\omega)$ is also a bounded solution of this equation. But the regularity of L implies uniqueness of the bounded solution and hence

$$x(t+\omega) \equiv x(t) \qquad (-\infty < t < \infty).$$

Thus the operator (4.6) maps any ω-periodic function $f(t)$ onto an ω-periodic function.

To find ω-periodic solutions of equation (4.44), it is convenient to rewrite it as

$$x(t) = Tf(t) = \sum_{j=-\infty}^{\infty} \int_0^{\omega} G(t, s + j\omega) f(s)\, ds$$

or equivalently,

$$Tf(t) = \int_0^{\omega} G_{\omega}(t, s) f(s)\, ds, \tag{4.45}$$

where

$$G_{\omega}(t, s) = \sum_{j=-\infty}^{\infty} G(t, s + j\omega) \qquad (-\infty < t < \infty;\ 0 \leqslant s \leqslant \omega). \tag{4.46}$$

The matrix-function (4.46) is called the *Green's function of the ω-periodic boundary-value problem.* It is readily shown that the Green's function is ω-periodic in t:

$$G_{\omega}(t + \omega, s) \equiv G_{\omega}(t, s). \tag{4.47}$$

§ 5. POSITIVITY OF THE GREEN'S FUNCTION

5.1. Wedges and cones

Let E be a real Banach space. A set $K \subset E$ is said to be a *wedge* if it is convex, closed and $x \in K$ implies that $\alpha x \in K$ for all $\alpha \geqslant 0$. A wedge K is said to be a *cone* if $x, -x \in K$ implies $x = 0$.

The simplest examples of cones in $C(R^1)$ and $B(R^1)$ are the sets K of all nonnegative functions. In R^n, the set of elements with nonnegative components (relative to some basis) is a cone.

Every cone $K \subset E$ generates a *semi-order* in E: we write $x \leqslant y$ if $y - x \in K$. It is clear that these inequalities may be added, multiplied by nonnegative numbers, and are preserved under passage to limits; moreover, $x \leqslant y$ and $y \leqslant z$ imply $x \leqslant z$.

A linear functional $l \in E^*$ is said to be *positive* if $l(x) \geqslant 0$ for $x \in K$. The set of all positive functionals forms a wedge, denoted by K^*.

A cone K is said to be solid if it contains interior elements. If $y-x$ is an interior element of K, we write $x \ll y$. This definition immediately implies the following propositions:

1. If $x \leqslant y$, $u \ll v$, then $x+u \ll y+v$.
2. If $x \ll y$ and $\alpha > 0$, then $\alpha x \ll \alpha y$.
3. If $x \leqslant y$, $y \ll z$ or $x \ll y$, $y \leqslant z$, then $x \ll z$.

The cones in the above examples are solid. Below, as a rule, we shall use only solid cones.

A cone K is said to be normal if $0 \leqslant x \leqslant y$ implies

$$\|x\| \leqslant N\|y\|, \tag{5.1}$$

where N does not depend on x, y. The cones K in the above examples are normal. In finite-dimensional (but not infinite-dimensional!) spaces any cone is normal.

A cone K is said to be minihedral if for any pair of elements $x, y \in E$ there exists an element u, with $x \leqslant u$, $y \leqslant u$, such that $x \leqslant v$, $y \leqslant v$ imply $u \leqslant v$. This element u is called the supremum of x, y and denoted by $u = \sup\{x, y\}$. We denote

$$x_+ = \sup\{0, x\}, \qquad x_- = \sup\{0, -x\}.$$

Obviously, $x = x_+ - x_-$. Note that in all our examples the cone K is minihedral.

In finite-dimensional spaces R^n the class of all minihedral cones consists of all cones whose elements have nonnegative components relative to some basis. In other words, a cone $K \subset R^n$ is minihedral if and only if it is the intersection of n half-spaces. Examples of nonminihedral cones are easily given.

The general definitions of this section are due to Krein (Krein and Rutman /1/, Krasnosel'skii /1/).

5.2. Functionals selecting cones

A continuous functional $q(x)$ defined on E is said to be sublinear if

$$q(\alpha x) \equiv \alpha q(x) \qquad (\alpha \geqslant 0, \quad x \in E) \tag{5.2}$$

and

$$q(x+y) \leqslant q(x) + q(y) \qquad (x, y \in E). \tag{5.3}$$

The simplest examples of sublinear functionals are the linear functionals $l(x)$ and the norm $\|x\|$.

Let $\{q_\gamma(x)\}$ be a family of sublinear functionals and

$$q_*(x) = \sup_\gamma q_\gamma(x) < \infty \qquad (x \in E). \tag{5.4}$$

It is easy to see that $q_*(x)$ is also sublinear. Denote by F the set of linear functionals satisfying the inequality

$$l(x) \leqslant q(x) \qquad (x \in E). \tag{5.5}$$

Lemma 5.1.

$$q(x) = \max_{l \in F} l(x) \qquad (x \in E). \tag{5.6}$$

Proof. Let $x_0 \in E$. It will suffice to construct a functional $l \in F$ such that $l(x_0) = q(x_0)$. We first define a functional l on the straight line $x = \alpha x_0 (-\infty < \alpha < \infty)$ by

$$l(x) = \alpha q(x_0).$$

Obviously,

$$l(\alpha x_0) \leqslant q(\alpha x_0) \qquad (-\infty < \alpha < \infty).$$

By the Hahn-Banach theorem (see Dunford and Schwartz /1/, Chapter 5), $l(x)$ can be extended to a functional of F which is continuous throughout E. Q. E. D.

Note that F is bicompact in the E-topology (Dunford and Schwartz /1/, Chapter 5) of the space E^*.

Let $q(x)$ be a sublinear functional. Set

$$K = \{x:\ q(x) \leqslant 0\} \tag{5.7}$$

and

$$K_{\text{int}} = \{x:\ q(x) < 0\}. \tag{5.8}$$

It is easily checked that K is a wedge. It is a cone if and only if the set F in (5.6) is total. We are interested in the sublinear functionals $q(x)$ such that K is a solid cone and K_{int} the set of the interior points of K. We shall say that these functionals *select* the cone K.

Any solid cone K may be selected by different sublinear functionals. We give here two basic constructions.

Let u_0 be a fixed interior point of the cone K. Then K is selected by the functional

$$q_0(x) = \min_{x+ru_0 \in K} r \qquad (x \in E). \tag{5.9}$$

The functional (5.9) can be defined by an equivalent formula

$$q_0(x) = -\min_{l \in S} l(x) \qquad (x \in E), \tag{5.10}$$

where S is the set of functionals $l \in K^*$ such that $l(u_0) = 1$. The proof that

$$\min_{x+ru_0 \in K} r = -\min_{l \in S} l(x) \qquad (x \in E)$$

is left to the reader. It follows from (5.10) that $q_0(x)$ is sublinear. The functional (5.9) can be described in another way.

Let Π be an arbitrary two-dimensional plane containing the straight line αu_0 $(-\infty < \alpha < \infty)$; this line divides Π into two half-planes; it is easy to see that (5.9) is linear on each of these half-planes; its values are easily determined, since $q_0(\alpha u_0) = -\alpha$ and $q_0(x)$ vanishes on the intersection of Π with the boundary of the cone K.

The functional (5.9) satisfies the inequality

$$|q_0(x)| \leqslant \frac{1}{r_0} \|x\| \qquad (x \in E),$$

where r_0 is the radius of the maximal ball with center at u_0 contained in K. It follows that

$$|q_0(x) - q_0(y)| \leqslant \frac{1}{r_0} \|x - y\| \qquad (x, y \in E).$$

A second example of a sublinear functional selecting the solid cone K is

$$q_1(x) = -\min_{l \in K^*, \|l\|=1} l(x) \qquad (x \in E). \tag{5.11}$$

It is easily checked that

$$q_1(x) = \begin{cases} \rho(x, K), & \text{if} \quad x \overline{\in} K, \\ -\rho(x, E \setminus K), & \text{if} \quad x \in K. \end{cases}$$

Let $q(x)$ be an arbitrary sublinear functional selecting a solid cone K. Since the element $q_0(x)u_0 + x$ (where $q_0(x)$ is the functional (5.9) and $x \in E$) lies on the boundary of K, it follows that

$$q[q_0(x)u_0 + x] \equiv 0 \qquad (x \in E)$$

Hence, by inequality (5.3), follow the estimates

$$q(-u_0)\,q_0(x) \leqslant q(x) \leqslant -q(u_0)\,q_0(x) \qquad (x \in K), \tag{5.12}$$

$$-q(u_0)\,q_0(x) \leqslant q(x) \leqslant q(-u_0)\,q_0(x) \qquad (x \overline{\in} K). \tag{5.13}$$

We now determine functionals defining a few special cones.

Let K be the cone of all nonnegative functions in $B(R^1)$. This cone is selected by the functional

$$q(x) = -\inf_{-\infty < t < \infty} x(t) \qquad (x \in B(R^1)).$$

Let $e_1, \ldots, e_n$ be a basis in R^n and K the cone

$$K = \{x\colon\ x = \xi_1 e_1 + \ldots + \xi_n e_n;\ \xi_j \geqslant 0\}.$$

This cone is selected by the functional

$$q(x) = -\min_j \xi_j \qquad (x \in R^n). \tag{5.14}$$

Finally, let K be the cone defined in R^n by

$$K = \{x\colon\ \sqrt{(Hx, x)} \leqslant (x, a)\}, \tag{5.15}$$

where a is a fixed vector, $(x, a) = 0$ or $(Hx, x) = 0$ implies that $x = 0$, and $(Hx, x) \geqslant 0$. The cone (5.15) is selected by the functional

$$q(x) = -(x, a) + \sqrt{(Hx, x)} \qquad (x \in R^n) \tag{5.16}$$

Functionals selecting a cone permit the use of analytical tools to describe certain geometric constructions. This method is apparently due to Kolesov /1/.

5.3. Space of cones

We denote by $\mathfrak{K}$ the set of all wedges in the space E. $\mathfrak{K}$ becomes a metric space when the distance between two elements $K_1, K_2 \in \mathfrak{K}$ is defined via the Hausdorff metric (see Hausdorff /1/):

$$\rho(K_1, K_2) = \max\left\{\sup_{x \in \Omega_1} \rho(x, \Omega_2), \sup_{y \in \Omega_2} \rho(y, \Omega_1)\right\}, \tag{5.17}$$

where

$$\Omega_1 = \{x\colon x \in K_1, \|x\| \leqslant 1\}, \quad \Omega_2 = \{y\colon y \in K_2, \|y\| \leqslant 1\}.$$

Formula (5.17) is equivalent to the formula

$$\rho(K_1, K_2) = \max\left\{\inf_{\Omega_1 \subset S_\delta(\Omega_2)} \delta, \inf_{\Omega_2 \subset S_\delta(\Omega_1)} \delta\right\}, \tag{5.18}$$

where $S_\delta(\Omega)$ denotes the set

$$S_\delta(\Omega) = \{x\colon \rho(x, \Omega) < \delta\}. \tag{5.19}$$

We claim that $\mathfrak{K}$ is a complete space. Let K_n be a fundamental sequence of wedges in $\mathfrak{K}$ and

$$\Omega_n = \{x\colon x \in K_n, \|x\| \leqslant 1\}.$$

Denote by Ω the set of limits of all fundamental sequences $x_{n_i} \in \Omega_{n_i}$ $(n_1 < n_2 < \ldots)$. Ω is obviously closed. Let $\varepsilon > 0$ be given and suppose that for $n, m \geqslant N(\varepsilon)$

$$\rho(\Omega_n, \Omega_m) < \varepsilon.$$

Let $x \in \Omega$, $x_{n_i} \in \Omega_{n_i}$ and $\rho(x_{n_i}, x) \to 0$; then for $m \geqslant N(\varepsilon)$

$$\rho(x, \Omega_m) = \lim_{i \to \infty} \rho(x_{n_i}, \Omega_m) \leqslant \varepsilon. \tag{5.20}$$

Now set $n_j = N(\varepsilon/2^{j+1})$. By the definition of n_j, it follows that for any fixed $x \in \Omega_n$, $n \geqslant N(\varepsilon/2)$, we can construct a sequence $x_{n_j} \in \Omega_{n_j}$ such that $\rho(x, x_{n_1}) < \varepsilon/2$ and $\rho(x_{n_j}, x_{n_{j+1}}) < \varepsilon/2^{j+1}$. Obviously, $\rho(x_{n_j}, x_{n_{j+l}}) < \varepsilon/2^j$ for any $l > 0$; hence the sequence x_{n_j} is fundamental. Now let $x_{n_j} \to x_* \in \Omega$ The inequality $\rho(x, x_{n_j}) < \varepsilon$ implies that

$$\rho(x, \Omega) \leqslant \rho(x, x_*) = \lim_{j \to \infty} \rho(x, x_{n_j}) \leqslant \varepsilon.$$

From this inequality and (5.20) it follows that

$$\lim_{n \to \infty} \max\left\{\sup_{x \in \Omega} \rho(x, \Omega_n), \sup_{y \in \Omega_n} \rho(y, \Omega)\right\} = 0.$$

To complete the proof we have to show that the set $\mathfrak{K}$ is the intersection of a wedge K and the unit ball. In other words, we must show that Ω is convex and $x \in \Omega$, $x \neq 0$ implies $x/\|x\| \in \Omega$. Both these properties follow from their validity for the sets Ω_n.

Note that the space $\mathfrak{K}$ has finite diameter: by (5.17),

$$\rho(K_1, K_2) \leqslant 1 \qquad (K_1, K_2 \in \mathfrak{K}).$$

We denote by $C(\mathfrak{K})$ the space of functions $K(t)$ continuous on $(-\infty, \infty)$ with values in $\mathfrak{K}$, with the usual metric

$$\rho_{C(\mathfrak{K})}[K_1(t), K_2(t)] = \sup_{-\infty < t < \infty} \rho_{\mathfrak{K}}[K_1(t), K_2(t)]. \tag{5.21}$$

As usual, a function $K(t) \in C(\mathfrak{K})$ is said to be almost periodic if the set of functions $K(t+h)$ $(-\infty < h < \infty)$ is compact in $C(K)$.

5.4. The cones $\hat{K}(t)$

We shall limit ourselves to the case $E = R^n$ (although many of the constructions are valid with slight changes for any Banach space E).

Let $K(t)$ be a fixed function in $C(\mathfrak{K})$. We denote by $\hat{K}(t)$ the set of functions $x(t) \in B(R^n)$ satisfying the condition

$$x(t) \in K(t) \qquad (-\infty < t < \infty). \tag{5.22}$$

The set $\hat{K}(t)$ is a wedge in $B(R^n)$. It is easily seen that $\hat{K}(t)$ is a cone if there is a set of values of t, dense on $(-\infty, \infty)$, such that the wedges $K(t)$ are cones in R^n.

The wedge $\hat{K}(t)$ may turn out to consist only of one point, the zero function. Investigation of the wedge $\hat{K}(t)$ in the general case is a complicated task. The situation is simpler if $K(t)$ is an almost periodic function.

We denote by $\mathfrak{N}$ the closure in $\mathfrak{K}$ of the range of the ap-function $K(t)$.

Lemma 5.2. Let $K(t)$ be an ap-function such that all the elements of $\mathfrak{N}$ are solid cones in R^n.

Then $\hat{K}(t)$ is a solid cone in $B(R^n)$.

Proof. We use the notation

$$\Omega(t) = \{x\colon\ x \in K(t),\ |x| \leqslant 1\} \qquad (-\infty < t < \infty). \tag{5.23}$$

Each set $\Omega(t)$ is convex, closed, bounded and solid (contains interior points). Denote by $x_0(t)$ the centroid of the set $\Omega(t)$; obviously,

$$x_0(t) = \left[\int_{\Omega(t)} dm(x) \right]^{-1} \int_{\Omega(t)} x \, dm(x), \tag{5.24}$$

where $m(x)$ is Lebesgue measure on R^n.

Formula (5.24) implies that $x_0(t) \in C(R^n)$. We now show that $x_0(t)$ is almost periodic.

Let h_i be a sequence of translations. Without loss of generality, we may assume that the sequence $K(t + h_i)$ converges in $C(\mathfrak{K})$ to an ap-function, which we denote by $K_*(t)$. Then the functions $x(t + h_i)$ obviously converge uniformly to the ap-function

$$x_*(t) = \left[\int_{\Omega_*(t)} dm(x) \right]^{-1} \int_{\Omega_*(t)} x \, dm(x).$$

Hence $x_0(t) \in B(R^n)$.

From (5.24) it follows that the values of $x_0(t)$ are interior points of the corresponding cones $K(t)$. Moreover, there exists $r_0 > 0$ such that, for any t, if $|x - x_0(t)| \leqslant r_0$ then $x \in K(t)$. To prove this we must show that the radii $r_0(t)$ of the maximal balls centered at $x_0(t)$ and lying in $K(t)$ are bounded away from zero. If not, there exists a sequence h_i such that $r_0(h_i) \to 0$. We may assume, without loss of generality that the cones $K(t)$ converge in $\mathfrak{K}$ to a cone $K_* \in \mathfrak{N}$. The points $x_0(h_i)$ then converge to

$$x_* = \left[\int_{\Omega_*} dm(x) \right]^{-1} \int_{\Omega_*} x \, dm(x),$$

which is an interior point of the solid cone K_* (by the assumptions of the lemma!). On the other hand, $r_0(h_i) \to 0$ implies that x_* is a boundary point of K_*. This is a contradiction.

It remains to note that any function $x(t) \in B(R^n)$ satisfying the inequality

$$\| x(t) - x_0(t) \|_{B(R^n)} \leqslant r_0,$$

belongs to $\hat{K}(t)$. Q. E. D.

Assume now that all values of an ap-function $K(t)$ are cones and that if $x,\ y - x \in K(t)$ then

$$|x| \leqslant N |y|, \tag{5.25}$$

where N does not depend on t. Then $\hat{K}(t)$ is a normal cone, since $x(t),\ y(t) - x(t) \in \hat{K}(t)$ obviously implies the inequality

$$\| x(t) \|_{B(R^n)} \leqslant N \| y(t) \|_{B(R^n)}. \tag{5.26}$$

It is easy to show that the existence of the constant N in inequalities (5.25) which does not depend on t is equivalent to the fact that all elements of $\mathfrak{R}$ are cones.

We now consider conditions for the cone $\hat{K}(t)$ to be minihedral.

Lemma 5.3. Let K_i be a sequence of minihedral cones converging in $\mathfrak{K}$ to a solid cone K. Let x_i be a sequence converging in R^n to a point x. Set

$$(x_i)_+ = \sup\{0, x_i\}, \tag{5.27}$$

where the supremum is in the sense of the semi-order generated by the cone K_i.

Then the cone K is minihedral and the sequence (5.27) converges in R^n to the point

$$x_+ = \sup\{0, x\}, \tag{5.28}$$

where the supremum is in the sense of the semi-order with respect to K.

Proof. As already noted, any minihedral cone in R^n is a cone of vectors with nonnegative components relative to some basis. Hence, there exists a sequence of normalized bases

$$g_1^i,\ g_2^i,\ \ldots,\ g_n^i \qquad (i = 1,\ 2,\ \ldots),$$

such that

$$K_i = \{x\colon\ x = \xi_1 g_1^i + \ \ldots\ + \xi_n g_n^i, \qquad \xi_j \geqslant 0\}. \tag{5.29}$$

Since the sequence (5.29) converges to a solid cone K, we may assume without loss of generality that each sequence of unit elements $g_j^i\ (i = 1,\ 2,\ \ldots)$ converges in R^n to an element g_j. Obviously, the vectors $g_1, \ldots, g_n$ are linearly independent and

$$K = \{x\colon\ x = \xi_1 g_1 + \ \ldots\ + \xi_n g_n,\ \ \xi_j \geqslant 0\}.$$

Hence the cone K is minihedral.

Now for the second part of our lemma.

Let

$$x_i = \xi_1^i g_1^i + \ \ldots\ + \xi_n^i g_n^i, \qquad x = \xi_1 g_1 + \ \ldots\ + \xi_n g_n.$$

By construction, the vectors $g_1, \ldots, g_n$ are such that $|x_i - x| \to 0$ implies

$$\lim_{i\to\infty} \xi_j^i = \xi_j \qquad (j = 1,\ \ldots,\ n).$$

Consequently, it follows from the relations

$$(x_i)_+ = \sum_{\xi_j^i \geqslant 0} \xi_j^i g_j^i, \qquad x_+ = \sum_{\xi_j \geqslant 0} \xi_j g_j$$

that $(x_i)_+ \to x_+$. Q. E. D.

Lemma 5.4. Under the assumptions of Lemma 5.2, suppose that every cone $K(t)$ is minihedral.

Then the cone $\hat{K}(t)$ is minihedral.

Proof. Let $x(t) \in B(R^n)$. For each fixed t we define an element

$$x_+(t) = \sup\{0, x(t)\}, \tag{5.30}$$

where the supremum is in the sense of the semi-order generated by $K(t)$. Obviously, the lemma will be proved if we can show that the function $x_+(t)$ is almost periodic.

The continuity of $x_+(t)$ follows immediately from Lemma 5.3.

From any sequence of translations h_k we can select a subsequence $h_{k(i)}$ such that the functions $K(t + h_{k(i)})$ converge in $C(\mathfrak{K})$ to some function $K_*(t)$, and the functions $x(t + h_{k(i)})$ converge in $C(R^n)$ to a function $x_*(t)$. Almost periodicity of $x_+(t)$ will be proved if we can show that the sequence of functions $x_+(t + h_{k(i)})$ converges in $C(R^n)$ to the function

$$[x_*(t)]_+ = \sup\{0, x_*(t)\} \qquad (-\infty < t < \infty).$$

If this is false, we can find a sequence t_j and a subsequence s_j of $h_{k(i)}$ such that

$$|x_+(t_j + s_j) - [x_*(t_j)]_+| \geqslant \varepsilon_0 \qquad (j = 1, 2, \ldots), \tag{5.31}$$

where ε_0 is a positive number. Since the functions $K(t)$ and $x(t)$ are almost periodic, we may assume that $K(t + t_j + s_j)$ and $K_*(t + t_j)$ converge to the same function $K_{**}(t)$ in $C(\mathfrak{K})$, while $x(t + t_j + s_j)$ and $x(t + t_j)$ converge in $C(R^n)$ to the same function $x_{**}(t)$. In particular, the minihedral cones $K(t_j + s_j)$ and $K_*(t_j)$ converge in $\mathfrak{K}$ to the cone $K_{**}(0)$, and the elements $x(t_j + s_j)$ and $x_*(t_j)$ converge in R^n to the point $x_{**}(0)$. By Lemma 5.4 it now follows that $x_+(t_j + s_j) \to [x_{**}(0)]_+$ and $[x_*(t_j)]_+ \to [x_{**}(0)]_+$, and this contradicts (5.31). Q. E. D.

5.5. Functionals selecting $K(t)$

Let $K(t)$ be a cone-valued ap-function satisfying the assumptions of Lemma 5.2. It follows from this lemma that there exist an ap-function $x_0(t) \in B(R^n)$ and a number $r_0 > 0$ such that, for every t, the point $x_0(t)$ belongs to $K(t)$ together with its spherical neighborhood of radius r_0.

Set

$$q_0(t, x) = \min_{x + r x_0(t) \in K(t)} r. \tag{5.32}$$

For every fixed t, sublinear functional (5.32) coincides with the functional (5.9) considered in Section 5.2. As we already know, the functional (5.32) satisfies a Lipschitz condition

$$|q_0(t, x) - q_0(t, y)| \leqslant \frac{1}{r_0} |x - y| \quad (x, y \in R^n). \tag{5.33}$$

For any function $x(t) \in B(R^n)$, the scalar function $q_0[t, x(t)]$ $(-\infty < t < \infty)$ is almost periodic (see Section 1.4) since all the functions $q_0(t, x)$ are almost periodic in t $(x \in R^n)$.

We note two more properties of the functional (5.32). First,

$$q_0[t, x_0(t)] = -1 \quad (-\infty < t < \infty). \tag{5.34}$$

Second,

$$\max\{q_0(t, x);\ q_0(t, -x)\} \geqslant \varepsilon_0 > 0 \quad (x \in R^n,\ |x| = 1). \tag{5.35}$$

We need prove only (5.35). Suppose this inequality false. Then there exist sequences t_i, x_i such that $|x_i| = 1$ and

$$q_0(t_i, x_i),\ q_0(t_i, -x_i) < \frac{1}{i} \quad (i = 1, 2, \ldots). \tag{5.36}$$

We may assume that the elements x_i converge in R^n to a point x_*, the sequence of cones $K(t_i)$ converges in $\mathfrak{K}$ to a cone K_*, and the numbers $q_0(t_i, x_i)$ and $q_0(t_i, -x_i)$ converge to certain numbers a and b. By (5.36), a and b are nonpositive; on the other hand, the inequalities

$$q_0(t_i, x_i) + q_0(t_i, -x_i) \geqslant 0 \quad (i = 1, 2, \ldots)$$

imply that $a + b \geqslant 0$. Hence $a = b = 0$. By the definition of $q_0(t, x)$, it follows that

$$q_0(t_i, x_i)\, x_0(t_i) + x_i,\ q_0(t_i, -x_i)\, x_0(t_i) - x_i \in K(t_i).$$

Therefore x_* and $-x_*$ both belong to K_*, i. e., $x_* = 0$. This contradicts the equality

$$|x_*| = \lim_{i\to\infty} |x_i| = 1.$$

Consider the following situation: given a functional $q(t, x)$, sublinear on $x \in R^n$ and almost periodic in t, we wish to ascertain whether the function

$$K(t) = \{x:\ q(t, x) \leqslant 0\} \qquad (-\infty < t < \infty) \tag{5.37}$$

satisfies the assumptions of Lemma 5.2.

Lemma 5.5. Let $q(t,x)$ be a functional, sublinear in $x \in R^n$ and almost periodic in t, satisfying the conditions:

1. Lipschitz condition:

$$|q(t, x) - q(t, y)| \leqslant a|x - y| \qquad (x, y \in R^n), \tag{5.38}$$

where a does not depend on t.

2. There exists a function $x_0(t) \in B(R^n)$ such that

$$q[t, x_0(t)] \leqslant -\varepsilon_0 < 0 \qquad (-\infty < t < \infty). \tag{5.39}$$

3. $$\max\{q(t, x),\ q(t, -x)\} \geqslant \varepsilon_1 > 0 \qquad (-\infty < t < \infty,\ |x| = 1). \tag{5.40}$$

Then (5.37) defines a cone-valued ap-function $K(t)$ satisfying the assumptions of Lemma 5.2.

There are no essential difficulties in the proof, which is therefore left to the reader.

5.6. Positive linear operators

Let E be a Banach space with a solid cone K. Let A be an operator in E.

The operator A is said to be positive (with respect to K) if $AK \subset K$, and strongly positive if $x \in K, x \neq 0$ imply $Ax \gg 0$ (i. e., Ax is an interior element of K). The operator A is said to be monotone if $x \leqslant y$ implies $Ax \leqslant Ay$. A linear operator A is monotone if and only if it is positive.

An essential role is played in the theory of positive operators by the so-called u_0-norm $\|x\|_{u_0}$, where u_0 is a fixed interior element of the cone K. This norm is defined by

$$\|x\|_{u_0} = \inf\{\alpha:\ -\alpha u_0 \leqslant x \leqslant \alpha u_0\} \qquad (x \in E). \tag{5.41}$$

It is easy to prove that if K is a normal cone, the norm (5.41) is equivalent to the original norm $\|x\|$.

An important characteristic of a linear operator A is its spectral radius $r(A)$:

$$r(A) = \lim_{n\to\infty} \| A^n \|^{\frac{1}{n}}. \tag{5.42}$$

Obviously,

$$r(A) \leqslant \|A\|. \tag{5.43}$$

Unlike the norm $\|A\|$ of A, the spectral radius is invariant under passage to equivalent norms. A useful estimate of the spectral radius is given in the following theorem, dating back to Frobenius and Urysohn.

Theorem 5.1. Let A be a positive linear operator. Let u_0 be an interior element of a normal cone K such that

$$Au_0 \ll u_0. \tag{5.44}$$

Then

$$r(A) < 1. \tag{5.45}$$

Proof. By (5.44), there exists $\eta > 0$ such that

$$Au_0 \leqslant (1-\eta)\, u_0. \tag{5.46}$$

Let $\| x \|_{u_0} \leqslant 1$, i. e., $-u_0 \leqslant x \leqslant u_0$. Then the positivity of the operator A implies the inequalities.

$$-Au_0 \leqslant Ax \leqslant Au_0,$$

whence, by virtue of (5.46),

$$-(1-\eta)\, u_0 \leqslant Ax \leqslant (1-\eta)\, u_0.$$

Consequently,

$$\| Ax \|_u \leqslant (1-\eta) \| x \|_{u_0} \qquad (x \in E).$$

By (5.43), we now get $r(A) \leqslant 1-\eta$. Q. E. D.

5.7. Green's functions of constant sign

We now return to our investigation of the differential ap-operator

$$Lx = \frac{d^m x}{dt^m} + A_1(t)\frac{d^{m-1}x}{dt^{m-1}} + \dots + A_m(t)x. \tag{5.47}$$

We assume the operator (5.47) to be regular (see Section 2.1). Then L has an inverse L^{-1}, defined on $C(R^n)$ with range in $C^m(R^n)$, and moreover $L^{-1}B(R^n) = B^m(R^n)$. It was proved in Section 4 that L^{-1} is representable in the form

$$L^{-1}f(t) = \int_{-\infty}^{\infty} G(t, s)f(s)\,ds, \tag{5.48}$$

where $G(t, s)$ is the Green's function.

To construct the Green's function, we first considered the equivalent ap-operator (4.8):

$$\tilde{L}u = \frac{du}{dt} + Q(t)u, \tag{5.49}$$

and then defined the Green's function of the latter by (4.9):

$$\tilde{G}(t, s) = \begin{cases} U(t)P_+U^{-1}(s), & \text{if} \quad t \geqslant s, \\ -U(t)P_-U^{-1}(s), & \text{if} \quad t < s, \end{cases} \tag{5.50}$$

where $U(t)$ (defined in R^{mn}) is the translation operator along trajectories of the equation $\tilde{L}u = 0$ from 0 to t, P_+ and $P_-(P_+ + P_- = I)$ are the projections onto the subspaces E_+ and E_- $(E_+ \dotplus E_- = R^{mn})$ of initial values of the solutions that decrease as $t \to \infty$ and as $t \to -\infty$. Finally, the Green's function was defined as the minor of order n in the upper right corner of the matrix $\tilde{G}(t, s)$ (see (4.12)).

Let $K(t)$ be a function with values in $\mathfrak{K}$. Recall (see Section 5.4) that $\hat{K}(t)$ denotes the set of functions $x(t) \in B(R^n)$ such that $x(t) \in K(t)$ $(-\infty < t < \infty)$. In Section 5.4 we discussed in detail conditions for $\hat{K}(t)$ to be a cone, and determined when this cone is solid. normal and minihedral.

The Green's function $G(t, s)$ is said to be *nonnegative with respect to the cone-valued function* $K(t)$ if

$$G(t, s)K(s) \subset K(t) \qquad (-\infty < t, s < \infty). \tag{5.51}$$

Similarly, the Green's function $G(t,s)$ is nonpositive if

$$-G(t, s) K(s) \subset K(t) \qquad (-\infty < t, s < \infty). \tag{5.52}$$

The following proposition is obvious.

Theorem 5.2. Let the Green's function $G(t,s)$ *be nonnegative with respect to the cone-valued function* $K(t)$.

Then the operator (5.48) is positive with respect to $\tilde{K}(t)$.

In many cases one has to consider operators

$$L_h x = \frac{d^m x}{dt^m} + A_1(t+h)\frac{d^{m-1}x}{dt^{m-1}} + \ldots + A_m(t+h)x, \tag{5.53}$$

together with the operator (5.47), where h is any number. As we have already remarked (see Section 2.4), the operators (5.53) form a compact set in the space of continuous operators acting from $C^m(R^n)$ to $C(R^n)$. The closure of this set was denoted by $H(L)$. It is easy to see that $H(L)$ is the set of regular ap-operators

$$L_* x = \frac{d^m x}{dt^m} + \overset{*}{A_1}(t)\frac{d^{m-1}x}{dt^{m-1}} + \ldots + \overset{*}{A_m}(t)x, \tag{5.54}$$

for each of which one can find a sequence h_j such that

$$\lim_{j\to\infty} \| A_i(t+h_j) - \overset{*}{A_i}(t) \|_{C(R^n)} = 0 \qquad (i = 1, \ldots, n). \tag{5.55}$$

A simple calculation shows that the Green's function $G_h(t,s)$ of the operator (5.53) is expressed in terms of the Green's function $G(t,s)$ of (5.47) by

$$G_h(t, s) = G(t+h, s+h). \tag{5.56}$$

It follows from Theorem 4.4 that if condition (5.55) holds the Green's function $G_*(t,s)$ of the operator (5.54) is given by

$$G_*(t, s) = \lim_{j\to\infty} G(t+h_j, s+h_j), \tag{5.57}$$

where the convergence is uniform in t and s. The set of Green's functions of all operators in $H(L)$ is denoted by $H[G(t,s)]$.

Let $K(t) \in B(\mathfrak{K})$ be a given ap-function. We denote by $H[K(t)]$ the closure in $B(\mathfrak{K})$ of the compact set of ap-functions $K(t+h)$ $(-\infty < h < \infty)$.

Assume that the Green's function $G(t, s)$ of the operator (5.47) is nonnegative with respect to the cone-valued function $K(t)$. Then the Green's functions $G(t+h, s+h)$ are nonnegative with respect to $K(t+h)$; the Green's function $G_*(t, s)$ of the operator (5.54) is nonegative with respect to the ap-function

$$K_*(t) = \lim_{j\to\infty} K(t+h_j)$$

if (5.55) is valid.

A cone-valued ap-function $K(t)$ is said to be *solid* if the values of all functions in $H[K(t)]$ are solid cones. As proved in Section 5.4, if an ap-function $K(t)$ is solid, then $\hat{K}(t)$ is a solid normal cone in $B(R^n)$.

We shall say the Green's function $G(t, s)$ of a regular ap-operator (5.47) is *strongly positive with respect to a solid ap-function* $K(t)$ if it is nonnegative and, for any Green's function $G_*(t, s) \in H[G(t, s)]$ and any t, there exists a set $\mathfrak{M}(t, G_*)$, dense in some infinite interval, such that for nonzero $x \in K_*(s)$ and $s \in \mathfrak{M}(t, G_*)$ the elements $G_*(t, s)x$ are interior points of the cone $K_*(t)$.

Similarly one defines *strongly negative* Green's functions.

Theorem 5.3. Let the Green's function $G(t, s)$ of a regular ap-operator (5.47) be strongly positive with respect to a solid cone-valued ap-function $K(t)$.

Then the operator (5.48) is strongly positive on the cone $\hat{K}(t)$.

Proof. Let $f(t)$ be a nonzero function in $\hat{K}(t)$. We must show that the function

$$x(t) = \int_{-\infty}^{\infty} G(t, s) f(s)\, ds$$

is an interior point of $\hat{K}(t)$ or, equivalently, that for any t the cone $K(t)$ contains $x(t)$ together with a spherical neighborhood of radius r, where r does not depend on t.

If not, there exist a sequence h_j and a sequence of elements y_j, lying on the boundaries $\Gamma(h_j)$ of the cones $K(h_j)$, such that

$$\lim_{j\to\infty} |x(h_j) - y_j| = 0. \tag{5.58}$$

We may assume without loss of generality that the operators L_{h_j} and the ap-functions $f(t+h_j)$ and $K(t+h_j)$ converge to a regular operator L_* and ap-functions $f_*(t)$ and $K_*(t)$, respectively. It follows from Theorem 4.4 that the ap-functions $x(t+h_j)$ converge uniformly to the function

$$x_*(t) = \int_{-\infty}^{\infty} G_*(t, s) f_*(s)\, ds,$$

where $G_*(t, s)$ is the Green's function of L_*. It then follows from (5.58) that $x_*(0)$ is on the boundary of the cone $K_*(0)$.

On the other hand, since the Green's function $G(t, s)$ is strongly positive, there exists s_* such that $G_*(0, s_*) f_*(s_*)$ is an interior point of $K_*(0)$. Since $G_*(0, s) f_*(s)$ is continuous in the neighborhood of s_*, it follows that $x_*(0)$ is an interior point of the cone $K_*(0)$. This is a contradiction. Q. E. D.

In Chapter 2 we shall indicate conditions for the Green's functions of various classes of regular ap-operators to be positive and strongly positive.

We leave it to the reader to prove the following proposition.

Theorem 5.4. Let $K(t)$ be an ap-function whose values are solid cones, for which one can find a regular ap-operator L whose Green's function $G(t, s)$ is nonegative with respect to $K(t)$.

Then the ap-function $K(t)$ is solid.

5.8. Converses of theorems on positivity of an integral operator

Theorem 5.2 has a natural converse:

Theorem 5.5. Let $K(t)$ be an almost periodic solid cone-valued function. Let the operator (5.48) be positive with respect to the cone $\hat{K}(t)$.

Then the Green's function $G(t, s)$ is nonnegative with respect to $K(t)$, i. e., $G(t, s)$ satisfies condition (5.51).

Proof. Let u_0 be a fixed interior element of the cone $K(s_0)$. We shall construct an ap-function $x(t) \in \hat{K}(t)$ such that

$$x(s_0) = u_0. \tag{5.59}$$

This function can be defined, for example, by a formula analogous to (5.24):

$$x(t) = \left[\int_{\Omega(t)} \rho(x)\, dm(x) \right]^{-1} \int_{\Omega(t)} x\rho(x)\, dm(x), \tag{5.60}$$

where the density $\rho(x)$ is so chosen that (5.59) holds. We now define a sequence of functions $z_n(t)$ as follows: $z_n(t)$ is periodic with period n, and defined on the interval $[s_0-\delta-1/n,\ s_0-\delta-1/n+n]$ by

$$z_n(t)=\begin{cases} 1, & \text{if} \quad |t-s_0|\leqslant\delta, \\ nt-ns_0+n\delta+1, & \text{if} \quad -\delta-1/n\leqslant t-s_0\leqslant-\delta, \\ -nt+ns_0+n\delta+1, & \text{if} \quad \delta\leqslant t-s_0\leqslant\delta+1/n, \\ 0, & \text{if} \quad \delta+1/n\leqslant t-s_0\leqslant n-\delta-1/n, \end{cases}$$

where δ is some positive number. The functions

$$x_n(t)=z_n(t)x(t)$$

are also almost periodic; they are uniformly bounded and converge pointwise to the function

$$y(t)=\begin{cases} x(t), & \text{if} \quad |t-s_0|\leqslant\delta, \\ 0, & \text{if} \quad |t-s_0|>\delta. \end{cases}$$

Therefore,

$$\int_{s_0-\delta}^{s_0+\delta} G(t,s)\,x(s)\,ds=\lim_{n\to\infty}\int_{-\infty}^{\infty} G(t,s)\,x_n(s)\,ds,$$

and, since the operator (5.48) is positive, we have

$$\int_{s_0-\delta}^{s_0+\delta} G(t,s)\,x(s)\,ds\in K(t) \qquad (-\infty<t<\infty;\ \delta>0). \tag{5.61}$$

Since $x(t)$ is continuous, it follows that for any fixed $t\neq s_0$

$$\lim_{\delta\to 0}\frac{1}{2\delta}\int_{s_0-\delta}^{s_0+\delta} G(t,s)\,x(s)\,ds=G(t,s_0)\,x(s_0)=G(t,s_0)\,u_0,$$

and from (5.61) we obtain

$$G(t,s_0)\,u_0\in K(t) \qquad (t\neq s_0).$$

Q. E. D.

The statement of this theorem is of course false if the cone-valued function $K(t)$ is not almost periodic.

The above construction of ap-functions in $\hat{K}(t)$ via equalities of type (5.60) does not carry over to ap-functions with values in infinite-dimensional spaces. A method constructing ap-functions in $\hat{K}(t)$ for this more complicated case has been developed by Pokrovskii /1/.

Chapter 2

ANALYSIS OF SPECIFIC TYPES OF AP-OPERATORS*

§ 6. FIRST-ORDER SYSTEMS OF EQUATIONS

6.1. General positivity conditions for the Green's function

In this section we shall study the ap-operator

$$Lx = \frac{dx}{dt} + A(t)\,x. \tag{6.1}$$

Its Green's function is denoted, as usual, by $G(t, s)$. Recall (see (4.9)) that

$$G(t, s) = \begin{cases} U(t)\,P_+U^{-1}(s) & \text{if} \quad t \geqslant s, \\ -U(t)\,P_-U^{-1}(s), & \text{if} \quad t < s, \end{cases} \tag{6.2}$$

where $U(t)$ is the fundamental matrix of solutions of the homogeneous system

$$\frac{dx}{dt} + A(t)\,x = 0 \tag{6.3}$$

satisfying the condition

$$U(0) = I, \tag{6.4}$$

P_+ and P_- are the projections onto the subspaces E_+ and E_- (of R^n) of initial values of the solutions of system (6.3), that decrease as $t \to \infty$ and as $t \to -\infty$, respectively (see Section 3).

Lemma 6.1. Suppose that the Green's function $G(t, s)$ of the operator (6.1) exists and is nonnegative with respect to a cone-valued function $K(t) \in C(\mathfrak{K})$ such that $K(0)$ is a solid cone.

* The main results of this chapter were published in Burd, Kolesov and Krasnosel'skii /1, 2/.

Then the trivial solution of system (6.3) is exponentially stable in the sense of Lyapunov, i.e., $P_- = 0$.

Proof. Suppose the statement false. Then one can find in the cone $K(0)$ an element $x_0 + y_0$ such that $x_0 \in E_+$, $y_0 \in E_-$ and $y_0 \neq 0$. Since the Green's function is nonnegative,

$$-U(t)P_-K(0) \subset K(t) \qquad (t < 0).$$

Letting $t \to -0$, we see that $-P_-K(0) \subset K(0)$. In particular, $-y_0 = -P_-(x_0 + y_0) \in K(0)$. It follows that $x_0 \in K(0)$ and so $x_0 - y_0 \in K(0)$. But then $y_0 = -P_-(x_0 - y_0) \in K(0)$. Therefore $y_0 = 0$ and we have derived a contradiction. Q. E. D.

Lemma 6.1 remains valid if instead of $K(0)$ some other cone $K(t_0)$ is solid. If none of the cones $K(t)$ is solid, the statement of the lemma fails in the general case.

It follows at once from Lemma 6.1 that the Green's function $G(t, s)$ of the operator (6.1) may be nonpositive with respect to a cone-valued function $K(t)$ (one of whose values is a solid cone) only if $P_+ = 0$.

We are interested in nonnegative Green's functions. It follows from Lemma 6.1 and from (6.2) that they can be written

$$G(t, s) = \begin{cases} U(t)U^{-1}(s), & \text{if } t \geqslant s, \\ 0, & \text{if } t < s. \end{cases} \tag{6.5}$$

The operator

$$U(t, s) = U(t)U^{-1}(s) \qquad (-\infty < t, s < \infty) \tag{6.6}$$

has a simple geometric meaning: a point moving along trajectories of equation (6.3) from time s to t goes from $x_0 \in R^n$ to $U(t, s)x_0$. The operator (6.6) is therefore known as the translation operator along the trajectories of system (6.3) [or the translation operator of the system].

When the system has constant coefficients, i. e., $A(t) \equiv A$, the translation operator is defined by

$$U(t, s) = e^{-(t-s)A} \qquad (-\infty < t, s < \infty).$$

In the general case, the operator $U(t, s)$ (often called the matriciant* of system (6.3)) can be defined by a series

$$U(t, s) = I - \int_s^t A(\tau)\,d\tau + \int_s^t \int_s^\tau A(\tau)A(\tau_1)\,d\tau_1\,d\tau - \ldots$$

* [This continental term seems to have no equivalent in English; see Gantmakher /1/.]

The translation operator $U(t, s)$ is said to be positive with respect to a cone-valued function $K(t)$ if

$$U(t, s)K(s) \subset K(t) \qquad (t > s).$$

Thus Lemma 6.1 implies the following

Theorem 6.1. The ap-operator (6.1) has a Green's function which is nonnegative with respect to a cone-valued function $K(t) \in C(\mathfrak{K})$, *one of whose values* $K(t_0)$ *is a solid cone, if and only if the following two conditions are satisfied:*

1. *The translation operator* $U(t, s)$ *is positive with respect to* $K(t)$.
2. *The translation operator* $U(t, s)$ *satisfies the inequalities*

$$|U(t, s)| \leqslant Me^{-\gamma(t-s)} \qquad (-\infty < s < t < \infty),$$

where $M, \gamma > 0$.

Analogous conditions can be formulated for the Green's function to be nonpositive.

Below we shall investigate the translation operator $U(t, s)$. Note that the basic constructions do not depend on the assumption that the matrix $A(t)$ is almost periodic.

6.2. Gâteaux derivatives

Let $q(x)$ be a sublinear functional (see Section 5.2). Then any function

$$\varphi(t) = q(x + th) \qquad (-\infty < t < \infty), \qquad (6.7)$$

where x and h are fixed elements of the Banach space E, is convex, since

$$\varphi\left(\frac{t_1 + t_2}{2}\right) = q\left[\frac{1}{2}(x + t_1 h) + \frac{1}{2}(x + t_2 h)\right] \leqslant$$
$$\leqslant q\left[\frac{1}{2}(x + t_1 h)\right] + q\left[\frac{1}{2}(x + t_2 h)\right] = \frac{1}{2}[\varphi(t_1) + \varphi(t_2)].$$

Thus the function (6.7) has right and left derivatives at any point. In other words, the limits

$$q'_+(x, h) = \lim_{t \to +0} \frac{q(x + th) - q(x)}{t} \qquad (x, h \in E) \qquad (6.8)$$

and

$$q'_{-}(x, h) = \lim_{t \to -0} \frac{q(x+th) - q(x)}{t} \qquad (x, h \in E) \tag{6.9}$$

exist. These limits will be called the right and left Gâteaux derivatives of the sublinear functional $q(x)$. Obviously, for any fixed x the functional $q'_{+}(x, h)$ $(h \in E)$ is sublinear. It satisfies the estimate

$$-q(-h) \leqslant q'_{+}(x, h) \leqslant q(h) \qquad (x, h \in E); \tag{6.10}$$

for any $\alpha \in (-\infty, \infty)$ we have the identity

$$q'_{+}(x, \alpha x + h) = \alpha q(x) + q'_{+}(x, h) \qquad (x, h \in E). \tag{6.11}$$

The right and left Gâteaux derivatives satisfy the equality

$$q'_{-}(x, h) = -q'_{+}(x, -h) \qquad (x, h \in E). \tag{6.12}$$

Note that since the function (6.7) is convex, we have

$$q'_{-}(x, h) \leqslant q'_{+}(x, h) \qquad (x, h \in E). \tag{6.13}$$

Evaluation of the Gâteaux derivatives is facilitated by the following lemma.

Lemma 6.2. Denote by $J(x)$ the set of linear functionals l such that

$$l(x) = q(x), \quad l(y) \leqslant q(y) \qquad (y \in E). \tag{6.14}$$

Then

$$q'_{+}(x, h) = \max_{l \in J(x)} l(h) \qquad (h \in E). \tag{6.15}$$

Proof. Let $l \in J(x)$. By Lemma 5.1, for $t > 0$

$$l(h) = \frac{l(x+th) - l(x)}{t} = \frac{l(x+th) - q(x)}{t} \leqslant \frac{q(x+th) - q(x)}{t},$$

whence follows the inequality

$$l(h) \leqslant q'_{+}(x, h) \qquad (l \in J(x), h \in E). \tag{6.16}$$

To complete the proof we must construct, for any nonzero $h \in E$, a functional $l_1 \in J(x)$ such that $l_1(h) = q'_+(x, h)$.

We define a functional l_0 on the elements $y = \gamma_1 x + \gamma_2 h$ $(-\infty < \gamma_1, \gamma_2 < \infty)$ by

$$l_0(y) = \gamma_1 q(x) + \gamma_2 q'_+(x, h). \tag{6.17}$$

Formula (6.17) remains valid if the elements x and h are collinear; this follows from (6.11). It is readily seen that

$$l_0(y) \leqslant q'_+(x, y) \qquad (y = \gamma_1 x + \gamma_2 h).$$

By the Hahn-Banach theorem (Dunford and Schwartz /1/) l_0 can be extended to a continuous linear functional l_1, defined throughout E and satisfying the inequality

$$l_1(y) \leqslant q'_+(x, y) \qquad (y \in E).$$

By construction, $l_1(h) = q'_+(x, h)$ and $l_1(x) = q(x)$. Q. E. D.

As proved in Section 5.2 (Lemma 5.1), any sublinear functional $q(x)$ can be represented as

$$q(x) = \max_{l \in F} l(x) \qquad (x \in E), \tag{6.18}$$

where F is the set of linear functionals $l \in E^*$ such that

$$l(x) \leqslant q(x) \qquad (x \in E). \tag{6.19}$$

It follows directly from the definition that F is convex and bicompact in the E-topology of the space E^*. We denote by F_{ex} the set of all extremal points (see Dunford and Schwartz /1/) of the set F. The sets $J(x)$ defined above are also convex and bicompact in the E-topology of E^*. Each of them is an extremal subset of F and, consequently, its extremal points are in F_{ex} (these points exist by virtue of the Krein-Mil'man theorem). It therefore follows from (6.18) that

$$q(x) = \max_{x \in F_{\text{ex}}} l(x) \qquad (x \in E). \tag{6.20}$$

Similarly, by (6.15),

$$q'_+(x, h) = \max_{l \in J_{\text{ex}}(x)} l(h) \qquad (h \in E), \tag{6.21}$$

where $J_{\text{ex}}(x)$ is the set of extremal points of $J(x)$.

Suppose now that the sublinear functional $q(x)$ is given by

$$q(x) = \max_{i=1,\dots,k} q_i(x) \qquad (x \in E), \tag{6.22}$$

where the functionals $q_i(x)$ are sublinear (not necessarily linear!). Denote by $i(x)$ the set of all i such that $q_i(x) = q(x)$. Obviously,

$$q'_+(x, h) = \max_{i \in i(x)} (q_i)'_+(x, h) \qquad (h \in E). \tag{6.23}$$

If the functional $q(x)$ is given by

$$q(x) = \sum_{i=1}^{k} q_i(x) \qquad (x \in E), \tag{6.24}$$

then

$$q'_+(x, h) = \sum_{i=1}^{k} (q_i)'_+(x, h) \qquad (h \in E). \tag{6.25}$$

The formulas cited above enable one to evaluate the right Gâteaux derivative $q'_+(x, h)$. They can also be used to evaluate the left derivative $q'_-(x, h)$, via (6.12).

We illustrate the evaluation of $q'_+(x, h)$ by a few examples.

a) Let $E = R^n$,

$$q(x) = |\xi_1| + \dots + |\xi_n| \qquad (x = \{\xi_1, \dots, \xi_n\} \in R^n). \tag{6.26}$$

Then by (6.25)

$$q'_+(x, h) = s(\xi_1, h_1) + \dots + s(\xi_n, h_n) \quad (h = \{h_1, \dots, h_n\} \in R^n), \tag{6.27}$$

where

$$s(\xi_i, h_i) = \begin{cases} h_i \operatorname{sign} \xi_i, & \text{if} \quad \xi_i \neq 0, \\ |h_i|, & \text{if} \quad \xi_i = 0. \end{cases} \tag{6.28}$$

b) Let $E = R^n$,

$$q(x) = \max_{i=1,\dots,n} |\xi_i| \qquad (x \in R^n). \tag{6.29}$$

Then by (6.23)

$$q'_+(x, h) = \max_{i \in i(x)} s(\xi_i, h_i). \tag{6.30}$$

c) Let E be the space C of all continuous scalar functions on $[0, 1]$,

$$q(x) = \|x\| = \max_{0 \leqslant t \leqslant 1} |x(t)| \qquad (x \in C). \tag{6.31}$$

The functional (6.31) can be written in the form (6.18), where F is the unit ball in C^*. The set F_{ex} of all extremal points consists (see Dunford and Schwartz /1/, p.441) of the functionals l such that $l(x)=x(\tau)$, where $\tau\in[0,1]$. Therefore, by (6.21),

$$q'_+(x,h)=\max\{\max_{\tau\in\Omega_+} h(\tau),\ -\min_{\tau\in\Omega_-} h(\tau)\}\qquad (h\in C), \tag{6.32}$$

where

$$\Omega_+=\{\tau:\ \tau\in[0,1],\ \ x(\tau)=\|x\|\} \tag{6.33}$$

and

$$\Omega_-=\{\tau:\ \tau\in[0,1],\ \ x(\tau)=-\|x\|\}. \tag{6.34}$$

d) The sublinear functional

$$q(x)=-\min_{i=1,\dots,n}\xi_i\qquad (x=\{\xi_1,\dots,\xi_n\}\in R^n) \tag{6.35}$$

selects in R^n the cone of all vectors with nonnegative components. By (6.23),

$$q'_+(x,h)=-\min_{i\in i(x)} h_i\qquad (h=\{h_1,\dots,h_n\}\in R^n), \tag{6.36}$$

where $i(x)$ is the set of all i such that $q(x)=-\xi_i$.

e) As a last example we consider the functional (5.9). To evaluate its Gâteaux derivatives, one uses (5.10) and (6.21).

6.3. Positivity of the translation operator

In order to study the position of curves $x=x(t)$ in the space E, it is convenient to consider scalar functions

$$q(t)=q[x(t)], \tag{6.37}$$

where $q(x)$ is a functional defined on E. The way in which the scalar functions (6.37) depend on t often provides information concerning the behavior of the vector-function $x(t)$. As is well known, this is the basic idea underlying the method of Lyapunov functions in stability theory, methods for investigating the dissipativity and convergence of systems of ordinary differential equations, methods for proving uniqueness and nonlocal continuability theorems, the method of direction functions in the theory of periodic solutions, and so on.

We are interested in the functions (6.37) when $q(x)$ is sublinear.

A function $x(t)$ with values in E is said to be (strongly) *differentiable* at a point t_0 if there exists an element $x'(t_0) \in E$ such that

$$\lim_{\Delta t \to 0} \left\| \frac{x(t_0 + \Delta t) - x(t_0)}{\Delta t} - x'(t_0) \right\| = 0.$$

When $E = R^n$, the function

$$x(t) = \{\xi_1(t), \ldots, \xi_n(t)\}$$

is differentiable if and only if all its components $\xi_i(t)$ are differentiable.

Lemma 6.3. *Let the function $x(t)$ be differentiable at the point t_0. Let the functional $q(x)$ be sublinear.*

Then the scalar function (6.37) has left and right derivatives $q'_+(t_0)$ and $q'_-(t_0)$ at t_0; moreover,

$$q'_+(t_0) = q'_+[x(t_0), x'(t_0)] \tag{6.38}$$

and

$$q'_-(t_0) = q'_-[x(t_0), x'(t_0)]. \tag{6.39}$$

Proof. We shall prove only (6.38). Let

$$x(t_0 + \Delta t) = x(t_0) + x'(t_0)\Delta t + \omega \Delta t,$$

where $\|\omega\| \to 0$ as $\Delta t \to 0$. Since $q(x)$ is sublinear, it follows that for $\Delta t > 0$

$$q[x(t_0) + x'(t_0)\Delta t] - \Delta t\, q(-\omega) \leqslant q[x(t_0) + x'(t_0)\Delta t + \omega \Delta t] \leqslant$$
$$\leqslant q[x(t_0) + x'(t_0)\Delta t] + \Delta t\, q(\omega),$$

whence

$$\frac{q[x(t_0) + x'(t_0)\Delta t] - q[x(t_0)]}{\Delta t} - q(-\omega) \leqslant \frac{q[x(t_0 + \Delta t)] - q[x(t_0)]}{\Delta t} \leqslant$$
$$\leqslant \frac{q[x(t_0) + x'(t_0)\Delta t] - q[x(t_0)]}{\Delta t} + q(\omega).$$

These inequalities imply (6.38).

Q. E. D.

It is sometimes convenient to characterize vector-functions $x(t)$ by scalar functions of the type

$$q(t) = q[t, x(t)], \tag{6.40}$$

where the functional $q(t, x)$ $(-\infty < t < \infty,\ x \in E)$ is sublinear in x for fixed t. If $q(t, x)$ is differentiable with respect to t and the derivative $q'_t(t, x)$ is jointly continuous in all variables, it is obvious that the function (6.40) has right and left derivatives $q'_+(t)$ and $q'_-(t)$, which are given by

$$q'_+(t_0) = b'_t[t_0, x(t_0)] + (q'_x)_+[t_0; x(t_0); x'(t_0)] \tag{6.41}$$

and

$$q'_-(t_0) = q'_t[t_0, x(t_0)] + (q'_x)_-[t_0; x(t_0); x'(t_0)]. \tag{6.42}$$

We now return to our investigation of equation (6.3) in R^n. The translation operator of equation (6.3) will be denoted, as before, by $U(t, s)$.

Let K be a solid cone in R^n. Recall that the translation operator is said to be positive if $U(t, s)K \subset K$ for $t \geqslant s$. The translation operator is said to be s t r o n g l y p o s i t i v e, if $U(t, s)x \gg 0$ for $t > s$, $x \in K$, $x \neq 0$. In our investigation of conditions for positivity and strong positivity of the translation operator along trajectories of the equation

$$\frac{dx}{dt} + A(t)x = 0 \tag{6.43}$$

we shall use conditions of the following types:

$$q'_-[x, -A(t)x] \leqslant 0 \qquad (x \in \Gamma,\ -\infty < t < \infty), \tag{6.44}$$

$$q'_+[x, -A(t)x] \leqslant 0 \qquad (x \in \Gamma,\ -\infty < t < \infty), \tag{6.45}$$

$$q'_-[x, -A(t)x] < 0 \qquad (x \in \Gamma;\ t \in \mathfrak{M};\ x \neq 0), \tag{6.46}$$

where $q(x)$ is a sublinear functional selecting the cone $K \subset R^n$, Γ is the boundary of K, $\mathfrak{M}$ is a set of values of t dense on $(-\infty, \infty)$.

L e m m a 6.4. A necessary and sufficient condition for any one of conditions (6.44) —(6.46) to hold is that it hold for the functional $q_0(x)$ defined by (5.9):

$$q_0(x) = \min_{x + ru_0 \in K} r \qquad (x \in R^n). \tag{6.47}$$

P r o o f. We shall show that (6.45) is equivalent to the condition

$$(q_0)'_+[x, -A(t)x] \leqslant 0 \qquad (x \in \Gamma,\ -\infty < t < \infty). \tag{6.48}$$

It follows from (5.12) and (5.13) that for any $x, h \in R^n$ $(-\infty < t < \infty)$

$$-q[q_0(x+th)u_0] \leqslant q(x+th) \leqslant q[-q_0(x+th)u_0].$$

Let $x \in \Gamma$; then for any $h \in R^n$ and $t > 0$

$$-q\left[\frac{q_0(x+th)-q_0(x)}{t}u_0\right] \leqslant \frac{q(x+th)-q(x)}{t} \leqslant$$
$$\leqslant q\left[-\frac{q_0(x+th)-q_0(x)}{t}u_0\right].$$

Letting $t \to 0$, we get the inequalities

$$-q[(q_0)'_+(x, h)u_0] \leqslant q'_+(x, h) \leqslant q[-(q_0)'_+(x, h)u_0].$$

In particular,

$$-q\{(q_0)'_+[x, -A(t)x]u_0\} \leqslant q'_+[x, -A(t)x] \leqslant$$
$$\leqslant q\{-(q_0)'_+[x, -A(t)x]u_0\}. \qquad (6.49)$$

Suppose inequality (6.48) is not valid for some t^* and $x^* \in \Gamma$. Then $(q_0)'_+[x^*, -A(t^*)x^*]u_0$ is an interior point of the cone K, and hence it follows from the left inequality of (6.48) that $q'_+[x^*, -A(t^*)x^*] > 0$. Therefore (6.45) implies (6.48).

If (6.48) holds, then (6.45) follows directly from the right inequality of (6.49).

Similarly one proves that (6.44) and (6.46) are equivalent respectively to

$$(q_0)'_-[x, -A(t)x] \leqslant 0 \qquad (x \in \Gamma;\ -\infty < t < \infty) \qquad (6.50)$$

and

$$(q_0)'_-[x, -A(t)x] < 0 \qquad (x \in \Gamma;\ x \neq 0;\ t \in \mathfrak{M}). \qquad (6.51)$$

Q. E. D.

Theorem 6.2. Let $q(x)$ be a sublinear functional selecting a solid cone K with boundary Γ.

Then (6.45) is a necessary condition and (6.44) a sufficient condition for the translation operator $U(t, s)$ to be positive. Condition (6.46) is sufficient for the operator $U(t, s)$ to be strongly positive.

Proof. The first assertion is trivial: if $U(t, s)$ is positive, then $q[U(t, s)x] \leqslant 0$ for $t \geqslant s$ and $x \in \Gamma$; consequently,

$$q'_+[x, -A(t)x] = \lim_{\Delta t \to +0} \frac{q[U(t+\Delta t, t)x] - q(x)}{\Delta t} =$$

$$= \lim_{\Delta t \to +0} \frac{q[U(t+\Delta t, t)x]}{\Delta t} \leqslant 0.$$

Now for the second assertion. Let (6.44) hold. Then, by Lemma 6.3, we have (6.50).

Consider the operator

$$Px = \begin{cases} x, & \text{if} \quad x \in K, \\ x + q_0(x) u_0, & \text{if} \quad x \,\overline{\in}\, K. \end{cases}$$

Obviously, P "projects" all elements $x \,\overline{\in}\, K$ onto the boundary Γ of K. A simple computation shows that

$$(q_0)'_-(x, h) = (q_0)'_-(Px, h) \qquad (x, h \in R^n). \tag{6.52}$$

Since the functional $q_0(x)$ satisfies a Lipschitz condition, the equation

$$\frac{dy}{dt} + A(t) Py = 0 \tag{6.53}$$

has a unique solution $y(t; s, y_0)$ for any initial value $y(s) = y_0$, defined for all $t \in (-\infty, \infty)$. Obviously, the second assertion will be proved, if we show that $y(t; s, y_0) \in K$ for $t \geqslant s$, $y_0 \in K$. Suppose this false. Then there exist t_1 and t_2 and an element $y_0 \in \Gamma$ such that $t_1 < t_2$ and $y(t; t_1, y_0) \,\overline{\in}\, K$ for $t_1 < t \leqslant t_2$. Consider the scalar function

$$\varphi(t) = q_0[y(t; t_1, y_0)] \qquad (t_1 \leqslant t \leqslant t_2).$$

By Lemma 6.3, the left derivative $\varphi'_-(t)$ is given by

$$\varphi'_-(t) = (q_0)'_-[y(t; t_1, y_0), y'_t(t; t_1, y_0)]$$

and consequently

$$\varphi'_-(t) = (q_0)'_-[y(t; t_1, y_0), \quad -A(t) Py(t; t_1, y_0)],$$

whence, by (6.52),

$$\varphi'_-(t) = (q_0)'_-[Py(t; t_1, y_0), \quad -A(t) Py(t; t_1, y_0)].$$

The elements $Py(t; t_1, y_0)$ belong to Γ. Hence (6.50) implies $\varphi'_-(t) \leqslant 0$ $(t_1 \leqslant t \leqslant t_2)$.

Since $\varphi(t_1)=0$, it follows from the last inequality that $\varphi(t)\leqslant 0$ for $t\in[t_1, t_2]$. Hence $y(t; t_1, y_0)\in K$ for $t\in[t_1, t_2]$, and this is a contradiction.

Before proving the third assertion of the theorem, we cite a general proposition, whose proof is left to the reader. If the translation operator is positive, then all points $U(t, s)x$ are interior points of cone K, provided $x\gg 0$ and $t\geqslant s$.

Let (6.46) hold. Then (6.45) also holds. Hence the translation operator is positive. It remains to prove that $U(t, s)x\gg 0$ for $x\in K$, $x\neq 0$ and $t>s$. If not, then $U(\tau, s)x\in\Gamma$ for $s\leqslant\tau\leqslant t$, and so

$$\varphi(\tau)=q[U(\tau, s)x]\equiv 0 \qquad (s\leqslant\tau\leqslant t),$$

which implies the identity

$$\varphi'_{-}(\tau)=q'_{-}[U(\tau, s)x, \ -A(\tau)U(\tau, s)x]\equiv 0 \qquad (s\leqslant\tau\leqslant t),$$

contradicting (6.46). Q. E. D.

Theorem 6.2 has a natural generalization to the case of a translation operator $U(t, s)$ which is positive or strongly positive not with respect to a constant cone K but with respect to a variable cone $K(t)$. Let the function $K(t)$ be selected by a functional depending on t (and smooth with respect to t) (see Section 5.5, Lemma 5.5). Then the assertions of Theorem 5.2 remain valid provided conditions (6.44) – (6.46) are replaced respectively by

$$q'_t(t, x)+(q'_x)_{-}[t, x, -A(t)x]\leqslant 0 \quad (x\in\Gamma(t); -\infty<t<\infty), \tag{6.54}$$

$$q'_t(t, x)+(q'_x)_{+}[t, x, -A(t)x]\leqslant 0 \quad (x\in\Gamma(t); -\infty<t<\infty) \tag{6.55}$$

and

$$q'_t(t, x)+(q'_x)_{-}[t, x, -A(t)x]<0 \quad (x\in\Gamma(t); x\neq 0; t\in\mathfrak{M}); \tag{6.56}$$

$\Gamma(t)$ denotes the boundary of $K(t)$.

6.4. Positivity of the translation operator with respect to a faceted cone

A cone $K\subset R^n$ is said to be *faceted* if it is the intersection of a finite number of half-spaces $l_i(x)=(x, y_i)\geqslant 0$ $(i=1, 2, \dots, k)$. A faceted cone can be selected by a sublinear functional

$$q(x)=-\min_{i=1,\dots,k}(x, y_i) \qquad (x\in R^n). \tag{6.57}$$

Obviously, we can select n linearly independent vectors from among $y_1, \ldots, y_h$. Henceforth we assume all faceted cones to be solid.

Consider the equation

$$\frac{dx}{dt} + Ax = 0, \tag{6.58}$$

where A is a constant matrix

$$A = \begin{Vmatrix} a_{11} & a_{12} & \ldots & a_{1n} \\ a_{21} & a_{22} & \ldots & a_{2n} \\ \cdot & \cdot & \cdot & \cdot \\ a_{n1} & a_{n2} & \ldots & a_{nn} \end{Vmatrix}. \tag{6.59}$$

It is useful to bear in mind that in this case (constant coefficients) conditions (6.46) are not only sufficient but also necessary for the translation operator of system (6.43) to be strongly positive (with respect to a faceted cone K). We shall not dwell on the proof.

We cite one more form of the conditions for the translation operator of equation (6.58) to be strongly positive: it is necessary and sufficient that the operator be positive and that the matrix (6.59) have in K an eigenvector, unique up to normalization, which is an interior element of K. Obviously, the last assertion is false if K is not a faceted cone.

We now consider in greater detail the case in which K is minihedral, i. e. K is a faceted cone which is the intersection of exactly n half-spaces. We may assume without loss of generality that K is the cone of vectors $x = \{\xi_1 \ldots, \xi_n\}$ with nonnegative components. It is easy to see that the translation operator of equation (6.58) is positive if and only if all the off-diagonal elements a_{ij} $(i \neq j)$ of the matrix (6.9) are nonpositive:

$$a_{ij} \leqslant 0 \qquad (i \neq j;\ i, j = 1, \ldots, n). \tag{6.60}$$

The sign of the diagonal elements a_{ii} has no significance. The condition for the translation operator to be strongly positive can be formulated conveniently with the help of an additional concept.

A sequence

$$a_{i_1 i_2},\ a_{i_2 i_3},\ \ldots,\ a_{i_{m-1} i_m},\ a_{i_m i_1} \tag{6.61}$$

is called a n o n d e g e n e r a c y p a t h of the matrix (6.59), if none of its terms vanish,

$$i_1 \neq i_2,\ i_2 \neq i_3,\ \ldots,\ i_{m-1} \neq i_m,\ i_m \neq i_1 \tag{6.62}$$

and lastly the sequence $i_1, i_2, \ldots, i_m$ includes all the integers $1, \ldots, n$ (some of which may appear several times). We then have the following proposition:

The translation operator is strongly positive with respect to the cone of vectors with nonnegative components if and only if (6.60) is valid and the matrix (6.59) has at least one nondegeneracy path.

For a variable matrix

$$A(t) = \begin{Vmatrix} a_{11}(t) & a_{12}(t) & \ldots & a_{1n}(t) \\ a_{21}(t) & a_{22}(t) & \ldots & a_{2n}(t) \\ \cdot & \cdot & \cdot & \cdot \\ a_{n1}(t) & a_{n2}(t) & \ldots & a_{nn}(t) \end{Vmatrix} \tag{6.63}$$

these conditions are only sufficient for the translation operator to be strongly positive. Conditions (6.60) remain unchanged: they must hold for all $t \in (-\infty, \infty)$; the assumption that there exists a nondegeneracy path means that the matrix (6.63) has nondegeneracy paths for all t (though they may be distinct for distinct t).

6.5. On construction of cone-valued functions $K(t)$

We have been discussing conditions for the translation operator to be positive or strongly positive with respect to a prescribed cone-valued function $K(t)$. These conditions determine classes of systems of differential equations which can be investigated by use of the cone-valued function $K(t)$.

A more important problem is the description and construction of a cone-valued function $K(t)$ with respect to which the translation operator of a prescribed system is positive or strongly positive. This function $K(t)$ must be such that the corresponding cone $\widehat{K}(t)$ in $B(R^n)$ is solid. Therefore, when constructing a cone-valued function $K(t)$ with respect to which the translation operator is positive or strongly positive, our aim should be to make the function periodic or almost periodic.

Suppose we are studying an equation

$$\frac{dx}{dt} + A(t)x = 0 \tag{6.64}$$

in R^n. Let K be an arbitrary cone in R^n and set

$$K(t) = U(t, 0)K \qquad (-\infty < t < \infty), \tag{6.65}$$

where $U(t, s)$ is the translation operator of equation (6.64). A simple check shows that

$$U(t, s) K(s) = K(t) \qquad (-\infty < t, s < \infty).$$

This means that $U(t, s)$ is positive with respect to the cone-valued function (6.65). Unfortunately, the function (6.65) is in general not necessarily almost periodic, even if $A(t)$ is independent of t. However, in the latter case, we can make $K(t)$ almost periodic by a suitable choice of K. The construction is not difficult and we leave it to the reader. Later, we shall construct cone-valued functions for some important special cases.

To end this subsection, we mention a special problem. Consider the equation

$$\frac{dy}{dt} + By = 0 \tag{6.66}$$

in R^n, with constant coefficients. Let K be some given solid cone in R^n, with respect to which the operator e^{-Bt} $(t>0)$ is strongly positive. Then (see Krasnosel'skii /1/) the matrix B has a simple eigenvalue λ_0 such that

$$\lambda_0 < \operatorname{Re}\lambda, \tag{6.67}$$

where λ is any other eigenvalue of B. It turns out that the converse is also true:

The operator e^{-Bt} $(t>0)$ *is strongly positive with respect to a solid cone* K *if inequality (6.67) holds.*

To prove this, we apply a linear substitution transforming (6.66) to the form

$$\frac{dy_1}{dt} + \lambda_0 y_1 = 0 \qquad (y_1 \text{ a scalar}),$$

$$\frac{dy_2}{dt} + Cy_2 = 0 \qquad (y_2 \in R^{n-1}),$$

where C is a matrix all of whose eigenvalues satisfy (6.67). Thus, the eigenvalues of the matrix $D = \lambda_0 I - C$ are in the left half-plane and hence (see, e. g., Lyapunov /1/, Malkin /1/) there exists a positive-definite matrix H such that $HD + D^*H$ is negative-definite. Now set

$$K = \{y\colon\ y = \{y_1, y_2\} \in R^n,\ -y_1 + \sqrt{(Hy_2, y_2)} \leqslant 0\}.$$

Our assertion now follows from the choice of H and Theorem 6.2.

Assume now that system (6.64) is reducible. Then there exists a continuously differentiable nonsingular ap-matrix $W(t)$, whose inverse is almost periodic, such that the matrix

$$B = W^{-1}(t)\,A(t)\,W(t) + W^{-1}(t)\,W'(t) \tag{6.68}$$

does not depend on t. The substitution

$$x(t) = W(t)\,y(t) \tag{6.69}$$

transforms equation (6.64) to equation (6.66). Obviously, the translation operator of equation (6.64) can be written

$$U(t, s) = W(t)\,e^{-B(t-s)}W^{-1}(s) \qquad (-\infty < t, s < \infty). \tag{6.70}$$

Thus we get the following proposition:

The operator $U(t, s)$ *is positive (strongly positive) with respect to the almost periodic cone-valued function*

$$K(t) = W(t)\,K, \tag{6.71}$$

if the translation operator $e^{-B(t-s)}$ *of equation (6.66) is positive (strongly positive) with respect to the constant cone* K.

6.6. On the stability of solutions of the homogeneous equation

We continue our investigation of the equation

$$Lx \equiv \frac{dx}{dt} + A(t)\,x = 0, \tag{6.72}$$

where $A(t)$ is a matrix with ap-elements. We shall assume that the translation operator $U(t, s)$ of equation (6.72) is positive with respect to some cone-valued function $K(t)$.

Lemma 6.5. Let $y_0(t)$ *be a vector-function, continuously differentiable on the whole real line, such that*

$$Ly_0(t) \geqslant 0 \qquad (-\infty < t < \infty). \tag{6.73}$$

Then

$$U(t, s)\,x \leqslant y_0(t) \qquad (t \geqslant s), \tag{6.74}$$

whenever $x \leqslant y_0(s)$.

Proof. Set

$$f(t) = Ly_0(t), \quad h(t) = y_0(t) - U(t, s)x. \tag{6.75}$$

The function $h(t)$ is the solution of the equation $Lh = f(t)$ with initial value $h(s) = y_0(s) - x \geqslant 0$. Obviously,

$$h(t) \equiv U(t, s)h(s) + \int_s^t U(t, \tau) f(\tau)\, d\tau. \tag{6.76}$$

Hence follows (6.74). Q. E. D.

Assume that the cone-valued function $K(t)$ is almost periodic and solid. Then (see Section 5.7) the closure $\mathfrak{R}$ of the range of $K(t)$ consists of solid cones. Moreover (see p. 42) there exists a constant N such that, for any cone $K \in \mathfrak{R}$, if $x, y - x \in K$, then

$$|x| \leqslant N|y|. \tag{6.77}$$

We write $x_0(t) \gg 0$, if, for every t, the vector $x_0(t)$ is in $K(t)$ together with a spherical neighborhood of radius r_0 independent of t.

Theorem 6.3. Suppose that the function $x_0(t)$ is bounded and has a derivative, bounded and continuous on the whole real line. Let

$$x_0(t) \gg 0, \quad Lx_0(t) \gg 0. \tag{6.78}$$

Then the trivial solution of equation (6.72) is exponentially stable in the sense of Lyapunov, i. e.

$$|U(t, s)| \leqslant Me^{-\gamma(t-s)} \qquad (t \geqslant s), \tag{6.79}$$

where $M, \gamma > 0$.

Proof. We introduce a family of norms in R^n (see (5.41)):

$$|x|_t = \inf\{\alpha\colon -\alpha x_0(t) \leqslant x \leqslant \alpha x_0(t)\} \quad (-\infty < t < \infty;\ x \in R^n). \tag{6.80}$$

Here and below in the proof, the sign $\leqslant$ in any inequality involving functions of t is defined by the semi-order generated by a suitable cone $K(t)$.

Obviously,

$$-|x|_t x_0(t) \leqslant x \leqslant |x|_t x_0(t) \qquad (-\infty < t < \infty),$$

so that

$$0 \leqslant x + |x|_t x_0(t) \leqslant 2|x|_t x_0(t) \qquad (-\infty < t < \infty),$$

and by (6.77)

$$|x| \leqslant (1+2N) \sup_\tau |x_0(\tau)| \cdot |x|_t \qquad (-\infty < t < \infty).$$

On the other hand, since $x_0(t)$ is contained in $K(t)$ together with a ball of radius r_0, the elements

$$x_0(t) + r_0 \frac{x}{|x|}, \qquad x_0(t) - r_0 \frac{x}{|x|}$$

are in $K(t)$, i. e.,

$$-\frac{|x|}{r_0} x_0(t) \leqslant x \leqslant \frac{|x|}{r_0} x_0(t) \qquad (-\infty < t < \infty),$$

whence it follows that

$$|x| \geqslant r_0 |x|_t \qquad (-\infty < t < \infty).$$

The inequalities just proved imply

$$m_1|x| \leqslant |x|_t \leqslant m_2|x| \qquad (x \in R^n, \ -\infty < t < \infty), \tag{6.81}$$

where m_1 and m_2 are positive constants.

The second condition in (6.78) implies the existence of a positive γ such that

$$Lx_0(t) \gg \gamma x_0(t). \tag{6.82}$$

Set

$$y_0(t) = e^{-\gamma(t-s)} x_0(t), \tag{6.83}$$

where $-\infty < t < \infty$ and s is a fixed number.
Obviously,

$$Ly_0(t) = e^{-\gamma(t-s)}[Lx_0(t) - \gamma x_0(t)] \geqslant 0. \tag{6.84}$$

Let $T(s)$ denote the ball $|x|_s \leqslant 1$, i. e., the set of $x \in R^n$ such that

$$-x_0(s) \leqslant x \leqslant x_0(s).$$

Since $y_0(s) = x_0(s)$, it follows from Lemma 6.5 that

$$-y_0(t) \leqslant U(t, s)x \leqslant y_0(t) \qquad (t \geqslant s,\ x \in T(s))$$

or, what is the same,

$$-e^{-\gamma(t-s)}x_0(t) \leqslant U(t, s)x \leqslant e^{-\gamma(t-s)}x_0(t) \quad (t \geqslant s,\ x \in T(s)).$$

From these inequalities it follows that

$$|U(t, s)x|_t \leqslant e^{-\gamma(t-s)} \qquad (t \geqslant s, \quad x \in T(s))$$

and by (6.81)

$$|U(t, s)x| \leqslant \frac{1}{m_1} e^{-\gamma(t-s)} \qquad (t \geqslant s,\ x \in T(s)),$$

which implies (6.79) with $M = m_1^{-1} m_2$. Q. E. D.

Theorem 6.3 has a converse, in the sense that if the Green's function of the ap-operator L is nonnegative there exists an ap-function $x_0(t)$ satisfying (6.78). This function $x_0(t)$ may be defined as an ap-solution of the equation $Lx = f(t)$, where $f(t) \in B(R^n)$ and $f(t) \gg 0$.

Theorem 6.3 leads on the one hand to various sufficient conditions for exponential stability, and on the other to effective conditions for the existence of a positive Green's function. Several specific applications of this theorem will be given below.

§ 7. SECOND-ORDER SYSTEMS

7.1. Equations with constant coefficients

Consider the ap-operator

$$Lx = \frac{d^2x}{dt^2} + Ax, \tag{7.1}$$

where A is a constant matrix.

The ap-operator (7.1) has a Green's function if and only if the matrix A has no nonnegative real eigenvalues.

Indeed, by Theorem 2.4, the Green's function $G(t, s)$ of the operator (7.1) exists if and only if the ap-operator

$$\tilde{L}u = \frac{du}{dt} + Qu, \tag{7.2}$$

where

$$Q = \left\| \begin{matrix} 0 & -I \\ A & 0 \end{matrix} \right\|, \tag{7.3}$$

is regular. In turn, the operator (7.2) is regular if and only if the matrix (7.3) has no pure imaginary eigenvalues, i. e., the matrix A has no nonnegative eigenvalues.

If the matrix A has no nonnegative eigenvalues, then (see Gantmakher /1/, Chap. 8) it can be expressed uniquely in the form

$$A = -B^2, \tag{7.4}$$

where B is a real matrix whose eigenvalues lie in the left half-plane. We leave it to the reader to prove that then the Green's function of the ap-operator (7.1) can be written

$$G(t, s) = \begin{cases} \frac{1}{2} B^{-1} e^{B(t-s)}, & \text{if} \quad t \geqslant s, \\ \frac{1}{2} B^{-1} e^{-B(t-s)}, & \text{if} \quad t \leqslant s. \end{cases} \tag{7.5}$$

The Green's function (7.5) cannot be nonnegative with respect to any constant cone $K \subset R^n$. Indeed, otherwise, for any $f_0 \in K$, the equation $Lx = f_0$ would have a solution $A^{-1}f_0$ in K, i. e., $A^{-1}K \subset K$ and, by well-known theorems on eigenvectors of positive linear operators (see Krein and Rutman /1/, Krasnosel'skii /1/), the matrix A would then have a positive eigenvalue, contrary to assumption.

Formula (7.5) implies that the Green's function $G(t, s)$ is nonpositive with respect to a constant cone K if and only if

$$e^{B\tau} K_0 \subset K \qquad (\tau \geqslant 0), \tag{7.6}$$

where $K_0 = -B^{-1}K$. Verification of (7.6) is a difficult task; even effective construction of a matrix B, given the matrix A, involves considerable difficulties. We shall therefore be interested in conditions for the Green's function to be negative, expressed in terms of the matrix A itself. In this respect only partial results have been obtained. The situation is even more complicated for ap-operators with variable coefficients.

7.2. The majorant test

In the sequel we shall need an elementary general proposition, which has been used in various forms by many authors.

Let K be a normal solid cone in a Banach space E. An additive and homogeneous operator which is positive with respect to this cone is also continuous.*

Lemma 7.1. Let D_0 be an additive and homogeneous operator acting in E, with domain of definition dense in E, which has an inverse D_0^{-1}, positive with respect to the cone K. Let B be a continuous linear operator such that $D_0^{-1}B$ is positive (this is the case for example, if B is positive). Let

$$D_0x_0 - Bx_0 \gg 0, \tag{7.7}$$

where x_0 is an interior element of K.

Then the operator

$$D_1 = D_0 - B \tag{7.8}$$

has a continuous positive inverse D_1^{-1}, defined on E, such that

$$D_1^{-1}f \geqslant D_0^{-1}f \qquad (f \in K). \tag{7.9}$$

Proof. By (7.7)

$$D_0^{-1}Bx_0 \ll x_0.$$

Therefore, it follows from Theorem 5.1 that the spectral radius $r(D_0^{-1}B)$ of the operator $D_0^{-1}B$ is less than 1. Hence the operator $(I - D_0^{-1}B)^{-1}$ exists and

$$(I - D_0^{-1}B)^{-1} = I + D_0^{-1}B + (D_0^{-1}B)^2 + \ldots \tag{7.10}$$

It follows from (7.10) that

$$(I - D_0^{-1}B)^{-1} f \geqslant f \qquad (f \in K). \tag{7.11}$$

Since $D_1 = D_0(I - D_0^{-1}B)$, D_1 is continuously invertible; at the same time, (7.11) implies the estimates (7.9). Q. E. D.

Similar lemmas have been proved by Collatz /1/.

It follows from (7.9) that, under the assumptions of Lemma 7.1, the operator D_1^{-1} is strongly positive if D_0^{-1} is strongly positive. Lemma 7.1 remains valid if (7.7) is replaced by the inequality $r(D_0^{-1}B) < 1$.

* Bakhtin, Krasnosel'skii, Stetsenko /1/.

Consider the ap-operators

$$L_0 x = \frac{d^2x}{dt^2} + A_0(t)x \qquad (7.12)$$

and

$$L_1 x = \frac{d^2x}{dt^2} + A_1(t)x. \qquad (7.13)$$

Let $K(t)$ be a solid almost periodic cone-valued function, and $\hat{K}(t)$ the solid cone (see Section 5.4) in $B(R^n)$ defined by $K(t)$.

Theorem 7.1. *Suppose that the Green's function $G_0(t, s)$ of the ap-operator (7.12) is nonpositive with respect to $K(t)$. Let*

$$A_1(t)x \geqslant A_0(t)x \qquad (x \in K(t), \ -\infty < t < \infty). \qquad (7.14)$$

Suppose that there exists an ap-function $x_0(t) \gg 0$, $x_0(t) \in B^2(R^n)$, such that

$$L_1 x_0(t) \ll 0. \qquad (7.15)$$

Then the ap-operator (7.13) is regular, its Green's function $G_1(t, s)$ is also nonpositive with respect to $K(t)$ and moreover

$$G_0(t, s)x - G_1(t, s)x \in K(t) \qquad (x \in K(s)). \qquad (7.16)$$

Proof. Set

$$D_0 x = -L_0 x, \quad D_1 x = -L_1 x \qquad (x \in B^2(R^n)).$$

The operators D_0 and D_1 satisfy the assumptions of Lemma 7.1. Therefore the ap-operator L_1 is regular. Inequalities (7.9) can then be written

$$\int_{-\infty}^{\infty} G_0(t, s)f(s)\,ds \geqslant \int_{-\infty}^{\infty} G_1(t, s)f(s)\,ds \qquad (f(t) \in \hat{K}(t))$$

or, equivalently,

$$\int_{-\infty}^{\infty} [G_0(t, s) - G_1(t, s)]f(s)\,ds \in K(t) \qquad (f(t) \in \hat{K}(t)). \qquad (7.17)$$

To derive (7.16) from (7.17), it is sufficient to repeat the reasoning of the proof of Theorem 5.5. Q. E. D.

Theorem 7.2. Under the assumptions of Theorem 7.1, let the Green's function $G_0(t, s)$ be strongly negative.

Then the Green's function $G_1(t, s)$ of the ap-operator (7.13) is strongly negative.

This theorem follows from (7.16).

In the analysis of specific operators L_1, the role of L_0 is conveniently assigned to ap-operators with constant coefficients.

7.3. Use of first-order systems

In this subsection we shall prove fairly general tests for positivity and strong positivity of the integral operator

$$Tf(t) = -\int_{-\infty}^{\infty} G(t, s) f(s)\, ds, \qquad (7.18)$$

where $G(t, s)$ is the Green's function of the operator

$$Lx = \frac{d^2x}{dt^2} + A(t)\, x. \qquad (7.19)$$

To simplify the exposition, we shall confine ourselves to a constant cone $\hat{K} \subset B(R^n)$ defined by a constant solid cone $K \subset R^n$. In investigation of the ap-operator (7.19) we shall need the translation operator $U(t, s)$ of the first-order linear system

$$\frac{dx}{dt} - A(t)\, x = 0 \qquad (7.20)$$

Theorem 7.3. Suppose there exists an ap-function $x_0(t) \in B^2(R^n)$ such that

$$x_0(t) \gg 0, \qquad Lx_0(t) \ll 0. \qquad (7.21)$$

Let the operator $U(t, s)$ be positive:

$$U(t, s) K \subset K \qquad (t \geqslant s). \qquad (7.22)$$

Then the operator (7.19) is regular and its Green's function $G(t, s)$ is nonpositive with respect to the cone K.

Before proceeding to the proof, we prove several lemmas.

Lemma 7.2. For any solid cone K in R^n, there exists a matrix B such that

$$Bx \gg 0 \qquad (x \in K,\ x \neq 0). \qquad (7.23)$$

A matrix B satisfying (7.23) can be constructed, for example, by setting

$$Bx = l_0(x)\, u_0 \qquad (x \in R^n),$$

where $u_0 \gg 0$ and $l_0(x)$ is a linear functional on R^n such that

$$l_0(x) \geqslant m\,|x| \qquad (x \in K;\ m > 0).$$

The existence of such functionals l_0 is geometrically evident (for a rigorous proof, see Krasnosel'skii /1/).

Lemma 7.3. Let (7.22) hold. Let B be a matrix satisfying condition (7.23).

Then for any $\varepsilon > 0$ there exists $a = a(\varepsilon) > 0$ such that

$$[A(t) + a^2 I + \varepsilon B]\,x \gg 0 \qquad (x \in K,\ x \neq 0,\ -\infty < t < \infty). \quad (7.24)$$

Proof. First let x be a fixed normalized element of the boundary Γ of K. We shall prove that there exists $a_0 = a_0(x) > 0$ such that for $a \geqslant a_0$

$$q\left[A(t)\,x + a^2 x + \frac{\varepsilon}{2}\,Bx\right] < 0 \qquad (-\infty < t < \infty), \quad (7.25)$$

where $q(x)$ is a sublinear functional selecting the cone K. If no such element exists, we can find a sequence t_k such that

$$q\left[A(t_k)\,x + k^2 x + \frac{\varepsilon}{2}\,Bx\right] \geqslant 0 \qquad (k = 1,\ 2,\ \ldots). \quad (7.26)$$

As usual, we may assume that the sequence of matrices $A(t_k)$ converges to a matrix A_0. Since $x \in \Gamma$, it follows that $q(x) = 0$ and, by (7.22) (via Theorem 6.2),

$$q'_+[x,\ A(t_k)\,x] \leqslant 0 \qquad (k = 1,\ 2,\ \ldots),$$

so that

$$q'_+(x,\ A_0 x) \leqslant 0.$$

Hence, by (6.10),

$$q'_+\left(x,\ A_0 x + \frac{\varepsilon}{2}\,Bx\right) \leqslant q'_+(x,\ A_0 x) + \frac{\varepsilon}{2}\, q'_+(x,\ Bx) \leqslant \frac{\varepsilon}{2}\, q(Bx),$$

and therefore,

$$q'_+\left(x,\ A_0 x + \frac{\varepsilon}{2}\,Bx\right) < 0. \quad (7.27)$$

By definition

$$q'_+\left(x,\ A_0x+\frac{\varepsilon}{2}Bx\right)=\lim_{\tau\to+0}\frac{q\left(x+\tau A_0x+\frac{\tau\varepsilon}{2}Bx\right)-q(x)}{\tau}=$$
$$=\lim_{\tau\to+0}q\left(A_0x+\frac{1}{\tau}x+\frac{\varepsilon}{2}Bx\right),$$

and (7.27) implies the existence of $k_0>0$ such that

$$q\left(A_0x+k_0^2x+\frac{\varepsilon}{2}Bx\right)<0.$$

Obviously, for $k\geqslant k_0$

$$q\left[A(t_k)x+k^2x+\frac{\varepsilon}{2}Bx\right]\leqslant q\left[A(t_k)x+k_0^2x+\frac{\varepsilon}{2}Bx\right],$$

and therefore

$$\overline{\lim_{k\to\infty}}\,q\left[A(t_k)x+k^2x+\frac{\varepsilon}{2}Bx\right]\leqslant$$
$$\leqslant\lim_{k\to\infty}q\left[A(t_k)x+k_0^2x+\frac{\varepsilon}{2}Bx\right]=q\left(A_0x+k_0^2x+\frac{\varepsilon}{2}Bx\right)<0,$$

contradicting (7.26).

Again, let x be a fixed normalized element of the boundary Γ of K. Then, for any $y\in K$ and $a\geqslant a_0(x)$,

$$q[A(t)y+a^2y+\varepsilon By]\leqslant q[A(t)y+a_0^2(x)y+\varepsilon By]\leqslant$$
$$\leqslant q\left[A(t)x+a_0^2(x)x+\frac{\varepsilon}{2}Bx\right]+\frac{\varepsilon}{2}q(Bx)+$$
$$+q[A(t)(y-x)+a_0^2(x)(y-x)+\varepsilon B(y-x)],$$

and it follows from (7.25) that

$$q[A(t)y+a^2y+\varepsilon By]\leqslant\frac{\varepsilon}{2}q(Bx)+b\,|y-x|\quad(-\infty<t<\infty),\tag{7.28}$$

where

$$b=[\sup_{-\infty<t<\infty}|A(t)|+a_0^2(x)+\varepsilon|B|]\max_{|z|\leqslant1}|q(z)|.$$

From (7.28) we get the inequality

$$q[A(t)y+a^2y+\varepsilon By]<0\tag{7.29}$$

for $a\geqslant a_0(x)$, $y\in K$ and

$$|y-x|\leqslant r(x)=\frac{\varepsilon}{4b}|q(Bx)|\tag{7.30}$$

(recall that $q(Bx) < 0$). Inequality (7.29) means that

$$A(t)y + a^2 y + \varepsilon By \gg 0 \quad (y \in K, |y - x| \leqslant r(x), \ a \geqslant a_0(x)). \quad (7.30)$$

Let Γ_0 be the intersection of Γ and the unit sphere $|x| = 1$. Select points $x_1, \ldots, x_s \in \Gamma_0$ such that the balls $|x - x_i| < r(x_i)$ cover Γ_0. Then it follows from (7.31) that

$$A(t)y + a^2 y + \varepsilon By \gg 0 \quad (y \in \Gamma_0, \ -\infty < t < \infty)$$

provided

$$a \geqslant \max_i a_0(x_i).$$

Consequently, for such values of a the operators $A(t) + a^2 I + \varepsilon B$ are strongly positive on the entire cone K. This implies the statement of Lemma 7.3.

Lemma 7.4. Under the assumptions of Theorem 7.3, suppose there exists a matrix B such that for sufficiently small positive ε the ap-operators

$$L_\varepsilon x = \frac{d^2 x}{dt^2} + A(t)x + \varepsilon Bx \quad (7.31)$$

are regular.

Then the ap-operator (7.19) is regular.

By Theorem 2.3, it is sufficient to show that the equation

$$\frac{d^2 x}{dt^2} + A(t)x = f(t) \quad (7.32)$$

with any function $f(t) \in B(R^n)$ has at least one solution in $C(R^n)$.

Let $\varepsilon_k \to 0$. Denote by $x_k(t)$ the ap-solutions of the equations

$$L_{\varepsilon_k} x = f(t) \qquad (k = 1, 2, \ldots).$$

Then

$$\frac{d^2 x_k(t)}{dt^2} + A(t)x_k(t) + \varepsilon_k B x_k(t) = f(t) \quad (k = 1, 2, \ldots). \quad (7.33)$$

We claim that the ap-functions $x_k(t)$ are uniformly bounded. If this is not so, there exists a sequence t_k such that

$$|x_k(t_k)| \geqslant \frac{1}{2} \| x_k(t) \|_{C(R^n)} \to \infty,$$

the matrix-functions $A(t+t_k)$ converge uniformly to an ap-matrix $A_*(t)$, and the ap-functions $x_0(t+t_k)$ and $Lx_0(t+t_k)$ converge respectively to ap-functions $x_*(t)$ and

$$L_*x_*(t) = \frac{d^2x_*(t)}{dt^2} + A_*(t)\,x_*(t).$$

Set

$$y_k(t) = \frac{x_k(t+t_k)}{\|x_k(\tau)\|_{C(R^n)}} \qquad (k = 1, 2, \ldots). \tag{7.34}$$

Obviously,

$$\frac{d^2y_k(t)}{dt^2} + A(t+t_k)\,y_k(t) + \varepsilon_k B y_k(t) = \frac{f(t+t_k)}{\|x_k(\tau)\|_{C(R^n)}}, \tag{7.35}$$

and therefore the second derivatives of the functions (7.34) are uniformly bounded. Hence the sequence (7.34) is compact in the sense of uniform convergence on any finite interval. Without loss of generality, we may assume that the sequence (7.34) itself converges uniformly on any finite interval to a function $y_0(t)$ $(-\infty < t < \infty)$.

Letting $k \to \infty$ (the details are left to the reader) in (7.35), we get

$$\frac{d^2y_0(t)}{dt^2} + A_*(t)\,y_0(t) = 0. \tag{7.36}$$

Since $|y_0(0)| \neq 0$, we may assume without loss of generality that $y_0(0) \overline{\in} K$. Obviously, $x_*(t) \gg 0$. Hence there exists a maximal α_0 such that the function

$$z_*(t) = x_*(t) + \alpha_0 y_0(t)$$

is in the cone $\hat{K} \subset C(R^n)$ of vector-functions with values in K.

Since $L_*x_*(t) \ll 0$, the function $z_*(t)$ is a solution, bounded on the whole real line, of the equation

$$\frac{d^2x}{dt^2} + A_*(t)\,x = -f_*(t),$$

where $f_*(t) \gg 0$.

Let $q(x)$ be a sublinear functional selecting K. Obviously,

$$\sup_{-\infty < t < \infty} q[z_*(t)] = 0.$$

Let t_j be a sequence of numbers such that the sequence $z_*(t_j)$ converges to a point $x_{**} \in \Gamma$. As before, we may assume that then the almost periodic functions $f_*(t+t_j)$ and the matrices $A_*(t+t_j)$ converge uniformly to an ap-function $f_{**}(t)$ and ap-matrix $A_{**}(t)$, respectively, and that the functions $z_{**}(t+t_j)$ converge uniformly on every finite interval to a function $z_{**}(t)$. The function $z_{**}(t)$ is a solution, bounded on the real line, of the equation

$$\frac{d^2x}{dt^2} + A_{**}(t)(x) = -f_{**}(t), \tag{7.37}$$

where $f_{**}(t) \gg 0$, and it satisfies the condition

$$0 = q[z_{**}(0)] = \sup_{-\infty < t < \infty} q[z_{**}(t)].$$

According to Theorem (6.2), condition (7.22) implies

$$q'_+[x_{**}, A_{**}(0)x_{**}] \leqslant 0.$$

Therefore, it follows from (6.15) that

$$\max_{l \in J(x_{**})} l[A_{**}(0)x_{**}] \leqslant 0, \tag{7.38}$$

where $J(x_{**})$ is the set of all linear functionals l such that

$$l(x_{**}) = 0, \quad l(x) \leqslant q(x) \quad (x \in R^n). \tag{7.39}$$

It follows from (7.39) that

$$l[z_{**}(0)] = 0, \quad l[z_{**}(t)] \leqslant 0 \quad (-\infty < t < \infty,\ l \in J(x_{**})).$$

Hence

$$q'_+\left[x_{**}, \frac{d^2z_{**}(0)}{dt^2}\right] = \max_{l \in J(x_{**})} l\left[\frac{d^2z_{**}(0)}{dt^2}\right] = \max_{l \in J(x_{**})} \frac{d^2l[z_{**}(0)]}{dt^2} \leqslant 0. \tag{7.40}$$

Since the derivative $q'_+(x_{**}, h)$ is a sublinear functional with respect to h, it follows from (7.37) that

$$q'_+\left[x_{**}, \frac{d^2z_{**}(0)}{dt^2}\right] + q'_+[x_{**}, A_{**}(0)x_{**}] + q'_+[x_{**}, f_{**}(0)] \geqslant 0.$$

It follows from (7.38) and (7.40) that the first two terms are nonpositive, and thus

$$q'_+[x_{**}, f_{**}(0)] \geqslant 0.$$

On the other hand, by (6.10),

$$q'_+[x_{**}, f_{**}(0)] \leqslant q[f_{**}(0)],$$

and since $f_{**}(0)$ is an interior element of the cone K,

$$q'_+[x_{**}, f_{**}(0)] < 0.$$

We have arrived at a contradiction.

Thus the ap-functions $x_k(t)$ satisfying conditions (7.33) are uniformly bounded. It follows from (7.33) that the sequence $x_k(t)$ is compact in the sense of uniform convergence on every finite interval.

Let $x_{k(i)}(t)$ be a subsequence converging uniformly on every finite interval to a function $x(t)$. Obviously, this function is a solution, bounded on the entire real line, of equation (7.32). Q. E. D.

We can now prove Theorem 3.2.

Let B be a matrix satisfying (7.23). Such a matrix exists by virtue of Lemma 7.2.

Let ε_0 be a positive such that

$$L_\varepsilon x_0(t) \ll 0 \qquad (0 < \varepsilon \leqslant \varepsilon_0), \tag{7.41}$$

where $x_0(t)$ is an ap-function that satisfies (7.21), and L_ε the operators (7.31). By Lemma 7.3, for any $\varepsilon \in (0, \varepsilon_0]$ there exists $a = a(\varepsilon) > 0$ such that (7.24) holds.

Set

$$G_a(t, s) = \begin{cases} -\frac{1}{2a} e^{-a(t-s)} I, & \text{if} \quad t \geqslant s, \\ -\frac{1}{2a} e^{a(t-s)} I, & \text{if} \quad t \leqslant s. \end{cases} \tag{7.42}$$

Obviously, the ap-solutions of the equations

$$L_\varepsilon x(t) = f(t) \qquad (f \in B(R^n))$$

coincide with those of the linear integral equation

$$x(t) = T_a x(t) + g(t),$$

where

$$T_a x(t) = -\int_{-\infty}^{\infty} G_a(t, s)[A(s) + a^2 I + \varepsilon B]\, x(s)\, ds$$

and

$$g(t) = \int_{-\infty}^{\infty} G_a(t, s) f(s)\, ds.$$

It follows from (7.24) that the operator T_a is positive with respect to the cone $\hat{K}$. Thus, (7.41) implies

$$x_0(t) - T_a x_0(t) \gg 0.$$

It follows from Lemma 7.1 (with D_0 the identity operator I and B the operator T_a) that the operator $I - T_a$ has a positive inverse $(I - T_a)^{-1}$. Hence the operators L_ε $(0 < \varepsilon \leqslant \varepsilon_0)$ are regular and their Green's functions $G(t, s; \varepsilon)$ are nonpositive with respect to the cone K.

Hence, by Lemma 7.4, the operator (7.19) is regular. By Theorem 4.3, its Green's function $G(t, s)$ is nonpositive because the Green's functions $G(t, s; \varepsilon)$ are nonpositive. This completes the proof of Theorem 7.3.

7.4. Strong positivity

Let $\mathfrak{A}$ denote the closure of the range of the ap-matrix $A(t)$.

Theorem 7.4. Suppose there exists a function $x_0(t)$ satisfying (7.21). Let

$$q'_-(x, Ax) < 0 \qquad (x \in \Gamma,\ x \neq 0,\ A \in \mathfrak{A}), \tag{7.43}$$

where $q(x)$ is a functional selecting the cone K, and Γ is the boundary of K.

Then the operator (7.18) is strongly positive with respect to the cone $\hat{K}$.

Proof. The positivity of the operator (7.18) follows from Theorem 7.3, since, by Theorem 6.2, (7.43) implies (7.22).

Suppose that the operator (7.18) is not strongly positive, i. e., there exists a nonzero ap-function $f(t) \in \hat{K}$ such that the ap-function $x(t) = Tf(t)$ lies on the boundary $\hat{\Gamma}$ of the cone $\hat{K}$. We may assume without loss of generality (see the proof of Lemma 7.41) that $z_0 = x(0) \in \Gamma$. We claim that $z_0 \neq 0$.

Obviously, for any $a > 0$,

$$x(0) = -\int_{-\infty}^{\infty} G_a(0, s)[A(s) + a^2 I]\, x(s)\, ds + \\ + \int_{-\infty}^{\infty} G_a(0, s) f(s)\, ds. \qquad (7.44)$$

Let $l(x)$ be a linear functional such that $l(x) \geqslant |x|$ $(x \in K)$. Then, for sufficiently large a,

$$l[A(s)x + a^2 x] \geqslant [a^2 - |A(s)| \cdot |l|]|x| \geqslant 0 \qquad (x \in K).$$

It therefore follows from (7.44) that

$$l[x(0)] \geqslant -\int_{-\infty}^{\infty} G_a(0, s)\, l[A(s)x(s) + a^2 x(s)]\, ds \geqslant 0,$$

and, if $x(0) = 0$, then

$$l[A(s)x(s) + a^2 x(s)] \equiv 0 \qquad (-\infty < s < \infty).$$

This identity implies that $x(s) \equiv 0$. But then

$$f(t) = -Lx(t) \equiv 0,$$

and this is a contradiction.

It follows from (6.12) and (6.15) that

$$q'_-[z_0, A(0)z_0] = \min_{l \in J(z_0)} l[A(0)z_0],$$

where $J(z_0)$ is the set of linear functionals such that

$$l(z_0) = 0, \quad l(x) \leqslant q(x) \qquad (x \in R^n).$$

Let l_0 be a functional in $J(z_0)$ such that

$$q'_-[z_0, A(0)z_0] = l_0[A(0)z_0].$$

Then

$$l_0[x''(0)] \leqslant 0, \quad l_0[f(0)] \leqslant 0, \quad l_0[A(0)z_0] < 0,$$

and thus

$$l_0[x''(0) + A(0)x(0) + f(0)] < 0.$$

On the other hand

$$l_0\left[\frac{d^2x(t)}{dt^2}+A(t)x(t)+f(t)\right]\equiv 0 \qquad (-\infty<t<\infty).$$

This contradiction completes the proof.

Recall that $H[A(t)]$ denotes the set of ap-matrices $A_*(t)$ which are uniform limits of sequences of the form $A(t+\tau_j)$.

Lemma 7.5. Let the ap-matrix $A(t)$ satisfy the conditions

$$q'_+[x, A(t)x]\leqslant 0 \qquad (x\in\Gamma,\ -\infty<t<\infty)$$

and

$$q'_-[x, A(t_0)x]<0 \qquad (x\in\Gamma,\ x\neq 0), \tag{7.45}$$

where t_0 is a fixed number and $q(x)$ a sublinear functional selecting a solid faceted cone K with boundary Γ.

Then, for any ap-matrix $A_*(t)\in H[A(t)]$, there exists t_* such that

$$q'_-[x, A_*(t_*)x]<0 \qquad (x\in\Gamma,\ x\neq 0). \tag{7.46}$$

Proof. Let $A_*(t)$ be the uniform limit of a sequence $A(t+\tau_j)$. Then the ap-matrix $A(t)$ is the uniform limit of the sequence $A_*(t-\tau_j)$. In particular, $A(t_0)$ is the limit of the sequence of matrices $A_*(t_0-\tau_j)$, whence it follows that

$$e^{A(t_0)}=\lim_{j\to\infty} e^{A_*(t_0-\tau_j)}.$$

By Theorem 6.2, it follows that the matrix $e^{A(t_0)}$ is strongly positive. Hence the matrices $e^{A_*(t_0-\tau_j)}$ are also strongly positive for large j.

Since K is a faceted cone, strong positivity of $e^{A_*(t_0-\tau_j)}$ implies strong positivity of all the matrices $e^{A_*(t_0-\tau_j)t}$ for $t>0$, or, equivalently, strong positivity of the translation operator of the equation

$$\frac{dx}{dt}=A_*(t_0-\tau_j)x$$

with constant coefficients. Hence (see Section 6.4), it follows that inequality (7.46) holds for $t_*=t_0-\tau_j$. Q. E. D.

Lemma 7.6. Let the translation operator $U(t,s)$ of the equation

$$\frac{dx}{dt}-A(t)x=0 \tag{7.47}$$

satisfy the condition

$$U(t, s)K \subset K \qquad (t \geqslant s), \tag{7.48}$$

where K is a faceted cone.

Then there exists $a > 0$ such that

$$A(t)x + a^2 x \in K \qquad (x \in K,\ -\infty < t < \infty). \tag{7.49}$$

Proof. It is clear that for any fixed point $x \in K$ there exists $a = a(x) > 0$ satisfying (7.49). Therefore, for any finite set of points $x_1, \ldots, x_m \in K$ there exists a common $a > 0$ such that

$$A(t)x_i + a^2 x_i \in K \qquad (i = 1, \ldots, m;\ -\infty < t < \infty). \tag{7.50}$$

Let $x_1, \ldots, x_m$ be normalized ($|x_i| = 1$) and lie on the one-dimensional edges of the cone K. Then every point $x \in K$ may be expressed as

$$x = \xi_1 x_1 + \ldots + \xi_m x_m,$$

where $\xi_i \geqslant 0$. Thus (7.49) follows from (7.50). Q. E. D.

Lemmas 7.5 and 7.6 make it possible to weaken the restrictive condition (7.43) for the case of faceted cones.

Theorem 7.5. Under the assumptions of Theorem 7.3, let K be a faceted cone. Let t_0 be a number such that

$$q'_-[x, A(t_0)x] < 0 \qquad (x \in \Gamma,\ x \neq 0). \tag{7.51}$$

Then the operator (7.18) is strongly positive with respect to the cone $\hat{K}$.

Proof. Theorem 7.3 implies that all the operators $-L_*^{-1}$ are positive with respect to $\hat{K}$, where

$$L_* x = \frac{d^2x}{dt^2} + A_*(t)x \qquad (A_*(t) \in H[A(t)]).$$

We claim that all the operators $-L_*^{-1}$ are strongly positive.

Otherwise, there exist an ap-matrix $A_*(t)$ and a nonzero ap-function $f(t) \in \hat{K}$ such that $x_*(0) \in \Gamma$, where $x_*(t)$ is an ap-solution of the equation

$$\frac{d^2x}{dt^2} + A_*(t)x + f(t) = 0 \tag{7.52}$$

It follows from (7.52) that

$$x_*(t) = -\int_{-\infty}^{\infty} G_a(t, s)[A_*(s)x_*(s) + a^2x_*(s)]\,ds - $$
$$- \int_{-\infty}^{\infty} G_a(t, s)f(s)\,ds, \qquad (7.53)$$

where $G_a(t, s)$ is the function (7.42).

Let l_0 be a linear functional, positive on K, such that $l_0[x_0(0)] = 0$. Then, by (7.53) and Lemma 7.6, for sufficiently large a we have the identity

$$l_0[A_*(s)x_*(s) + a^2x_*(s)] \equiv 0 \qquad (-\infty < s < \infty),$$

and this implies

$$l_0[x_*(s)] \equiv 0 \qquad (-\infty < s < \infty) \qquad (7.54)$$

and

$$l_0[A_*(s)x_*(s)] \equiv 0 \qquad (-\infty < s < \infty). \qquad (7.55)$$

As in the proof of Theorem 7.4, it is easy to show that the ap-function $x_*(s)$ never vanishes. By (7.54) all its values lie on the boundary of the cone K. Thus it follows from Lemma 7.5 that there exists t_* such that

$$q'_-[x_*(t_*),\ A_*(t_*)x_*(t_*)] < 0. \qquad (7.56)$$

Denote by J_* the set of normalized linear functionals, positive on K, such that $l[x_*(t_*)] = 0$. On the one hand, by Lemma 6.2 and (7.56),

$$\max_{l \in J_*} l[A_*(t_*)x_*(t_*)] > 0.$$

Let $l_* \in J_*$ and $l_*[A_*(t_*)x_*(t_*)] > 0$. On the other hand, repeating the reasoning in the proof of identity (7.55), one easily shows that

$$l_*[A_*(t_*)x_*(t_*)] = 0.$$

This is a contradiction. Q. E. D.

Theorem 7.5 remains valid for unfaceted cones, provided one assumes a priori that there exists a satisfying condition (7.49) and replaces condition (7.51) by the stronger inequality

220733

$$q'_-[x, A_*(t_*)x] < 0 \qquad (x \in \Gamma,\ x \neq 0;\ A_*(t) \in H(A)),$$

where t_* depends on the ap-matrix $A_*(t)$.

On the other hand, Lemmas 7.5 and 7.6 fail to hold for unfaceted cones.

7.5. More about systems with constant coefficients

In the case of ap-operators

$$Lx = \frac{d^2x}{dt^2} + Ax \tag{7.57}$$

with constant coefficients, the theorems of the preceding subsections can be formulated more conveniently in other terms.

T h e o r e m 7.6. *Let the real parts of the eigenvalues of the matrix A be negative. Let*

$$e^{At}K \subset K \qquad (t > 0), \tag{7.58}$$

where K is a solid cone in R^n.

Then the Green's function $G(t, s)$ of the ap-operator (7.57) exists and is nonpositive with respect to K.

T h e o r e m 7.7. *Let the real parts of the eigenvalues of the matrix A be negative. Let*

$$q'_-(x, Ax) < 0 \qquad (x \in \Gamma,\ x \neq 0), \tag{7.59}$$

where Γ is the boundary of a solid cone K in R^n.

Then the Green's function $G(t, s)$ of the ap-operator (7.57) exists and the operator

$$Tf(t) = -\int_{-\infty}^{\infty} G(t, s)f(s)\,ds \tag{7.60}$$

is strongly positive with respect to $\hat{K}$.

T h e o r e m 7.8. *Let the real parts of the eigenvalues of the matrix A be negative. Let K be a solid faceted cone. Let*

$$e^{At}x \gg 0 \qquad (x \in K,\ \ x \neq 0,\ \ t > 0). \tag{7.61}$$

Then the Green's function $G(t, s)$ of the operator (7.57) exists and the operator (7.60) is strongly positive with respect to $\hat{K}$.

It is interesting to note that for $n = 2$ (when R^2 is the plane), the conditions of Theorems 76 and 78 are not only sufficient but also necessary for the operator (7.60) to be positive and strongly positive, respectively. Let us prove this statement.

Choose a basis in R^2 such that K is the set of vectors with nonnegative components. Suppose that, relative to this basis,

$$A = \begin{Vmatrix} a_{11} & a_{12} \\ a_{21} & a_{22} \end{Vmatrix}. \tag{7.62}$$

If $-L^{-1}$ is positive, then $-A^{-1}K \subset K$. Hence the matrix $-A^{-1}$ has a positive eigenvalue. Consequently, the matrix A has a negative eigenvalue λ_1. Since the operator (7.51) has a Green's function, the second eigenvalue is also negative. Thus the first condition of Theorems 7.6 and 7.8 holds.

Obviously,

$$-A^{-1} = \frac{1}{\Delta}\begin{Vmatrix} -a_{22} & a_{12} \\ a_{21} & -a_{11} \end{Vmatrix}, \tag{7.63}$$

where

$$\Delta = a_{11}a_{22} - a_{21}a_{12} = \lambda_1\lambda_2 > 0, \tag{7.64}$$

and so

$$a_{12} \geqslant 0, \quad a_{21} \geqslant 0. \tag{7.65}$$

The last inequalities mean (see Section 6.4) that condition (7.58) holds.

If the operator $-L^{-1}$ is strongly positive, then $-A^{-1}$ is also strongly positive. Hence

$$a_{12} > 0, \quad a_{21} > 0,$$

so that condition (7.61) is satisfied.

When $n \geqslant 3$ the conditions of Theorems 7.6 and 7.8 are no longer necessary.

Note, in conclusion, that under the conditions of Theorem 7.8 not only is the operator (7.60) strongly positive with respect to the cone $\widehat{K}$, but moreover the Green's function $G(t, s)$ is strongly negative with respect to the cone K.

7.6. Remark on first-order systems

Let the first-order ap-operator

$$L_1 x = \frac{dx}{dt} - A(t)\,x$$

be regular and suppose that its Green's function $G(t,s)$ is nonnegative with respect to a solid faceted cone K. Suppose in addition that inequality (7.51) is valid. Reasoning as in the proof of Theorem 7.5, one readily sees that the integral operator

$$T_1 f(t) = \int_{-\infty}^{\infty} G(t, s) f(s)\,ds$$

is strongly positive with respect to the cone $\hat{K}$.

§ 8. HIGHER-ORDER SCALAR EQUATIONS

8.1. Statement of the problem

In this section we shall investigate ap-operators

$$Lx = \frac{d^m x}{dt^m} + a_1(t)\frac{d^{m-1}x}{dt^{m-1}} + \dots + a_m(t)\,x, \tag{8.1}$$

where $a_i(t)$ are scalar ap-functions. The Green's function $G(t,s)$ of a regular ap-operator (8.1) is a scalar function. We are interested in conditions for this function to be nonpositive or nonnegative.

When investigating the operator (8.1), it is convenient to use the notation and terminology of the previous sections. Let K_+ denote the positive half-line in R^1; then $\hat{K}_+$ is the cone of nonnegative scalar functions in $B(R^1)$. $G(t,s)$ is nonnegative as a scalar function if and only if it is nonnegative with respect to the cone K_+. If $G(t,s)$ is strongly positive with respect to the cone K_+, it is nonnegative and for every fixed t one of the infinite intervals $(-\infty, t)$ or (t, ∞) contains no interval $\alpha < s < \beta$ on which $G(t,s) \equiv 0$.

A convenient aid to investigation of the operator (8.1) (see Section 2.1) is the operator

$$\tilde{L}u = \frac{du}{dt} + Q(t)\,u, \tag{8.2}$$

where $u \in R^m$, and

$$Q(t) = \begin{Vmatrix} 0 & -1 & 0 & \dots & 0 \\ 0 & 0 & -1 & \dots & 0 \\ \cdot & \cdot & \cdot & \cdot & \cdot \\ 0 & 0 & 0 & \dots & -1 \\ a_m(t) & a_{m-1}(t) & a_{m-2}(t) & \dots & a_1(t) \end{Vmatrix}. \tag{8.3}$$

By Theorem 2.4, the operator (8.1) is regular if and only if the operator (8.2) is regular.

If (8.2) is regular, then (see Section 4.2) its Green's function can be written

$$\tilde{G}(t, s) = \begin{cases} U(t) P_+ U^{-1}(s), & \text{if} \quad t \geqslant s, \\ -U(t) P_- U^{-1}(s), & \text{if} \quad t < s, \end{cases} \tag{8.4}$$

where $U(t)$ $(U(0) = I)$ is the fundamental matrix of solutions of the system

$$\tilde{L} u = 0, \tag{8.5}$$

and P_+, P_- are the projections onto the subspaces $E_+, E_- \subset R^m$ of initial values of the solutions of equation (8.5) that decrease as $t \to -\infty$ and $t \to \infty$, respectively (see Section 3.2). Set

$$e_1 = \{1, 0, \dots, 0\},\ e_2 = \{0, 1, \dots, 0\}, \dots \\ \dots, e_m = \{0, 0, \dots, 1\}. \tag{8.6}$$

The vectors (8.6) form a basis in R^m. It follows from Theorem 4.2 and from (8.4) that the Green's function $G(t, s)$ of the operator (8.1) has the form

$$G(t, s) = \begin{cases} (U(t) P_+ U^{-1}(s) e_m, e_1), & \text{if} \quad t \geqslant s, \\ -(U(t) P_- U^{-1}(s) e_m, e_1), & \text{if} \quad t < s. \end{cases} \tag{8.7}$$

Theorem 8.1. If the Green's function $G(t, s)$ of a regular ap-operator (8.1) is nonnegative with respect to the cone $K_+ \subset R^1$, then it is strongly positive with respect to K_+.

Proof. If this assertion is false, we can select numbers $\alpha_1 < \beta_1 < t_0 < \alpha_2 < \beta_2$ such that $G(t_0, s)$ vanishes for $s \in (\alpha_1, \beta_1)$ and for $s \in (\alpha_2, \beta_2)$. By (8.7), this means that

$$(U(t_0) P_+ U^{-1}(s) e_m, e_1) = 0 \qquad (\alpha_1 < s < \beta_1) \tag{8.8}$$

and

$$(U(t_0) P_- U^{-1}(s) e_m, e_1) = 0 \qquad (\alpha_2 < s < \beta_2). \tag{8.9}$$

Obviously,

$$\frac{d}{ds} U^{-1}(s) = U^{-1}(s) Q(s),$$

so that

$$\left(U(t_0) P_+ U^{-1}(s) Q(s) e_m, e_1\right) = 0 \qquad (\alpha_1 < s < \beta_1)$$

and

$$\left(U(t_0) P_- U^{-1}(s) Q(s) e_m, e_1\right) = 0 \qquad (\alpha_2 < s < \beta_2)$$

or, what is the same,

$$-\left(U(t_0) P_+ U^{-1}(s) e_{m-1}, e_1\right) + \\ + a_1(s)\left(U(t_0) P_+ U^{-1}(s) e_m, e_1\right) = 0 \qquad (\alpha_1 < s < \beta_1)$$

and

$$-\left(U(t_0) P_- U^{-1}(s) e_{m-1}, e_1\right) + \\ + a_1(s)\left(U(t_0) P_- U^{-1}(s) e_m, e_1\right) = 0 \qquad (\alpha_2 < s < \beta_2),$$

whence, by (8.8) and (8.9),

$$\left(U(t_0) P_+ U^{-1}(s) e_{m-1}, e_1\right) = 0 \qquad (\alpha_1 < s < \beta_1)$$

and

$$\left(U(t_0) P_- U^{-1}(s) e_{m-1}, e_1\right) = 0 \qquad (\alpha_2 < s < \beta_2).$$

Repeating this argument, we get the equalities

$$\left(U(t_0) P_+ U^{-1}(s) e_i, e_1\right) = 0 \quad (i = 1, \ldots, m;\ \alpha_1 < s < \beta_1)$$

and

$$\left(U(t_0) P_- U^{-1}(s) e_i, e_1\right) = 0 \quad (i = 1, \ldots, m;\ \alpha_2 < s < \beta_2).$$

Since for any s the vectors $U^{-1}(s) e_1, \ldots, U^{-1}(s) e_m$ constitute a basis in R^m, the last equalities imply that for any $u \in R^m$

$$(U(t_0) P_+ u, e_1) = (U(t_0) P_- u, e_1) = 0,$$

and thus

$$(U(t_0) u, e_1) = 0 \qquad (u \in R^m).$$

On the other hand, $(U(t_0)u, e_1) = 1$ for $u = U^{-1}(t_0)e_1$. This is a contradiction. Q. E. D.

One proves similarly that if the Green's function of the ap-operator (8.1) is nonpositive with respect to $K_+ \subset R^1$ then it is strongly negative with respect to K_+.

In connection with Theorem 8.1, we are confining ourselves in this section to conditions for the existence and nonnegativity or nonpositivity of the Green's function. For operators (8.1) with variable coefficients only partial results are available. More complete results have been obtained for operators

$$Lx = \frac{d^m x}{dt^m} + a_1 \frac{d^{m-1}x}{dt^{m-1}} + \dots + a_m x \tag{8.10}$$

with constant coefficients.

8.2. A necessary condition for the Green's function of the ap-operator (8.10) to be of constant sign

The following proposition is easily verified:

The ap-operator (8.10) is regular if and only if its characteristic equation

$$\chi(\lambda) \equiv \lambda^m + a_1\lambda^{m-1} + \dots + a_m = 0 \tag{8.11}$$

has no zero or pure imaginary roots.

Henceforth we shall assume that this condition is satisfied.

The roots of equation (8.11) fall naturally into two groups: the roots in the left half-plane and those in the right half-plane. Let α_- be the maximum real part of the roots in the first group, and α_+ the minimum real part of the roots in the second. If the first group is empty, we set $\alpha_- = -\infty$; if the second is empty, we set $\alpha_+ = \infty$. Equation (8.11) has no roots in the strip $\alpha_- < \mathrm{Re}\,\lambda < \alpha_+$. Let Λ_- and Λ_+ denote the sets of roots λ such that $\mathrm{Re}\,\lambda = \alpha_-$ and $\mathrm{Re}\,\lambda = \alpha_+$, respectively. If all the roots of equation (8.11) are in one half-plane, one of sets Λ_- or Λ_+ is empty.

Let $\alpha_- \in \Lambda_-$, if α_- is finite and $\alpha_+ \in \Lambda_+$ if α_+ is finite. Then we shall say that the characteristic polynomial $\chi(\lambda)$ is t a m e .*

T h e o r e m 8.2. Let the Green's function $G(t, s)$ of the regular ap-operator (8.10) be of constant sign.

Then the characteristic polynomial $\chi(\lambda)$ is tame.

* [Russian original: p r a v i l ' n y i . This is apparently an ad hoc term and we therefore prefer "tame" to the rather overworked "regular."]

Proof. Note first that in the case of constant coefficients the formula (8.7) for the Green's function can be written

$$G(t,s)=\begin{cases} (e^{-Q(t-s)}P_+e_m, e_1), & \text{if} \quad t\geqslant s, \\ -(e^{-Q(t-s)}P_-e_m, e_1), & \text{if} \quad t<s. \end{cases} \tag{8.12}$$

Here

$$Q=\begin{Vmatrix} 0 & -1 & 0 & \dots & 0 \\ 0 & 0 & -1 & \dots & 0 \\ \cdot & \cdot & \cdot & \cdot & \cdot \\ 0 & 0 & 0 & \dots & -1 \\ a_m & a_{m-1} & a_{m-2} & \dots & a_1 \end{Vmatrix}. \tag{8.13}$$

Let $\lambda_1, \dots, \lambda_k$ $(k\leqslant m)$ be the distinct roots of the characteristic equation (8.11). Denote their multiplicities by $r_1, \dots, r_k$ $(r_1+\dots+r_k=m)$. Consider the vector-function

$$g(z)=\{1, z, \dots, z^{m-1}\}$$

of the complex variable z and define vectors

$$g_i^j=\frac{d^{j-1}g(\lambda_i)}{dz^{j-1}} \qquad (i=1, \dots, k;\ j=1, \dots, r_i). \tag{8.14}$$

A simple computation shows that for any fixed i

$$-Qg_i^1=\lambda_i g_i^1, \quad -Qg_i^2=\lambda_i g_i^2+g_i^1, \dots, \quad -Qg_i^{r_i}=\lambda_i g_i^{r_i}+g_i^{r_i-1}.$$

This means that the vectors g_i^1 are the eigenvectors of the matrix $-Q$ corresponding to the eigenvalues λ_i, while $g_i^2, \dots, g_i^{r_i}$ are the associated eigenvectors corresponding to the same eigenvalues. It follows from general theorems of linear algebra that the vectors (8.14) form a basis for complex m-space E. Thus the vector $e_m=\{0, 0, \dots, 1\}$ can be expressed as

$$e_m=\sum_{i=1}^{k}\sum_{j=1}^{r_i}\alpha_i^j g_i^j. \tag{8.15}$$

The linear span E_i of the vectors $g_i^1, \dots, g_i^{r_i}$ is the root subspace of the matrix $-Q$ corresponding to the eigenvalue λ_i. Every $x\in E$ can be expressed uniquely as

$$x=x_1+\dots+x_k \qquad (x_i\in E_i;\quad i=1, \dots, k).$$

This representation defines projections

$$P(\lambda_i)x = x_i \qquad (x \in E;\ i = 1, \ldots, k)$$

onto the root subspaces E_i. Obviously,

$$P(\lambda_i)\, e_m = \sum_{j=1}^{r_i} \alpha_i^j g_i^j \qquad (i = 1, \ldots, k).$$

We claim that none of the vectors $P(\lambda_i)e_m$ $(i = 1, \ldots, k)$ vanish. Otherwise, for some i,

$$e^{-Qt}P(\lambda_i)e_m = 0 \qquad (-\infty < t < \infty).$$

Differentiating this identity, we get

$$e^{-Qt}Q^l P(\lambda_i)e_m = 0 \qquad (l = 1, 2, \ldots, m-1),$$

whence (since Q and $P(\lambda_i)$ commute)

$$P(\lambda_i)Q^l e_m = 0 \qquad (l = 0, 1, 2, \ldots, m-1),$$

and therefore

$$P(\lambda_i)\, e_1 = P(\lambda_i)e_2 = \ldots = P(\lambda_i)e_m = 0,$$

i. e.,

$$P(\lambda_i)x = 0 \qquad (x \in E).$$

This is a contradiction.

By the general theory of ordinary linear differential equations (see, e. g., Coddington and Levinson /1/),

$$\left(e^{-Qt}P(\lambda_i)e_m, e_1\right) = M(t, \lambda_i)\, e^{\lambda_i t}, \tag{8.16}$$

where

$$M(t, \lambda_i) = \alpha_i^1 + \alpha_i^2 t + \ldots + \frac{\alpha_i^{r_i}}{(r_i - 1)!}\, t^{r_i - 1}. \tag{8.17}$$

The left-hand side of (8.16) is the solution of the equation $Lx = 0$ with initial value $P(\lambda_i)e_m \neq 0$. Therefore $M(t, \lambda_i)$ are nonzero polynomials. Obviously,

$$M(t, \bar{\lambda}) = \overline{M(t, \lambda)}$$

for any complex eigenvalue λ of $-Q$.

Obviously,

$$P_+ = \sum_{\mathrm{Re}\,\lambda_i < 0} P(\lambda_i), \qquad P_- = \sum_{\mathrm{Re}\,\lambda_i > 0} P(\lambda_i).$$

Hence it follows from (8.16) and (8.12) that

$$G(t, s) = \begin{cases} \sum_{\mathrm{Re}\,\lambda_i < 0} M(t-s, \lambda_i) e^{\lambda_i (t-s)}, & \text{if} \quad t \geqslant s, \\ -\sum_{\mathrm{Re}\,\lambda_i > 0} M(t-s, \lambda_i) e^{\lambda_i (t-s)}, & \text{if} \quad t < s. \end{cases} \tag{8.18}$$

Suppose that Λ_- is nonempty and $\alpha_- \in \Lambda_-$. Then for large t the function $G(t, 0)$ is of variable sign. Similarly, if Λ_+ is nonempty and $\alpha_+ \in \Lambda_+$, then $G(0, s)$ is of variable sign for large s. Thus, if the Green's function of the ap-operator (8.10) is of constant sign the characteristic polynomial $\chi(\lambda)$ is tame. Q. E. D.

8.3. Sufficient conditions for the Green's function of the ap-operator (8.10) to be of constant sign

The Cauchy function $K(t, s)$ of the differential operator (8.1) is defined as the solution of the equation $Lx = 0$ satisfying initial conditions

$$K(s, s) = \ldots = K^{(m-2)}(s, s) = 0, \quad K^{(m-1)}(s, s) = 1. \tag{8.19}$$

In the case of differential operators (8.10) with constant coefficients, the Cauchy function $K(t, s)$ depends on the difference of the arguments t, s:

$$K(t, s) = k(t-s) = (e^{-Q(t-s)} e_m, e_1), \tag{8.20}$$

where Q is the matrix (8.13).

If all the roots of the polynomial (8.11) have negative real parts, then by (8.12) the Green's function of (8.10) is defined by

$$G(t, s) = \begin{cases} k(t-s), & \text{if} \quad t \geqslant s, \\ 0, & \text{if} \quad t < s. \end{cases} \tag{8.21}$$

On the other hand, if all the roots of (8.11) have positive real parts, then

$$G(t, s) = \begin{cases} 0, & \text{if} \quad t \geqslant s, \\ -k(t-s), & \text{if} \quad t < s. \end{cases} \tag{8.22}$$

Thus, if the roots of the polynomial (8.11) lie in one half-plane, then the Green's function $G(t,s)$ of the operator (8.10) is of constant sign if and only if its Cauchy function $k(t)$ is of constant sign on the corresponding interval $(0, \infty)$ or $(-\infty, 0)$. By (8.19), we have the following proposition:

The Cauchy function $k(t)$ *is of constant sign on* $(0, \infty)$ *if it is nonnegative there; it is of constant sign on* $(-\infty, 0)$ *if for odd* m *it is nonnegative there, and for even* m *it is nonpositive.*

Any regular ap-operator (8.10) with constant coefficients can be expressed as the superposition

$$L = L_1 L_2 \tag{8.23}$$

of two regular ap-operators with constant coefficients all of whose characteristic roots lie in the left and right half-planes of the complex plane, respectively. Obviously,

$$L^{-1} = L_2^{-1} L_1^{-1} = L_1^{-1} L_2^{-1},$$

whence it follows that the Green's function $G(t,s)$ of the operator (8.10) is the superposition of the Green's functions $G_1(t,s)$, $G_2(t,s)$ of L_1 and L_2:

$$G(t, s) = \int_{-\infty}^{\infty} G_2(t, \sigma) G_1(\sigma, s) \, d\sigma. \tag{8.24}$$

Thus, a sufficient condition for the Green's function $G(t,s)$ to be of constant sign is that the Green's functions of the operators L_1 and L_2 both be of constant sign. As mentioned, the question whether the Green's functions of two operators L_1 and L_2, each having characteristic roots only in one of the half-plane, are of constant sign reduces to the analogous question (for $t \geqslant 0$ and $t \leqslant 0$) for their Cauchy functions.

A polynomial is called a Hurwitz polynomial if all its roots are situated in the left half-plane.

We confine ourselves in this section to analysis of an operator L_1 whose characteristic polynomial is a Hurwitz polynomial. If all the roots of the characteristic polynomial lie in the right half-plane, we need only apply a "time reversal"; the characteristic polynomial of the new operator will then be a Hurwitz polynomial.

A solid cone $K \subset R^m$ is said to be admissible if it is situated in the half-space $(u, e_1) \geqslant 0$ and $e_m \in K$. Our subsequent investigation will be based on the following theorem.

Theorem 8.3. The Cauchy function $k(t)$ of the differential operator (8.10) is nonnegative for $t \geqslant 0$ if and only if there exists an admissible cone K such that

$$e^{-Qt}K \subset K \qquad (0 \leqslant t < \infty). \tag{8.25}$$

Proof. Sufficiency follows from (8.20). To prove necessity, we need only consider the set

$$K = \{u : (e^{-Qt}u, e_1) \geqslant 0 \quad \text{for} \quad t \geqslant 0\}$$

and check that it is an admissible cone satisfying (8.25) (see proof of Theorem 8.1). Q. E. D.

To use Theorem 8.3, we must know how to construct admissible cones in the phase space R^m. In this subsection we shall describe a fairly general method to this end, based on the following theorem of Yakubovich /1/ and Kalman /1/.

Consider the following problem.

Let A be a given matrix with Hurwitz characteristic polynomial, a and b two nonzero vectors. It is required to find conditions for the existence of a positive-definite matrix H such that

$$HA + A^*H < 0 \tag{8.26}$$

and

$$Ha + b = 0. \tag{8.27}$$

Inequality (8.26) means that the matrix $HA + A^*H$ is negative-definite.

Theorem 8.4. There exists a symmetric positive-definite matrix H satisfying (8.26) and (8.27) if and only if

$$(a, b) < 0, \qquad (Aa, b) > 0 \tag{8.28}$$

and for all real ω,

$$\mathrm{Re}\left(A_{i\omega}^{-1}a,\, b\right) < 0, \tag{8.29}$$

where $A_{i\omega} = i\omega I - A$, a $i = \sqrt{-1}$.

Let the Hurwitz polynomial (8.11) be tame, and let λ_0 be its largest real root. We shall call the polynomial (8.11) **completely tame** if the following conditions hold:

1. λ_0 is a simple root, and all other roots λ of the characteristic polynomial (8.11) satisfy the inequality $\operatorname{Re}\lambda < \lambda_0$.
2. There exists a polynomial

$$\varkappa(\lambda) = (t_1 + \lambda) \ldots (t_{m-2} + \lambda)$$

of degree $m-2$ such that

$$t_1, t_2, \ldots, t_{m-2} > 0 \qquad (8.30)$$

and

$$\operatorname{Re}\frac{\varkappa(i\omega)}{\chi(\lambda_0 + i\omega)} < 0 \qquad (-\infty < \omega < \infty). \qquad (8.31)$$

Set

$$\chi_0(\lambda) = \frac{\chi(\lambda_0 + \lambda)}{\lambda}. \qquad (8.32)$$

Then

$$\frac{\varkappa(i\omega)}{\chi(\lambda_0 + i\omega)} = \frac{S_1(\omega^2) + i\omega S_2(\omega^2)}{i\omega\,|\chi_0(i\omega)|^2},$$

where

$$S_1(\omega^2) = \operatorname{Re}[\varkappa(i\omega)\,\chi_0(-i\omega)]$$

and

$$\omega S_2(\omega^2) = \operatorname{Im}[\varkappa(i\omega)\,\chi_0(-i\omega)],$$

so that inequality (8.31) is equivalent to

$$S_2(\lambda) < 0 \qquad (0 \leqslant \lambda < \infty).$$

Note that $S_2(\lambda)$ is a polynomial of degree $m-2$.

Lemma 8.1. Let the characteristic polynomial (8.11) be completely tame.

Then the Cauchy function $k(t)$ of the ap-operator (8.10) is nonnegative for $t \geqslant 0$.

Proof. Consider the ap-operator

$$L_0 x = \frac{d}{dt}\chi_0\left(\frac{d}{dt}\right)x, \qquad (8.33)$$

where

$$\chi_0(\lambda) = \lambda^{m-1} + b_1\lambda^{m-2} + \dots + b_{m-1}$$

is the polynomial (8.32). Note that $\chi_0(\lambda)$ is a Hurwitz polynomial and therefore $b_{m-1} > 0$.

A simple computation shows that the Cauchy function $k(t)$ of the operator (8.10) is related to the Cauchy function $k_0(t)$ of (8.33) by

$$k(t) = e^{\lambda_0 t} k_0(t).$$

Thus, in order to prove the lemma it will suffice to show that the function $k_0(t)$ is nonnegative (for $t \geqslant 0$).

By Theorem 8.3, in order to prove that the Cauchy function $k_0(t)$ of the ap-operator (8.33) is nonnegative (for $t \geqslant 0$) one must construct an admissible cone K in the phase space R^m, with respect to which the translation operator $e^{-Q_0 t}$ of the equation

$$\frac{du}{dt} + Q_0 u = 0 \tag{8.34}$$

is positive, where

$$Q_0 = \begin{Vmatrix} 0 & -1 & 0 & \dots & 0 \\ 0 & 0 & -1 & \dots & 0 \\ \cdot & \cdot & \cdot & \cdot & \cdot \\ 0 & 0 & 0 & \dots & -1 \\ 0 & b_{m-1} & b_{m-2} & \dots & b_1 \end{Vmatrix}$$

To construct the cone K it is convenient to introduce a new basis.

The polynomial $\lambda\chi_0(\lambda)$ is the characteristic polynomial of the matrix $-Q_0$. Consequently, this matrix has a zero eigenvalue, the corresponding eigenvector being e_1. Denote by

$$g_0 = \{\eta_1, \dots, \eta_m\}$$

the eigenvector of the matrix $-Q_0^*$ corresponding to the zero eigenvalue. We shall assume g_0 normalized so that $(g_0, e_1) = \eta_1 = 1$. Then every $u \in R^m$ can be expressed uniquely in the form

$$u = (u, g_0)e_1 + u_1.$$

The set of all vectors u_1 forms a certain $(m-1)$-dimensional subspace E_0. Let $f_1, \dots, f_{m-1}$ be a basis for E_0, and denote by B the matrix whose column vectors are $e_1, f_1, \dots, f_{m-1}$, respectively.

We now perform the substitution

$$u = Bv.$$

Equation (8.34) is then transformed into a system of two equations

$$\left.\begin{aligned}\frac{dv_0}{dt} &= 0,\\ \frac{dv_1}{dt} &= C_1 v_1,\end{aligned}\right\} \tag{8.35}$$

where v_1 is an $(m-1)$-vector and v_0 a scalar.

Set

$$a = B^{-1} e_m = \{a_0, a_1\}, \qquad b = B^* e_1 = \{b_0, b_1\},$$

where a_1 and b_1 are $(m-1)$-vectors. Obviously,

$$a_0 = (a, e_1) = (e_m, g_0) = \eta_m = \frac{1}{b_{m-1}} > 0$$

and

$$b_0 = (b, e_1) = (e_1, e_1) = 1.$$

Let

$$C = \begin{Vmatrix} 0 & 0 \\ 0 & C_1 \end{Vmatrix}.$$

Then

$$(a, \varkappa(C^*) b) = (e_m, \varkappa(-Q_0^*) e_1) = 0.$$

Thus it follows from (8.30) and the inequality $a_0 > 0$ that

$$(a_1, \varkappa(C_1^*) b_1) < 0.$$

It is easily checked that

$$(C_1 a_1, \varkappa(C_1^*) b_1) = (e_m, -Q_0^* \varkappa(-Q_0^*) e_1) = 1.$$

It therefore follows from inequality (8.31) and Theorem 8.4 that there exists a positive-definite matrix H such that

$$Ha_1 + \varkappa(C_1^*) b_1 = 0 \tag{8.36}$$

and

$$HC_1 + C_1^*H < 0.$$

Set

$$q(v) = -\min\{v_0 - \sqrt{(H_1v_1, v_1)}, \quad v_0 + (b_1, v_1),$$
$$t_1v_0 + ((C_1^* + t_1I)b_1, v_1), \ldots, t_1 \ldots t_{m-3}v_0 +$$
$$+ ((C_1^* + t_1I) \ldots (C_1^* + t_{m-3})b_1, v_1)\},$$

where

$$v = \{v_0, v_1\}, \qquad H_1 = \frac{a_0\varkappa(0)}{(Ha_1, a_1)} H.$$

The functional $q(v)$ is so constructed that the cone that it selects is admissible relative to the old basis. Moreover, the translation operator of system (8.34) is positive with respect to this cone if the linear functional

$$l(v) = \varkappa(0)v_0 + (\varkappa(C_1^*)b_1, v_1)$$

is a support functional at a for the circular cone

$$K_0 = \{v: v_0 \geqslant \sqrt{(H_1v_1, v_1)}\}.$$

But the last condition is equivalent to (8.36). Q. E. D.

We return to an operator

$$Lx = \frac{d^mx}{dt^m} + a_1\frac{d^{m-1}x}{dt^{m-1}} + \ldots + a_mx, \tag{8.37}$$

whose characteristic polynomial $\chi(\lambda)$ may have roots in both half-planes.

We are interested in conditions for its Green's function to be of constant sign. By Theorem 8.2, we must assume the polynomial $\chi(\lambda)$ to be tame. Consequently, $\chi(\lambda)$ can be expressed in the form

$$\chi(\lambda) = \chi_1(\lambda)\chi_2(-\lambda), \tag{8.38}$$

where $\chi_1(\lambda)$, $\chi_2(\lambda)$ are tame Hurwitz polynomials. The decomposition (8.38) induces a decomposition of the operator (8.37) as the superposition (8.23) of two operators. If the Green's functions of these two operators are of constant sign then the Green's function $G(t, s)$ is also of constant sign. Its sign is $(-1)^{m_2}$, where m_2 is the degree of the polynomial $\chi_2(\lambda)$. Hence, *if the Green's function* $G(t, s)$ *is of constant sign, then*

$$a_mG(t, s) \geqslant 0. \tag{8.39}$$

A Hurwitz polynomial $\chi(\lambda)$ is said to be completely tame in the limit if there exists a sequence of completely tame polynomials converging to $\chi(\lambda)$. From Lemma 8.1 we get

Theorem 8.5. Let the characteristic polynomial $\chi(\lambda)$ of the operator (8.7) be expressible in the form (8.38), where $\chi_1(\lambda)$ and $\chi_2(\lambda)$ are Hurwitz polynomials, completely tame in the limit.

Then the Green's function of the operator (8.37) is of constant sign and inequality (8.39) holds.

8.4. Special classes of operators with constant coefficients

Theorem 8.5 implies that the Green's function of the ap-operator (8.37) is of constant sign if each of the polynomials $\chi_1(\lambda)$ and $\chi_2(\lambda)$ in (8.38) has at most one pair of simple complex roots λ and $\bar{\lambda}$. Hence it follows that the conditions of Theorem 8.2 are not only necessary but also sufficient if both polynomials $\chi_1(\lambda)$ and $\chi_2(\lambda)$ are of degree at most four. Unfortunately, in general the conditions of Theorem 8.2 are only necessary for constancy of sign of the Green's function. For example, the Green's function of a fifth-order operator with tame characteristic polynomial $(\lambda+1)(\lambda^2+2\lambda+2)^2$ is of variable sign.

In simple cases, the condition that the characteristic polynomial be tame may be expressed directly in terms of its coefficients. For example, for third-order operators the characteristic polynomial

$$\chi(\lambda) = \lambda^3 + a\lambda^2 + b\lambda + c \tag{8.40}$$

is tame if either

$$c \neq 0, \qquad (2a^3 - 9ab + 27c)^2 + 4(3b - a^2)^3 \leqslant 0 \tag{8.41}$$

or

$$ac > 0, \qquad a(2a^3 - 9ab + 27c) \leqslant 0. \tag{8.42}$$

The first condition means that all the roots of polynomial (8.41) are real and nonvanishing. The second means that the roots $\lambda_1, \lambda_2, \lambda_3$ are in the same half-plane,

$$0 < |\lambda_1| \leqslant |\operatorname{Re}\lambda_2| = |\operatorname{Re}\lambda_3|,$$

if λ_1 is real and λ_2 and λ_3 are complex conjugates.

It would be interesting to find similar conditions for the characteristic polynomial of fourth-order operators to be tame.

8.5. Operators with variable coefficients

In this subsection we shall study the problem of existence and constancy of sign of the Green's function of an ap-operator

$$Lx = \frac{d^m x}{dt^m} + a_1(t) \frac{d^{m-1}x}{dt^{m-1}} + \dots + a_m(t)\, x \tag{8.43}$$

with variable coefficients. Simultaneously, we shall establish comparison theorems for the Green's function of the ap-operator (8.43) and the Green's function of a certain regular ap-operator

$$L_0 x = \frac{d^m x}{dt^m} + b_1(t) \frac{d^{m-1}x}{dt^{m-1}} + \dots + b_m(t)\, x. \tag{8.44}$$

The basic idea of the proofs presented below was already known to Lyapunov. It has been used in different forms by many authors investigating the existence and constancy of sign of Green's functions of boundary-value problems.

Theorem 8.6. Let the Green's function $G_0(t, s)$ *of the regular operator (8.44) be of constant sign. Suppose there exists an ap-function* $u_0(t) \in B^m(R^1)$ *such that*

$$u_0(t) \gg 0, \quad \varepsilon L_0 u_0(t) \gg 0, \quad \varepsilon L u_0(t) \gg 0, \tag{8.45}$$

where $\varepsilon = 1$ *if* $G_0(t, s) \geqslant 0$ *and* $\varepsilon = -1$ *if* $G_0(t, s) \leqslant 0$. *Suppose that for* $t \neq s$

$$LG_0(t, s) \leqslant 0. \tag{8.46}$$

Then the ap-operator (8.43) is regular and its Green's function satisfies the inequality

$$\varepsilon G(t, s) \geqslant \varepsilon G_0(t, s) \qquad (-\infty < t, s < \infty). \tag{8.47}$$

Proof. To fix ideas, we assume that $G_0(t, s) \geqslant 0$. The reasoning in the case $G_0(t, s) \leqslant 0$ is analogous.

The equation

$$Lx = f(t) \qquad (f(t) \in B(R^1)) \tag{8.48}$$

may be written

$$L_0x - Dx = f(t), \tag{8.49}$$

where

$$Dx = \sum_{k=1}^{m} [b_k(t) - a_k(t)] \frac{d^{m-k}x}{dt^{m-k}}. \tag{8.50}$$

Under the change of variables

$$x = L_0^{-1}y \tag{8.51}$$

equation (8.49) becomes

$$y - DL_0^{-1}y = f(t). \tag{8.52}$$

The operator DL_0^{-1} is continuous in the space of ap-functions. Moreover, by (8.46) the operator DL_0^{-1} is positive with respect to the cone $\hat{K}_+$ of nonnegative functions in $B(R^1)$. Set

$$y_0(t) = L_0u_0(t).$$

It follows from (8.45) that

$$DL_0^{-1}y_0(t) \ll y_0(t).$$

Therefore (see Theorem 5.1) the spectral radius of the operator DL_0^{-1} is less than one, and so

$$y(t) = f(t) + DL_0^{-1}f(t) + (DL_0^{-1})^2 f(t) + \ldots \tag{8.53}$$

Together with (8.51), this implies that the ap-operator (8.43) is regular.

To complete the proof, note that (8.51) and (8.52) imply the inequality

$$L_0x(t) \geqslant f(t) \qquad (-\infty < t < \infty). \tag{8.54}$$

This inequality can be written

$$L^{-1}f(t) \geqslant L_0^{-1}f(t) \qquad (-\infty < t < \infty). \tag{8.55}$$

This in turn means that

$$G(t, s) \geqslant G_0(t, s) \qquad (-\infty < t, s < \infty)$$

Q. E. D.

Consider the family of operators

$$L(\mu)x = \frac{d^m x}{dt^m} + [\mu a_1(t) + (1-\mu) b_1(t)] \frac{d^{m-1}x}{dt^{m-1}} + \dots$$
$$\dots + [\mu a_m(t) + (1-\mu) b_m(t)]\, x, \qquad (8.56)$$

depending on a scalar parameter $\mu \in [0, 1]$. Obviously,

$$L(\mu)x = L_0 x - \mu D x. \qquad (8.57)$$

Therefore, under the assumptions of Theorem 8.6, each of the operators $L(\mu)$ $(0 \leqslant \mu \leqslant 1)$ is regular. Denote the Green's function of the ap-operator (8.57) by $G(t, s; \mu)$. From (8.7) it follows that

$$G(t, s; \mu) = \begin{cases} (U(t; \mu) P_+(\mu) U^{-1}(s; \mu) e_m, e_1), & \text{if } t \geqslant s, \\ -(U(t; \mu) P_-(\mu) U^{-1}(s; \mu) e_m, e_1), & \text{if } t < s. \end{cases}$$

By Theorem 4.3, the projections $P_+(\mu)$ and $P_-(\mu)$ depend continuously on $\mu \in [0, 1]$, and hence the dimensions of the subspaces $P_+(\mu)R^m$ and $P_-(\mu)R^m$ are the same for all $\mu \in [0, 1]$. In particular, this implies the following proposition:

The trivial solution of the equation $Lx = 0$ *is exponentially stable if the trivial solution of the equation* $L_0 x = 0$ *is exponentially stable.*

We now cite a few corollaries of Theorem 8.6.

Set

$$L_0 x = \left(\frac{d}{dt} - \lambda_1\right) \dots \left(\frac{d}{dt} - \lambda_m\right) x, \qquad (8.58)$$

where

$$\lambda_1 > \lambda_2 > \dots > \lambda_k > 0 > \lambda_{k+1} > \dots > \lambda_m. \qquad (8.59)$$

The ap-operator (8.58) is regular and its Green's function $G_0(t, s)$ *is nonnegative if* k *is even, nonpositive if* k *is odd.*

This function can be expressed in the explicit form

$$G_0(t, s) = \begin{cases} \displaystyle\sum_{j=k+1}^{m} \frac{e^{\lambda_j (t-s)}}{\chi_0'(\lambda_j)}, & \text{if } t \geqslant s, \\ \displaystyle -\sum_{j=1}^{k} \frac{e^{\lambda_j (t-s)}}{\chi_0'(\lambda_j)}, & \text{if } t < s, \end{cases} \qquad (8.60)$$

where

$$\chi_0(\lambda) = (\lambda - \lambda_1) \dots (\lambda - \lambda_m). \qquad (8.61)$$

Set

$$H(t, s)=\begin{cases} -\sum\limits_{j=k+1}^{m}\dfrac{\chi(t;\lambda_j)}{\chi_0'(\lambda_j)}e^{\lambda_j(t-s)}, & \text{if}\quad t\geqslant s,\\ \sum\limits_{j=1}^{k}\dfrac{\chi(t;\lambda_j)}{\chi_0'(\lambda_j)}e^{\lambda_j(t-s)}, & \text{if}\quad t<s.\end{cases}$$

Here $\chi(t;\lambda)$ is the "characteristic polynomial"

$$\chi(t;\lambda)=\lambda^m+a_1(t)\lambda^{m-1}+\ldots+a_m(t) \qquad (8.62)$$

of the ap-operator (8.43)

Note that

$$DL_0^{-1}x(t)=\int_{-\infty}^{\infty}H(t,s)x(s)\,ds.$$

Theorem 8.7. Let

$$H(t,s)\geqslant 0 \qquad (-\infty<t,\ s<\infty) \qquad (8.63)$$

and

$$(-1)^k a_m(t)\gg 0. \qquad (8.64)$$

Then the ap-operator (8.43) is regular and its Green's function satisfies the inequality

$$a_m(t)G(t,s)\geqslant 0 \qquad (-\infty<t,\ s<\infty). \qquad (8.65)$$

Proof. Inequality (8.46) follows from (8.63). Now, inequality (8.64) means that (8.45) is valid with $u_0(t)\equiv 1$. Thus, by Theorem 8.6, the ap-operator (8.43) is regular and its Green's function satisfies (8.65). Q. E. D.

To apply Theorem 8.7, we must know how to select numbers $\lambda_1,\ldots,\lambda_m$ satisfying inequalities (8.59) for which (8.63) is valid. We know of no general prescription, but in certain simple cases, the procedure is self-evident.

Let $\lambda_1(t),\ldots,\lambda_m(t)$ denote the roots of the equation

$$\chi(t;\lambda)=0. \qquad (8.66)$$

We shall assume that the roots are simple and real and that

$$\alpha_j\leqslant\lambda_j(t)\leqslant\beta_j \qquad (j=1,\ldots,m), \qquad (8.67)$$

where

$$\beta_1 > \alpha_1 > \ \ldots \ > \beta_k > \alpha_k > 0 > \beta_{k+1} > \alpha_{k+1} > \ \ldots \ > \beta_m > \alpha_m.$$

That (8.64) holds is obvious; inequality (8.63) will also hold if we set

$$\lambda_1 = \beta_1, \ldots, \lambda_k = \beta_k; \ \ \lambda_{k+1} = \alpha_{k+1}, \ldots, \lambda_m = \alpha_m.$$

We have thus proved

Theorem 8.8. Let the roots $\lambda_j(t)$ $(j = 1, \ldots, m)$ *of equation (8.66) be simple and real and satisfy (8.67).*

Then the ap-operator (8.43) is regular and its Green's function $G(t, s)$ *satisfies the inequality*

$$(-1)^k G(t, s) \geqslant 0 \qquad (-\infty < t, s < \infty). \tag{8.68}$$

The role of conditions (8.67) in investigation of the oscillatory properties of solutions of equations with variable coefficients was noticed in a different situation by Levin /2/.

In conclusion, we cite one more regularity test for the ap-operator (8.43), in terms of the properties of its "characteristic polynomial" (8.62).

Let $\lambda_1, \ldots, \lambda_m$ be arbitrary numbers. Consider the divided differences

$$\Delta^0(t; \lambda_1) = \chi(t; \lambda_1),$$

$$\Delta^1(t; \lambda_1, \lambda_2) = \frac{\Delta^0(t; \lambda_1) - \Delta^0(t; \lambda_2)}{\lambda_1 - \lambda_2},$$

$$\cdots\cdots\cdots\cdots\cdots\cdots\cdots$$

$$\Delta^{m-1}(t; \lambda_1, \ldots, \lambda_m) = \frac{\Delta^{m-2}(t; \lambda_1, \ldots, \lambda_{m-1}) - \Delta^{m-2}(t; \lambda_2, \ldots, \lambda_m)}{\lambda_1 - \lambda_m},$$

of the "characteristic polynomial" $\chi(t; \lambda)$. If some of the numbers λ_j are equal, the divided differences are understood as limits, i. e., the numbers λ_j are first given small perturbations and the limits of the quotients are then considered.

Denote by L_j $(j = 0, 1, \ldots, m)$ the ap-operators $L_0 = I$ and

$$L_j = \left(\frac{d}{dt} - \lambda_1\right) \ldots \left(\frac{d}{dt} - \lambda_j\right).$$

The corresponding characteristic polynomials are denoted by $\chi_j(\lambda)$ $(j = 0, 1, \ldots, m)$.

We need the following lemma.

Lemma 8.2. Let $\chi(\lambda)$ be an arbitrary polynomial of degree m $(m \geqslant 2)$.
Then

$$\chi(\lambda) = \sum_{j=0}^{m-1} \Delta^j(\lambda_1, \ldots, \lambda_{j+1}) \chi_j(\lambda) + \chi_m(\lambda), \tag{8.69}$$

where $\Delta^j(\lambda_1, \ldots, \lambda_{j+1})$ are the divided differences of $\chi(\lambda)$.

Proof. We prove (8.69) by induction on m.

Identity (8.69) is easily checked for $m = 2$. Let (8.69) be valid for any polynomial of degree $m-1$ $(m > 2)$; we prove that it is valid for polynomials of degree m.

Set

$$\theta(\lambda) = \frac{\chi(\lambda) - \chi(\lambda_1)}{\lambda - \lambda_1}, \qquad \theta_j(\lambda) = \frac{\chi_{j+1}(\lambda)}{\lambda - \lambda_1} \quad (j = 0, 1, \ldots, m-2).$$

In this notation, identity (8.69) is equivalent to

$$\theta(\lambda) = \sum_{j=0}^{m-2} \Delta^j(\lambda_2, \ldots, \lambda_{j+2}) \theta_j(\lambda) + \theta_{m-1}(\lambda), \tag{8.70}$$

where $\Delta^j(\lambda_2, \ldots, \lambda_{j+2})$ are the divided differences of the polynomial $\theta(\lambda)$ at the points $\lambda_2, \ldots, \lambda_m$. By the induction hypothesis, we have (8.70) and hence also (8.69). Q. E. D.

Theorem 8.9. Suppose there exist constants $\lambda_1 < 0, \ldots, \lambda_{m-1} < 0$ such that $\Delta^j(t; \lambda_1, \ldots, \lambda_{j+1}) \leqslant 0$

$$(j = 0, 1, \ldots, m-2; \ -\infty < t < \infty). \tag{8.71}$$

Let $u_0(t) \in B^m(R^1)$ be an ap-function such that

$$u_0(t) \gg 0, \qquad Lu_0(t) \gg 0 \tag{8.72}$$

and

$$L_{m-1}u_0(t) \gg 0. \tag{8.73}$$

Then the ap-operator (8.43) is regular, its Green's function is nonnegative and the trivial solution of the equation $Lx = 0$ is stable.

Proof. By induction on m, one can prove that

$$\Delta^{m-1}(t; \lambda_1, \ldots, \lambda_m) = a_1(t) + \lambda_1 + \ldots + \lambda_m. \tag{8.74}$$

The proof of this simple proposition is omitted.

Henceforth we shall assume that the negative number λ_m is so large in absolute value that

$$\Delta^{m-1}(t; \lambda_1, \ldots, \lambda_m) \leqslant 0 \qquad (-\infty < t < \infty) \tag{8.75}$$

and

$$L_m u_0(t) = \left(\frac{d}{dt} - \lambda_m\right) L_{m-1} u_0(t) \gg 0.$$

By Lemma 8.2, the ap-operator (8.43) can be written

$$Lx = \sum_{j=0}^{m-1} \Delta^j (t; \lambda_1, \ldots, \lambda_{j+1}) L_j x + L_m x. \tag{8.76}$$

Consequently, by the change of variables $x = L_m^{-1} y$ the equation

$$Lx = f(t)$$

becomes

$$y - Dy = f(t),$$

where

$$Dy(t) = -\sum_{j=0}^{m-1} \Delta^j (t; \lambda_1, \ldots, \lambda_{j+1}) L_j L_m^{-1}.$$

It follows from (8.71) and (8.75) that the continuous operator D is positive with respect to the cone $\hat{K}_+$ of nonnegative functions in $B(R^1)$. The proof is now completed in the same way as that of Theorem 8.6 (the operator L_m plays the same role here as L_0 in the assumptions of that theorem). Q. E. D.

A proposition similar to Theorem 8.9 may be found in Berezhnoi and Kolesov /1/.

§ 9. SECOND-ORDER SCALAR AP-OPERATORS

9.1. Connection with oscillatory properties of solutions of the homogeneous equation

This section is devoted to an investigation of the Green's function of the ap-operator

$$Lx = \frac{d^2x}{dt^2} + p(t)\frac{dx}{dt} + q(t)x. \tag{9.1}$$

The operator (9.1) is a particular case of the scalar operators investigated in § 8. Hence, all the propositions of § 8 are valid for this operator. In particular, by Theorem 8.1, if the Green's function $G(t, s)$ of the operator (9.1) is nonnegative it is strongly positive, and if it is nonpositive, it is strongly negative.

Below we establish several theorems concerning the existence and constancy of sign of the Green's function of the ap-operator (9.1) which do not follow from the results of the preceding section.

The solutions of the homogeneous equation

$$\frac{d^2x}{dt^2} + p(t)\frac{dx}{dt} + q(t)x = 0 \tag{9.2}$$

are said to be *nonoscillatory* on $(-\infty, \infty)$ if every nontrivial solution $x(t)$ of this equation vanishes at most once.

Theorem 9.1. Let the operator (9.1) be regular.

Then its Green's function $G(t, s)$ is of constant sign if and only if the solutions of the homogeneous equation are nonoscillatory.

This subsection presents only part of the proof. Let us show first that nonoscillation of the solutions is a necessary condition for the Green's function to be of constant sign. Indeed, if a nontrivial solution $x_*(t)$ of equation (9.2) vanishes at two points t_1 and t_2 $(t_1 < t_2)$ then, by Sturm theory (see, e.g., Petrovskii /1/), every solution $x(t)$ of equation (9.2) other than $x_*(t)$ changes sign on the interval (t_1, t_2). In particular, the Green's function $G(t, s)$ changes sign on (t_1, t_2) either for $s < t_1$ or for $s > t_2$, so that it cannot be of constant sign.

The converse is easy to prove if the trivial solution of equation (9.2) is stable in Lyapunov's sense, for either increasing or decreasing t. To fix ideas, let the trivial solution be stable as $t \to \infty$. Then

$$G(t, s) = \begin{cases} K(t, s), & \text{if} \quad t \geqslant s, \\ 0, & \text{if} \quad t < s, \end{cases} \tag{9.3}$$

where $K(t, s)$ is the Cauchy function of equation (9.2), and the Cauchy function is positive for $t > s$.

The case in which the trivial solution of equation (9.2) is unstable both as $t \to \infty$ and as $t \to -\infty$ will be examined in Section 9.7.

We shall now investigate in detail the properties of solutions of equation (9.2). Theorem 9.1 allows us to confine ourselves to the case in which the solutions of equation (9.2) are nonoscillatory.

In subsequent constructions we shall assume that

$$M(p)=\lim_{T\to\infty}\frac{1}{2T}\int_{-T}^{T}p(t)\,dt\geqslant 0. \tag{9.4}$$

This involves no loss of generality, since if necessary we need only substitute $t=-\tau$ in equation (9.2).

9.2. The Riccati equation

The properties of solutions of equations (9.2) are closely bound up with the properties of solutions of the Riccati equation

$$\frac{dz}{dt}=z^2-p(t)z+q(t), \tag{9.5}$$

which is obtained from (9.2) by setting

$$z=-x'/x. \tag{9.6}$$

Therefore, both here and in Section 9.3—9.5 we shall devote close attention to the properties of solutions of equation (9.5), on the assumption that the solutions of equation (9.2) are nonoscillatory on $(-\infty, \infty)$.

Obviously, there exists $m>0$ such that every solution $z(t)$ of equation (9.5) with initial value $z(t_0)\geqslant m$ escapes to ∞ in a finite time with increasing t, and every solution $z(t)$ with initial value $z(t_0)\leqslant -m$ escapes to $-\infty$ in a finite time with decreasing t. On the other hand, since the solutions of equation (9.2) are nonoscillatory on $(-\infty, \infty)$, it follows that for any t_0 we can find initial values $z(t_0)$ for which the solutions are globally continuable in the direction of increasing time, and also initial values for which the solutions are globally continuable in the direction of decreasing time. Let Z^+ denote the set of values at $t=t_0$ of the first group of solutions, and Z^- the analogous set for solutions in the second group. Let $z_1(t)$ be the solution of equation (9.5) with initial value

$$z_1(t_0)=\inf\{z: z\in Z^-\},$$

and $z_2(t)$ the solution of this equation with initial value

$$z_2(t_0)=\sup\{z: z\in Z^+\}.$$

Then $z_1(t)$ is defined for all $t \leqslant t_0$, and $z_2(t)$ for all $t \geqslant t_0$.

We claim that

$$z_1(t_0) \leqslant z_2(t_0). \tag{9.7}$$

If this is false, the solution $z(t)$ of equation (9.5) with initial value $z(t_0)$ such that

$$z_1(t_0) > z(t_0) > z_2(t_0)$$

goes to ∞ in a finite time with increasing t and to $-\infty$ with decreasing t. This means that the solution $x(t)$ of equation (9.2) with initial values

$$x(t_0) = 1, \quad x'(t_0) = -z(t_0)$$

vanishes at least twice. But this contradicts the assumption that the solutions of equation (9.2) are nonoscillatory on $(-\infty, \infty)$.

Inequality (9.7) implies that $z_1(t)$ and $z_2(t)$ are the solutions of equation (9.5) bounded on the whole real line. The solutions $z_1(t)$ and $z_2(t)$ do not depend on the choice of the initial time t_0.

We shall call $z_1(t)$ and $z_2(t)$ the extremal solutions of the Riccati equation (9.5). They may, of course, coincide.

Lemma 9.1. Let equation (9.5) have two distinct extremal solutions $z_1(t)$ and $z_2(t)$.

Then for any solution $z(t)$ with initial value $z(t_0)$ such that

$$z_1(t_0) < z(t_0) < z_2(t_0),$$

we have the limiting equalities

$$\lim_{t\to\infty} [z(t) - z_1(t)] = 0 \tag{9.8}$$

and

$$\lim_{t\to-\infty} [z_2(t) - z(t)] = 0. \tag{9.9}$$

Proof. Equalities (9.8) and (9.9) are proved in the same way, so we shall prove only (9.8).

Set

$$h(t) = z(t) - z_1(t) \qquad (t \geqslant t_0).$$

It is easy to see that

$$h'(t) = h^2(t) - [p(t) - 2z_1(t)]\,h(t),$$

and therefore

$$h(t) = \frac{h(t_0)\exp\left\{-\int\limits_{t_0}^{t}[p(\tau)-2z_1(\tau)]\,d\tau\right\}}{1-h(t_0)\int\limits_{t_0}^{t}\left\{\exp\left[-\int\limits_{t_0}^{s}[p(\tau)-2z_1(\tau)]\,d\tau\right]\right\}ds}.$$

The function $h(t)$ is defined and positive for all $t \geqslant t_0$. Hence

$$\int\limits_{t_0}^{t}\left\{\exp\left[-\int\limits_{t_0}^{s}[p(\tau)-2z_1(\tau)]\,d\tau\right]\right\}ds \leqslant \frac{1}{z(t_0)-z_1(t_0)} \quad (t \geqslant t_0)$$

and, since $z(t_0)$ may be arbitrarily near $z_2(t_0)$,

$$\int\limits_{t_0}^{t}\left\{\exp\left[-\int\limits_{t_0}^{s}[p(\tau)-2z_1(\tau)]\,d\tau\right]\right\}ds \leqslant \frac{1}{z_2(t_0)-z_1(t_0)} \quad (t \geqslant t_0).$$

From the last two inequalities it follows that

$$\lim_{t\to\infty}\exp\left\{-\int\limits_{t_0}^{t}[p(\tau)-2z_1(\tau)]\,d\tau\right\} = 0.$$

Equality (9.8) now follows from the inequalities

$$0 < h(t) \leqslant \frac{h(t_0)\exp\left\{-\int\limits_{t_0}^{t}[p(\tau)-2z_1(\tau)]\,d\tau\right\}}{[z_2(t_0)-z(t_0)]\int\limits_{t_0}^{t}\left\{\exp\left[-\int\limits_{t_0}^{s}[p(\tau)-2z_1(\tau)]\,d\tau\right]\right\}ds}.$$

Q. E. D.

Propositions of this type were first noted, apparently, by Sobol' /1/.

Later on, the inequalities*

$$\varlimsup_{T\to\infty}\frac{1}{T}\int\limits_{t_0}^{T} z_1(t)\,dt + \varliminf_{T\to\infty}\frac{1}{T}\int\limits_{t_0}^{T} z(t)\,dt \leqslant M(p), \tag{9.10}$$

* To prove them one must express $z_1(t)+z(t)$ and $z_2(t)+z(t)$ in terms of the corresponding solutions of equation (9.2) and then use Liouville's well-known formula for the Wronskians.

$$\overline{\lim_{T\to-\infty}} \frac{1}{T}\int_{t_0}^{T} z_2(t)\,dt + \overline{\lim_{T\to-\infty}} \frac{1}{T}\int_{t_0}^{T} z(t)\,dt \geqslant M(p), \tag{9.11}$$

will be useful; in (9.10), $z(t)$ denotes a solution of equation (9.5) with initial value $z(t_0) < z_1(t_0)$, while in (9.11) $z(t)$ is a solution of (9.5) with initial value $z(t_0) > z_2(t_0)$.

Let equation (9.5) have two distinct extremal solutions which are almost periodic functions. The reader is invited to prove that in this case

$$\inf_{-\infty<t<\infty} [z_2(t) - z_1(t)] > 0, \tag{9.12}$$

whence it follows that

$$M(z_1) + M(z_2) = M(p). \tag{9.13}$$

9.3. Differential inequalities for the Riccati equation

The subsequent investigation will rely essentially on the following theorems on differential inequalities.

Theorem 9.2. Let $z_0(t)$ be continuous on the entire real line and satisfy the differential inequality

$$D^*z_0(t) \geqslant z_0^2(t) - p(t)z_0(t) + q(t) \qquad (-\infty < t < \infty), \tag{9.14}$$

where $D^*z_0(t)$ is the right upper derivative of the function $z_0(t)$.

Then the solutions of equation (9.2) are nonoscillatory on $(-\infty, \infty)$ and

$$z_1(t) \leqslant z_0(t) \leqslant z_2(t) \qquad (-\infty < t < \infty). \tag{9.15}$$

Proof. By (9.14),

$$z(t) \leqslant z_0(t) \qquad (t_0 \leqslant t < \infty),$$

where $z(t)$ is the solution of (9.5) with initial value $z(t_0) = z_0(t_0)$. Therefore, the solution

$$x(t) = \exp\left[-\int_{t_0}^{t} z(\tau)\,d\tau\right]$$

of equation (9.2) is defined and positive for all $t \geqslant t_0$. It follows from Sturm's theory that no nontrivial solution of equation (9.2) can vanish twice on the interval (t_0, ∞). Since t_0 is arbitrary, the solutions of equation (9.2) are nonoscillatory on the whole real line.

It remains to prove inequalities (9.15).

We show, for example, that

$$z_0(t) \leqslant z_2(t) \qquad (-\infty < t < \infty).$$

Suppose this is false, so that at some time t_0

$$z_0(t_0) > z_2(t_0). \tag{9.16}$$

Then it follows from (9.14) and (9.16) that

$$z_0(t) \geqslant z(t) \qquad (t \geqslant t_0), \tag{9.17}$$

where $z(t)$ is a solution of equation (9.5) with initial value $z(t_0) > z_2(t_0)$. It follows from (9.17) that any solution $z(t)$ with initial value $z(t_0)$ greater than $z_2(t_0)$ is globally continuable in the direction of increasing time. This is a contradiction.

The inequality

$$z_0(t) \geqslant z_1(t) \qquad (-\infty < t < \infty)$$

is proved in a similar manner. Q. E. D.

The assertion of Theorem 9.2 has been noted by many authors.

Theorem 9.3. Let $\varphi(t)$ *be an ap-function with uniformly continuous derivative* $\varphi'(t)$, *satisfying the differential inequality*

$$\varphi'(t) \leqslant \varphi^2(t) - p(t)\varphi(t) + q(t), \tag{9.18}$$

and suppose $\varphi(t)$ *is not a solution of equation (9.5).*

Then either

$$\inf_{-\infty < t < \infty} [\varphi(t) - z_2(t)] > 0 \tag{9.19}$$

or

$$\inf_{-\infty < t < \infty} [z_1(t) - \varphi(t)] > 0. \tag{9.20}$$

Proof. Suppose that at some time

$$z_1(t_0) \leqslant \varphi(t_0) \leqslant z_2(t_0). \tag{9.21}$$

We shall show that the inequality

$$\varphi(t) \leqslant z_2(t) \tag{9.22}$$

cannot hold for all $t \leqslant t_0$.

Otherwise, by virtue of Lemma 9.1,

$$\lim_{t \to -\infty} [z_2(t) - \varphi(t)] = 0. \tag{9.23}$$

Select a sequence $t_n \to -\infty$ such that the limits

$$\varphi_*(t) = \lim_{n\to\infty} \varphi(t + t_n), \qquad \varphi'_*(t) = \lim_{n\to\infty} \varphi'(t + t_n),$$
$$p^*(t) = \lim_{n\to\infty} p(t + t_n), \qquad q^*(t) = \lim_{n\to\infty} q(t + t_n)$$

exist uniformly in $t \in (-\infty, \infty)$, and the sequence of functions $z_2(t + t_n)$ converges uniformly on every finite interval to a function $z_2^*(t)$. Obviously, $z_2^*(t)$ is a solution of the equation

$$z' = z^2 - p^*(t)z + q^*(t) \tag{9.24}$$

bounded on the whole real line. The function $\varphi_*(t)$ satisfies the differential inequality

$$\varphi'_*(t) \leqslant \varphi_*^2(t) - p^*(t)\varphi_*(t) + q^*(t) \tag{9.25}$$

and is not a solution of equation (9.24). Now, it follows from inequality (9.22) that

$$\varphi_*(t) \leqslant z_2^*(t) \tag{9.26}$$

for all $t \in (-\infty, \infty)$, and from (9.23) that

$$\varphi_*(0) = z_2^*(0). \tag{9.27}$$

It follows from (9.25)—(9.27) that the function $\varphi_*(t)$ satisfies the equation (9.24) for $-\infty < t \leqslant 0$, and since $\varphi_*(t)$ is almost periodic it is a solution of equation (9.24) for $0 \leqslant t < \infty$ also. This is a contradiction.

Similarly, one proves that the inequality

$$\varphi(t) \geqslant z_1(t)$$

cannot hold for all $t \geqslant t_0$. Hence there exist t_1 and t_2 such that

$$\varphi(t_1) > z_2(t_1) \tag{9.28}$$

and

$$\varphi(t_2) < z_1(t_2). \tag{9.29}$$

It follows from (9.28), (9.29) and (9,18) that

$$\varphi(t) > z_2(t) \qquad (-\infty < t \leqslant t_1)$$

and

$$\varphi(t) < z_1(t) \qquad (t_2 \leqslant t < \infty).$$

Reasoning as in the proof that inequality (9.22) cannot hold, we see that in fact

$$\varphi(t) \geqslant z_2(t) + \delta \qquad (-\infty < t \leqslant t_1) \tag{9.30}$$

and

$$\varphi(t) \leqslant z_1(t) - \delta \qquad (t_2 \leqslant t < \infty), \tag{9.31}$$

where δ is positive. By (9.30), (9.31) and (9.10), (9.11) we have

$$2M(\varphi) \geqslant M(p) + \delta$$

and

$$2M(\varphi) \leqslant M(p) - \delta,$$

But these inequalities are contradictory.

Thus, inequalities (9.21) cannot hold for any t_0 and so we necessarily have one of the inequalities

$$\varphi(t) < z_1(t) \qquad (-\infty < t < \infty)$$

or

$$\varphi(t) > z_2(t) \qquad (-\infty < t < \infty).$$

Hence (using arguments similar to those used to disprove (9.22)), we see that either (9.19) or (9.20) must hold. Q. E. D.

9.4. Almost periodic solutions of the Riccati equation

We shall say that equation (9.2) is coarsely nonoscillatory on $(-\infty, \infty)$ if there is a nonnegative nonzero ap-function $b(t)$ such that the solutions of the equation

$$Lx + b(t)x = 0 \tag{9.32}$$

are nonoscillatory.

Theorem 9.4. The Riccati equation (9.5) has two distinct almost periodic extremal solutions $z_1(t)$ and $z_2(t)$ if and only if equation (9.2) is coarsely nonoscillatory on $(-\infty, \infty)$.

Proof. Sufficiency. Consider the auxiliary equation

$$u'(t) = u^2(t) - p(t)u(t) + q(t) + b(t), \tag{9.33}$$

where $b(t)$ is a nonnegative nonzero ap-function such that the solutions of equation (9.2) are nonoscillatory on $(-\infty, \infty)$. Let the extremal solutions of the Riccati equation (9.33) be $u_1(t)$ and $u_2(t)$. These functions satisfy the inequalities

$$\begin{aligned} u_1'(t) &\geqslant u_1^2(t) - p(t)u_1(t) + q(t), \\ u_2'(t) &\geqslant u_2^2(t) - p(t)u_2(t) + q(t). \end{aligned} \tag{9.34}$$

It follows from (9.34) and Theorem 9.2 that

$$z_1(t) \leqslant u_1(t) \leqslant u_2(t) \leqslant z_2(t) \qquad (-\infty < t < \infty) \tag{9.35}$$

and, since $b(t)$ is a nonzero function,

$$z_1(t) < z_2(t) \qquad (-\infty < t < \infty). \tag{9.36}$$

We shall show that in fact a stronger inequality

$$\inf_{-\infty < t < \infty} [z_2(t) - z_1(t)] > 0 \tag{9.37}$$

is valid.

If inequality (9.37) is false, there exists a sequence t_j $(j = 1, 2, \ldots)$ such that

$$\lim_{j \to \infty} [z_2(t_j) - z_1(t_j)] = 0. \tag{9.38}$$

Without loss of generality, we may assume that the sequences of ap-functions $p(t + t_j)$, $q(t + t_j)$ and $b(t + t_j)$ converge uniformly on the whole real line to ap-functions $p^*(t)$, $q^*(t)$ and $b^*(t)$, and the sequences $z_1(t+t_j)$, $z_2(t+t_j)$, $u_1(t+t_j)$, $u_2(t+t_j)$ converge uniformly on any finite interval to functions $z_1^*(t)$, $z_2^*(t)$, $u_1^*(t)$, $u_2^*(t)$.

The functions $z_1^*(t)$, $z_2^*(t)$ and $u_1^*(t)$, $u_2^*(t)$ are solutions, bounded on the whole real line, of the equations

$$\frac{dz}{dt}=z^2-p^*(t)z+q^*(t) \tag{9.39}$$

and

$$\frac{du}{dt}=u^2-p^*(t)u+q^*(t)+b^*(t), \tag{9.40}$$

respectively. Inequalities (9.35) imply

$$z_1^*(t)\leqslant u_1^*(t)\leqslant u_2^*(t)\leqslant z_2^*(t) \qquad (-\infty<t<\infty).$$

Hence, by (9.38), it follows that

$$z_1^*(t)\equiv u_1^*(t)\equiv u_2^*(t)\equiv z_2^*(t).$$

But this is impossible, for $b^*(t)$ is not identically zero. Note that $z_1^*(t)$ and $z_2^*(t)$ are the extremal solutions of equation (9.39). To prove this, one uses Lemma 9.1 and repeats the arguments used previously.

Let the sequence t_j $(j=1, 2, \ldots)$ be such that the sequences of ap-functions $p(t+t_j)$ and $q(t+t_j)$ converge uniformly on the whole real line to ap-functions $p^*(t)$ and $q^*(t)$, and the sequence $z_1(t+t_j)$ converges uniformly on any finite interval to the extremal solution $z_1^*(t)$ of equation (9.39). We claim that the sequence $z_1(t+t_j)$ converges uniformly on the entire real line to $z_1^*(t)$. Otherwise, there exist a sequence of numbers s_k, a sequence of indices j_k and a positive number ε_0 such that

$$\left|z_1\left(t_{j_k}+s_k\right)-z_1^*(s_k)\right|>\varepsilon_0 \qquad (k=1, 2, \ldots). \tag{9.41}$$

We may assume here that the limits $t\in(-\infty, \infty)$

$$p^{**}(t)=\lim_{k\to\infty}p\left(t+t_{j_k}+s_k\right)=\lim_{k\to\infty}p^*(t+s_k),$$
$$q^{**}(t)=\lim_{k\to\infty}q\left(t+t_{j_k}+s_k\right)=\lim_{k\to\infty}q^*(t+s_k)$$

exist uniformly in $t\in(-\infty,\infty)$. We may also assume that the sequences $z_1(t+t_{j_k}+s_k)$ and $z_1^*(t+s_k)$ converge uniformly on any finite interval to solutions, bounded on the real line, of the equation

$$\frac{dz}{dt}=z^2-p^{**}(t)z+q^{**}(t).$$

As noted above, in this situation extremal solutions may converge only to extremal solutions, with upper (lower) extremal solutions converging to an upper (lower) extremal solution. Thus

$$\lim_{k\to\infty} z_1(t_{j_k}+s_k)=\lim_{k\to\infty} z_1^*(s_k).$$

This contradicts (9.41).

We have thus proved that the upper extremal solution $z_1(t)$ is almost periodic. One proves similarly that the lower extremal solution $z_2(t)$ is also almost periodic.

Necessity. Let $z_1(t)$ and $z_2(t)$ be distinct extremal solutions of equation (9.5) which are ap-functions. We shall show that for sufficiently small μ the solutions of the equation

$$Lx+\mu x=0 \tag{9.42}$$

are nonoscillatory on $(-\infty, \infty)$. This will complete the proof.

Let $z = h + z_2(t)$, and transform the Riccati equation

$$\frac{dz}{dt}=z^2-p(t)z+q(t)+\mu \tag{9.43}$$

to the form

$$\frac{dh}{dt}=h^2+[2z_2(t)-p(t)]h+\mu. \tag{9.44}$$

It follows from (9.13) that $2M(z_2) - M(p) > 0$. Thus, the existence of almost periodic solutions of equation (9.44) is equivalent to solvability of the integral equation

$$h(t)=-\int_t^\infty \exp\left\{\int_s^t [2z_2(\tau)-p(\tau)]\,d\tau\right\}h^2(s)\,ds-$$
$$-\mu\int_t^\infty \exp\left\{\int_s^t [2z_2(\tau)-p(\tau)]\,d\tau\right\}ds \tag{9.45}$$

in the space $B(R^1)$. For small μ, the operator on the right of (9.45) maps a ball of sufficiently small radius into itself, acting there as a contraction operator. Hence, for small μ the Riccati equation (9.43) has solutions bounded on the real line. It then follows from Theorem 9.2 that the solutions of equation (9.42) are nonoscillatory on $(-\infty, \infty)$. Q. E. D.

A direct consequence of the proof is the following proposition:

Equation (9.2) is coarsely nonoscillatory on $(-\infty,\infty)$ *if and only if its solutions are nonoscillatory on* $(-\infty,\infty)$ *and the extremal solutions* $z_1(t)$ *and* $z_2(t)$ *of the Riccati equation (9.5) satisfy inequality (9.37).*

We mention one more result.

Theorem 9.5. Suppose there exists an ap-function $\varphi(t)$, *whose derivative is uniformly continuous, such that*

$$\frac{d\varphi(t)}{dt} \geqslant \varphi^2(t) - p(t)\varphi(t) + q(t) \qquad (-\infty < t < \infty), \tag{9.46}$$

but $\varphi(t)$ *is not a solution of equation (9.5).*

Then the Riccati equation (9.5) has two distinct almost periodic extremal solutions $z_1(t)$ *and* $z_2(t)$ *and*

$$z_1(t) < \varphi(t) < z_2(t) \qquad (-\infty < t < \infty). \tag{9.47}$$

To prove the first part of this theorem, it is sufficient to note that the solutions of equation (9.32) are nonoscillatory on $(-\infty, \infty)$ if

$$b(t) = \varphi'(t) - \varphi^2(t) + p(t)\varphi(t) - q(t).$$

Inequalities (9.47) follow from Theorem 9.2 and from the fact that $\varphi(t)$ is not a solution of (9.5).

Propositions similar to Theorems 9.4 and 9.5 may be found in Halanay /1/, Markus and Moore /1/.

9.5. Stability of solutions of equations (9.2)

The second-order equation (9.2) is equivalent to a system of two differential equations

$$\left.\begin{aligned} \frac{du_1}{dt} &= u_2, \\ \frac{du_2}{dt} &= -q(t)u_1 - p(t)u_2. \end{aligned}\right\} \tag{9.48}$$

Let $U(t, s)$ denote the fundamental matrix of this system. In this subsection we shall investigate conditions for the validity of the estimate

$$|U(t, s)| \leqslant Me^{-\gamma(t-s)} \qquad (\gamma > 0;\ t \geqslant s). \tag{9.49}$$

Theorem 9.6. Let the solutions of equation (9.2) be nonoscillatory on $(-\infty, \infty)$.

Then inequality (9.49) is valid if and only if there exists an ap-function $\varphi(t)$ *with uniformly continuous derivative such that*

1. $$\frac{d\varphi(t)}{dt} \leqslant \varphi^2(t) - p(t)\varphi(t) + q(t) \qquad (-\infty < t < \infty), \tag{9.50}$$

but $\varphi(t)$ *is not a solution of the Riccati equation (9.5);*

2. $$0 \leqslant M(\varphi) \leqslant \frac{1}{2} M(p). \tag{9.51}$$

Proof. Sufficiency. From inequalities (9.11), (9.51) and Theorem 9.3, we obtain the inequality

$$z_1(t) \geqslant \varphi(t) + \delta \qquad (-\infty < t < \infty), \tag{9.52}$$

where δ is a positive number.

Denote by $z(t)$ the solution of equation (9.5) satisfying the initial condition

$$z(s) = z_1(s) - \delta/2.$$

The functions

$$x_1(t, s) = \exp\left[-\int_s^t z_1(\tau)\, d\tau\right]$$

and

$$x_2(t, s) = \exp\left[-\int_s^t z(\tau)\, d\tau\right]$$

are solutions of equation (9.2). The matrix

$$\begin{Vmatrix} 1 & 1 \\ -z_1(s) & -z_1(s) + \delta/2 \end{Vmatrix}$$

and its inverse are uniformly bounded in the norm for all $s \in (-\infty, \infty)$. Using this and inequality (9.52), we see that the conditions of our theorem are indeed sufficient.

Necessity. Consider the auxiliary equation

$$Lx - \mu x = 0, \tag{9.53}$$

where μ is a positive number. The regularity of operator L implies that for sufficiently small μ the norm of the fundamental matrix $U(t, s; \mu)$ of the system

$$\frac{du_1}{dt} = u_2,$$
$$\frac{du_2}{dt} = -[q(t) - \mu]\, u_1 - p(t)\, u_2$$

satisfies the inequality

$$|U(t, s; \mu)| \leqslant M_1 e^{-\gamma_1 (t-s)} \qquad (\gamma_1 > 0;\ t \geqslant s). \tag{9.54}$$

It follows from Theorem (9.4) that the extremal solutions $z_1(t; \mu)$ and $z_2(t; \mu)$ of the Riccati equation

$$\frac{dz}{dt} = z^2 - p(t)z + q(t) - \mu$$

are ap-functions. Now, it follows from (9.54) that

$$0 < M[z_1(t; \mu)] < M[z_2(t; \mu)].$$

Finally, by (9.13),

$$M[z_1(t; \mu)] < \frac{1}{2} M(p).$$

Thus it suffices to set $\varphi(t)$ equal to $z_1(t; \mu)$. Q. E. D.

Note the following corollary of Theorem (9.6):

Let the solutions of equation (9.2) be nonoscillatory on the whole real line. Suppose there exists an ap-function $x_0(t)$ *with uniformly bounded second derivative such that*

$$x_0(t) \gg 0, \quad Lx_0(t) \geqslant 0 \qquad (Lx_0(t) \not\equiv 0). \tag{9.55}$$

Then inequality (9.49) is valid.

This proposition is obtained from Theorem 9.6 by setting

$$\varphi(t) = -\frac{x_0'(t)}{x_0(t)}.$$

Another stability test for the solutions of equation (9.2) is the following.

Theorem 9.7. Let the solutions of equation (9.2) be non-oscillatory on $(-\infty, \infty)$. *Assume that*

$$M(p) > 0 \tag{9.56}$$

and

$$M\left\{ \int_{-\infty}^{t} \exp\left[-\int_{s}^{t} p(\tau)\, d\tau\right] q(s)\, ds \right\} \geqslant 0, \quad q(t) \not\equiv 0. \tag{9.57}$$

Then inequality (9.49) is valid.

Proof. The function

$$\varphi(t) = \int_{-\infty}^{t} \exp\left[-\int_{s}^{t} p(\tau)\,d\tau\right] q(s)\,ds,$$

as an ap-solution of the linear equation

$$\frac{dx}{dt} = -p(t)x + q(t), \tag{9.58}$$

satisfies the differential inequality (9.50). By Theorem 9.3, this implies the validity of one of inequalities (9.19) and (9.20). If we show that (9.20) is satisfied, the proof will be complete.

Suppose the assertion false, so that (9.19) is valid. Then there exists $\delta > 0$ such that

$$\varphi(t) - z_2(t) \geqslant \delta \qquad (-\infty < t < \infty). \tag{9.59}$$

Denote by $x_1(t)$ the solution of equation (9.58) with initial value $x_1(0) = z_2(0)$. The differential inequality

$$\frac{dx_1(t)}{dt} \leqslant x_1^2(t) - p(t)x_1(t) + q(t) \qquad (0 \leqslant t < \infty)$$

implies that

$$x_1(t) \leqslant z_2(t) \qquad (0 \leqslant t < \infty).$$

Hence, using (9.59), we obtain

$$\varphi(t) - x_1(t) \geqslant \delta \qquad (0 \leqslant t < \infty). \tag{9.60}$$

On the other hand, it follows from (9.56) that the ap-solution $\varphi(t)$ is asymptotically stable in the large. This means that inequality (9.69) cannot hold. Q. E. D.

9.6. Necessary and sufficient conditions for existence and constancy of sign of the Green's function

In the above subsections, we made a detailed study of the properties of solutions of the linear homogeneous equation (9.2) and the Riccati equation (9.5). The results will enable us to formulate necessary and sufficient conditions for the existence and constancy of sign of the Green's function of the operator (9.1).

Theorem 9.8. The ap-operator (9.1) is regular and its Green's function is nonnegative if and only if

1. the solutions of equation (9.2) are nonoscillatory on $(-\infty, \infty)$;

2. there exists an ap-function $x_0(t)$*, with uniformly continuous second derivative, satisfying inequalities (9.55).*

Proof. Necessity is obvious.

To prove sufficiency, note that by the corollary to Theorem 9.6 the fundamental matrix $U(t, s)$ of system (9.48) satisfies inequality (9.49). Thus the ap-operator (9.1) is regular and we have

$$G(t, s)=\begin{cases} K(t, s), & \text{if} \quad t \geqslant s, \\ 0, & \text{if} \quad t<s. \end{cases} \tag{9.61}$$

It now follows from (9.61) and the assumption that the solutions of equation (9.2) are nonoscillatory on $(-\infty, \infty)$ that the Green's function $G(t, s)$ is nonnegative. Q. E. D.

Theorem 9.9. The ap-operator (9.1) is regular and its Green's function is nonpositive if and only if there exists an ap-function $x_0(t)$*, with uniformly continuous second derivative, such that*

$$x_0(t) \gg 0, \quad L x_0(t) \leqslant 0 \qquad (L x_0(t) \not\equiv 0). \tag{9.62}$$

To prove sufficiency, set $\varphi(t)=-\dfrac{x_0'(t)}{x_0(t)}$. Then the second of inequalities (9.62) is simply inequality (9.46). Thus, by Theorem 9.5, it follows that the Riccati equation (9.5) has two distinct extremal solutions $z_1(t)$ and $z_2(t)$ which are almost periodic functions. Next, it follows from (9.47) that

$$M(z_2)>0 \tag{9.63}$$

and

$$M(z_1)<0. \tag{9.64}$$

Inequalities (9.63) and (9.64) imply that the ap-operator (8.1) is regular; its Green's function is given by

$$=\begin{cases} \dfrac{1}{z_1(s)-z_2(s)} \exp\left[-\displaystyle\int_s^t z_2(\tau)\, d\tau\right] & \text{for} \quad t \geqslant s, \\ \dfrac{1}{z_1(s)-z_2(s)} \exp\left[-\displaystyle\int_s^t z_1(\tau)\, d\tau\right] & \text{for} \quad t<s. \end{cases} \tag{9.65}$$

This formula at once implies that $G(t, s)$ is nonpositive.
The necessity of the conditions is obvious. Q. E. D.

9.7. Completion of proof of Theorem 9.1

Let the ap-operator (9.1) be regular, and suppose the solutions of equation (9.2) are nonoscillatory on $(-\infty, \infty)$. Since dichotomy of solutions is preserved under small perturbations, it follows that for sufficiently small μ the ap-operator

$$L(\mu)x = Lx - \mu x \tag{9.66}$$

is regular. According to Theorem 9.4, the Riccati equation

$$\frac{dz}{dt} = z^2 - p(t)z + q(t) - \mu$$

has two distinct extremal solutions $z_1(t; \mu)$ and $z_2(t; \mu)$ which are almost periodic. Now, since the trivial solutions of the equation $L(\mu)x = 0$ is unstable and there is an exponential dichotomy of solutions, it follows that

$$M[z_1(t; \mu)] < 0, \quad \text{and} \quad M[z_2(t; \mu)] > 0.$$

Hence, by (9.65), it follows that the Green's function $G(t, s; \mu)$ of the ap-operator (9.66) is nonpositive. It remains to note that

$$G(t, s) = \lim_{\mu \to 0} G(t, s; \mu)$$

uniformly in $t, s \in (-\infty, \infty)$. This completes the proof of Theorem 9.1.

9.8. Integral test for nonoscillation

According to Theorem 9.1, the Green's function of a regular ap-operator (9.1) is of constant sign if and only if all solutions of equation (9.2) are nonoscillating on $(-\infty, \infty)$. Thus various tests for nonoscillation are of interest. One of these is stated in Theorem 9.2. However, it is difficult to find a function satisfying the differential inequality (9.14). We now present an integral test for nonoscillation $(-\infty, \infty)$ of solutions of equation (9.2), due to Levin /1/.

Theorem 9.10. Suppose there exists a function $x_1(t)$, positive for $-\infty < t < \infty$, such that

$$\int_t^\infty \frac{1}{x_1^2(s)} \exp\left[-\int_0^s p(\tau)\,d\tau\right] ds < \infty.$$

Suppose that for any $-\infty < t_1 \leqslant t_2 < \infty$,

$$\int_{t_1}^{t_2} x_2(t)\, Lx_1(t)\, dt \leqslant 1, \tag{9.67}$$

where

$$x_2(t) = x_1(t) \int_t^\infty \frac{1}{x_1^2(s)} \exp\left[\int_s^t p(\tau)\,d\tau\right] ds.$$

Then the solutions of equation (9.2) are nonoscillatory on $(-\infty, \infty)$

Proof. Consider the function

$$w(t) = x_1'(t)\, x_2(t) + x_1(t)\, x_2(t)\, z(t), \tag{9.68}$$

where $z(t)$ is any solution of the Riccati equation (9.5). Suppose that

$$w(t_0) \leqslant 0. \tag{9.69}$$

We claim that then, for all $t \geqslant t_0$,

$$w(t) \leqslant 1. \tag{9.70}$$

If this is not true, there exist t_1 and t_2, $t_0 \leqslant t_1 < t_2$, such that

$$w(t_1) = 0, \qquad w(t_2) = 1, \tag{9.71}$$

and

$$0 < w(t) < 1 \qquad (t_1 < t < t_2). \tag{9.72}$$

Obviously,

$$w'(t) = \frac{w^2(t) - w(t)}{x_1(t)\, x_2(t)} + x_2(t)\, Lx_1(t). \tag{9.73}$$

Integrating (9.73) from t_1 to t_2 and using (9.71), we obtain

$$1 = \int_{t_1}^{t_2} \frac{w^2(t) - w(t)}{x_1(t)\, x_2(t)}\, dt + \int_{t_1}^{t_2} x_2(t)\, Lx_1(t)\, dt. \tag{9.74}$$

Thus, by (9.72),

$$\int_{t_1}^{t_2} x_2(t)\, Lx_1(t)\, dt > 1. \tag{9.75}$$

Inequality (9.75) contradicts (9.67).

Thus, for any $t_0 \in (-\infty, \infty)$, the Riccati equation has a solution defined on (t_0, ∞). This implies the statement of our theorem.

It is easy to show that under the assumptions of Theorem 9.10 the Riccati equation (9.5) has two distinct extremal solutions. The converse is also true:

If the Riccati equation (9.5) has two distinct extremal solutions, there exists a function $x_1(t)$ *satisfying the assumptions of Theorem 9.10.*

Note, in addition, that under the assumptions of Theorem 9.10 equation (9.2) is coarsely nonoscillatory on $(-\infty, \infty)$ if inequality (9.67) is replaced by the stronger inequality

$$\int_{t_1}^{t_2} x_2(t)\, Lx_1(t)\, dt \leqslant 1 - \varepsilon \qquad (-\infty < t_1 \leqslant t_2 < \infty),$$

where ε is a positive number, and moreover

$$\sup_{-\infty < t < \infty} [x_1(t)\, x_2(t)] < \infty.$$

There is an extensive literature dedicated to investigating oscillation properties of solutions of linear differential equations. The reader is referred to the interesting review of Levin /2/, which includes a fairly exhaustive bibliography.

Chapter 3

*GLOBAL THEOREMS ON AP-SOLUTIONS OF NONLINEAR EQUATIONS**

In this and the following chapter we investigate ap-solutions of nonlinear differential equations: we prove existence theorems, estimate the number of such solutions, and study their stability. Toward this end, we use the method of integral equations. The methods of the theory of cones play an essential role.

§ 10. GENERAL EXISTENCE THEOREMS

10.1. Passage to the integral equation

In the first two chapters we examined linear differential operators with ap-coefficients. In this chapter, the theory developed above is applied to the investigation of ap-solutions of nonlinear equations.

Let

$$Lx = \frac{d^m x}{dt^m} + A_1(t)\frac{d^{m-1}x}{dt^{m-1}} + \dots + A_m(t)x \tag{10.1}$$

be a regular ap-operator. This means (see Section 2) that every linear equation

$$Lx = f(t) \tag{10.2}$$

with an almost periodic right-hand side $f(t)$ has a unique ap-solution

$$x(t) = Tf(t). \tag{10.3}$$

* The main results of this chapter were reported by the authors at the Third All-Union Conference on Theoretical and Applied Mechanics (Moscow, 1968).

Here T is the integral operator

$$Tf(t) = \int_{-\infty}^{\infty} G(t, s) f(s)\, ds, \tag{10.4}$$

whose kernel is the Green's function of the ap-operator (10.1).

Now consider the nonlinear equation

$$Lx = f(t, x), \tag{10.5}$$

where the vector-function $f(t, x)$ is jointly continuous in both variables, and almost periodic in t uniformly with respect to x in any ball $|x| \leqslant r$.

It may be readily checked that the ap-function $x(t)$ is a solution of equation (10.5) if and only if it is a solution of the integral equation

$$x(t) = \int_{-\infty}^{\infty} G(t, s) f[s, x(s)]\, ds. \tag{10.6}$$

The right-hand side of equation (10.6) defines a nonlinear integral operator

$$\Pi x(t) = \int_{-\infty}^{\infty} G(t, s) f[s, x(s)]\, ds, \tag{10.7}$$

which may be written

$$\Pi = T\mathfrak{f}, \tag{10.8}$$

where T is the operator (10.4) and $\mathfrak{f}$ the superposition operator

$$\mathfrak{f}x(t) = f[t, x(t)]. \tag{10.9}$$

The properties of $f(t, x)$ (see Section 1) entail that the nonlinear operator $\mathfrak{f}$ acts in $B(R^n)$ and is continuous and bounded there. Since this is also true of the linear operator T, it follows that Π acts in $B(R^n)$ and is bounded and continuous there.

Recall that a nonlinear operator is bounded when the norms of its values are bounded on any ball.

The first problem arising in the investigation of equation (10.5) is the existence of ap-solutions. Various fixed-point principles are naturally used here.

The most elementary of these principles is the contracting-mapping principle. It is applicable, for example, when $f(t, x)$ satisfies a Lipschitz condition

$$|f(t, x) - f(t, y)| \leqslant a(r)|x - y| \quad (-\infty < t < \infty; |x|, |y| \leqslant r) \tag{10.10}$$

in the space variables. The operator Π then satisfies the inequality

$$\| \Pi x - \Pi y \|_{B(R^n)} \leqslant \| T \| a(r) \cdot \| x - y \|_{B(R^n)}.$$

on every ball $\| x \|_{B(R^n)} \leqslant r$ in $B(R^n)$. It is obvious that

$$\| \Pi x \|_{B(R^n)} \leqslant \| T \| \cdot \| fx \|_{B(R^n)} \leqslant \leqslant \| T \| [\| f(t, 0) \|_{B(R^n)} + ra(r)] (\| x \|_{B(R^n)} \leqslant r).$$

Hence equation (10.5) has a unique solution in the ball $\| x \|_{B(R^n)} \leqslant r$ if

$$a(r) \| T \| < 1, \quad \| T \| [\| f(t, 0) \|_{B(R^n)} + ra(r)] \leqslant r. \tag{10.11}$$

These inequalities are the simplest and coarsest conditions for the existence of ap-solutions. In the general case, their actual application is hampered by the lack of good methods to evaluate (or even estimate) the norm of the linear operator T.

Existence theorems for ap-solutions arising from the contracting-mapping principle have been established by Friedrichs, Stoker, Demidovich, Corduneanu and others.

Several fixed-point principles (Schauder principle, Browder principle, the theory of completely continuous vector fields and others) use the complete continuity of the relevant operators. Unfortunately, however, this theory cannot be applied in the general case to the problem of the existence of ap-solutions. The difficulty lies in that the superposition operator (10.9) is completely continuous in $B(R^n)$ only if $f(t, x)$ does not depend on x, and the linear operator T is also not completely continuous.

Thus, in order to investigate equation (10.6) one should use only fixed-point principles not involving complete continuity of the operators. Below we shall apply a generalized contracting-mapping principle, together with certain fixed point-principles which are applicable in the case of operators with an invariant cone in Banach space.

10.2. Generalized contracting-mapping principle

Let $\mathfrak{R}$ be a complete metric space with metric $\rho(x, y)$. An operator Π acting in $\mathfrak{R}$ is said to be a *generalized contraction* if

$$\rho(\Pi x, \Pi y) \leqslant q(\alpha, \beta)\rho(x, y) \qquad (\alpha \leqslant \rho(x, y) \leqslant \beta), \tag{10.12}$$

where, for $0 < \alpha \leqslant \beta < \infty$,

$$q(\alpha, \beta) < 1. \tag{10.13}$$

Π is a generalized contraction if, for example,

$$\rho(\Pi x, \Pi y) \leqslant \rho(x, y) - \gamma[\rho(x, y)],$$

where the function $\gamma(u)$ is positive and continuous for $u > 0$.

Theorem 10.1. Let Π be an operator which maps the complete metric space $\mathfrak{R}$ into itself and is a generalized contraction.

Then the equation

$$x = \Pi x \tag{10.14}$$

has a unique solution $x^* \in \mathfrak{R}$. The successive approximations

$$x_{n+1} = \Pi x_n \quad (n = 0, 1, 2, \ldots) \tag{10.15}$$

converge to this solution for any initial approximation $x_0 \in \mathfrak{R}$.

Proof. Consider the sequence

$$\alpha_n = \rho(x_n, x_{n-1}) \qquad (n = 1, 2, \ldots)$$

By (10.12), this sequence is nonincreasing; let α^* be its limit.

If $\alpha^* > 0$, then, for sufficiently large k and all $m = 1, 2, \ldots$, it follows from (10.12) that

$$\alpha_{k+m} \leqslant [q(\alpha^*, \alpha^* + 1)]^m (\alpha^* + 1),$$

and this is a contradiction, so that $\alpha^* = 0$.

Let $\varepsilon > 0$ be given. We choose n such that

$$\alpha_n \leqslant \frac{\varepsilon}{2}\left[1 - q\left(\frac{\varepsilon}{2}, \varepsilon\right)\right],$$

and show that Π maps the ball $\rho(x, x_n) \leqslant \varepsilon$ into itself. In fact, if $\rho(x, x_n) < \varepsilon/2$ then

$$\rho(\Pi x, x_n) \leqslant \rho(\Pi x, \Pi x_n) + \rho(\Pi x_n, x_n) \leqslant \\ \leqslant \rho(x, x_n) + \alpha_n \leqslant \varepsilon/2 + \alpha_n < \varepsilon.$$

If $\varepsilon/2 \leqslant \rho(x, x_n) \leqslant \varepsilon$, then again

$$\rho(\Pi x, x_n) \leqslant \rho(\Pi x, \Pi x_n) + \rho(\Pi x_n, x_n) \leqslant \\ \leqslant q(\varepsilon/2, \varepsilon)\, \rho(x, x_n) + \alpha_n < \varepsilon.$$

The invariance of the ball $\rho(x, x_n) \leqslant \varepsilon$ implies

$$\rho(x_{n+m}, x_n) \leqslant \varepsilon \qquad (m = 1, 2, \ldots).$$

Hence the sequence (10.15) is fundamental. Its limit x^* is a solution of equation (10.14). The uniqueness of this solution is obvious.

Note that under the assumptions of Theorem 10.1 the successive approximations converge uniformly with respect to initial values in any ball. Theorem 10.1 is due to Krasnosel'skii.

Under the assumptions of the generalized contracting-mapping principle, an equivalent metric may be defined in $\mathfrak{R}$, with respect to which Π satisfies the conditions of the ordinary contracting-mapping principle. This follows from an interesting theorem of Meyers /1/ on the converse of the contracting-mapping principle.

10.3. Equations with monotone operators

Let K be a cone in a Banach space E.

A nonlinear operator Π acting in E is said to be positive if $\Pi K \subset K$. Π is monotone on a subset $\mathfrak{M} \subset E$ if $x \leqslant y$ $(x, y \in \mathfrak{M})$ implies $\Pi x \leqslant \Pi y$ (i. e., $y - x \in K$ implies $\Pi y - \Pi x \in K$). Note that in contradistinction to the case of linear operators, *the properties of positivity and monotonicity are independent for nonlinear operators*.

We shall often have occasion to consider special sets in E, "conic intervals." If $u, v \in E$ and $u \leqslant v$, the set of all $x \in E$ such that $u \leqslant x \leqslant v$ is called the conic interval $\langle u, v \rangle$. We recall that a cone K is normal if $0 \leqslant x \leqslant y$ implies $\|x\| \leqslant N \|y\|$, where N is a constant. Let K be normal; then $x \in \langle u, v \rangle$ implies

$$\| x \| \leqslant \| u \| + \| x - u \| \leqslant \| u \| + N \| v - u \|.$$

Thus:

If K is normal, a conic interval is a bounded set.

Theorem 10.2. Let K be a normal cone. Let the conic interval $\langle u, v\rangle$ be invariant with respect to an operator Π, monotone on $\langle u, v\rangle$. Further, let Π satisfy the condition

$$\Pi y - \Pi x \leqslant S(y - x) \qquad (u \leqslant x \leqslant y \leqslant v), \tag{10.16}$$

where S is a positive linear operator whose spectral radius is less than 1.

Then equation (10.14) has a unique solution x^* on $\langle u, v\rangle$ and the successive approximations (10.15) converge to x^* for any initial approximation $x_0 \in \langle u, v\rangle$.

Proof. Consider the two sequences

$$u_n = \Pi u_{n-1} \qquad (n = 1,\ 2,\ \ldots;\ u_0 = u),$$
$$v_n = \Pi v_{n-1} \qquad (n = 1,\ 2,\ \ldots;\ v_0 = v).$$

It follows from the monotonicity of Π and the invariance of the cone interval that

$$u_0 \leqslant u_1 \leqslant u_2 \leqslant \ldots \leqslant u_n \leqslant \ldots \leqslant v_n \leqslant \ldots \leqslant v_1 \leqslant v_0. \tag{10.17}$$

By (10.16), we have

$$0 \leqslant v_n - u_n \leqslant S^n(v - u) \qquad (n = 1,\ 2,\ \ldots),$$

and as K is normal,

$$\| v_n - u_n \| \leqslant N \| S^n \| \cdot \| v - u \|.$$

By assumption, $\|S^n\| \to 0$; hence,

$$\lim_{n\to\infty} \| v_n - u_n \| = 0.$$

From (10.17) and the normality of K it follows that u_n and v_n are fundamental sequences. Their common limit x^* is a solution of equation (10.14).

As to the second part of the theorem, from the monotonicity of Π we have that the elements x_n in (10.15) satisfy the inequalities

$$u_n \leqslant x_n \leqslant v_n \qquad (n = 1,\ 2,\ \ldots),$$

and so

$$\| x_n - x^* \| \leqslant \| x_n - u_n \| + \| u_n - x^* \| \leqslant N \| v_n - u_n \| + \| u_n - x^* \|,$$

whence $\| x_n - x^* \| \to 0$. Q. E. D.

10.4. Uniformly concave operators

Let K be a solid normal cone in a Banach space E. Denote by K_{in} the set of interior elements of K and define a metric on K via

$$\rho(x, y) = \min\{\alpha:\ x \leqslant e^{\alpha} y,\ y \leqslant e^{\alpha} x\}. \tag{10.18}$$

It can easily be verified that ρ is a metric and it can be shown as well that K_{in} is complete with respect to this metric.

Consider a conic interval $\langle u, v\rangle$, where $u, v \in K_{\text{in}}$. Obviously,

$$\rho(x, y) \leqslant \rho(u, v) \qquad (u \leqslant x,\ y \leqslant v), \tag{10.19}$$

i. e., $\langle u, v\rangle$ is a bounded set in the metric (10.18).

The inequalities

$$x \leqslant e^{\rho(x, y)} y, \quad y \leqslant e^{\rho(x, y)} x$$

imply that

$$-[e^{\rho(x, y)} - 1]\, x \leqslant x - y \leqslant [e^{\rho(x, y)} - 1]\, y$$

and moreover,

$$-[e^{\rho(x, y)} - 1]\, v \leqslant x - y \leqslant [e^{\rho(x, y)} - 1]\, v.$$

By (10.19),

$$e^{\rho(x, y)} - 1 \leqslant e^{\rho(u, v)} \rho(x, y) \qquad (u \leqslant x,\ y \leqslant v),$$

so that

$$-e^{\rho(u, v)} \rho(x, y)\, v \leqslant x - y \leqslant e^{\rho(u, v)} \rho(x, y)\, v.$$

From the normality of the cone K it follows that

$$\| x - y \| \leqslant m_1 \rho(x, y) \qquad (u \leqslant x,\ y \leqslant v), \tag{10.20}$$

where $m_1 = (1 + 2N)\, e^{\rho(u, v)} \| v \|$.

It turns out that the converse inequality

$$m_2 \rho(x, y) \leqslant \| x - y \| \qquad (u \leqslant x,\ y \leqslant v), \tag{10.21}$$

where m_2 is a positive constant, is also valid.

In fact, since u is an interior element of K, it has a spherical neighborhood of radius δ contained in K. Thus, each element of the conic interval is contained in K with its spherical neighborhood of radius δ.

Let $u \leqslant x,\ y \leqslant v;\ x \neq y$. Then

$$y - \delta \frac{x-y}{\|x-y\|}, \quad x - \delta \frac{y-x}{\|y-x\|} \in K,$$

i. e.,

$$x \leqslant \left(1 + \frac{\|x-y\|}{\delta}\right) y, \qquad y \leqslant \left(1 + \frac{\|x-y\|}{\delta}\right) x.$$

Hence, we have

$$\rho(x, y) \leqslant \ln\left(1 + \frac{\|x-y\|}{\delta}\right),$$

yielding (10.21).

Inequalities (10.20) and (10.21) signify that the metric (10.18) is equivalent to the metric $\|x-y\|$ on any conic interval $\langle u, v\rangle$ $(u, v \in K_{\text{in}})$. Hence each of these conic intervals is a complete metric space.

An operator Π is said to be *uniformly concave on a conic interval* $\langle u, v\rangle$ $(u,\ v \in K_{\text{in}})$ if it is monotone on $\langle 0, v\rangle$, $\Pi u \in K_{\text{in}}$, and for any interval $[a, b] \subset (0, 1)$ there exists $\eta = \eta(a, b) > 0$, such that

$$\Pi(\tau x) \geqslant (1+\eta)\,\tau \Pi x \qquad (a \leqslant \tau \leqslant b,\ x \in \langle u, v\rangle). \tag{10.22}$$

Let Π be uniformly concave on $\langle u, v\rangle$ and let $x, y \in \langle u, v\rangle$, $x \neq y$. Then the inequalities

$$e^{-\rho(x, y)} x \leqslant y, \quad e^{-\rho(x, y)} y \leqslant x$$

and the monotonicity of Π imply

$$\Pi[e^{-\rho(x, y)} x] \leqslant \Pi y, \quad \Pi[e^{-\rho(x, y)} y] \leqslant \Pi x.$$

Thus, according to (10.22), it follows that

$$(1+\eta)\, e^{-\rho(x, y)} \Pi x \leqslant \Pi y, \quad (1+\eta)\, e^{-\rho(x, y)} \Pi y \leqslant \Pi x,$$

where $\eta = \eta[e^{-\rho(u, v)}, e^{-\rho(x, y)}]$. Hence

$$\rho(\Pi x, \Pi y) \leqslant \rho(x, y) - \ln\{1 + \eta[e^{-\rho(u, v)}, e^{-\rho(x, y)}]\}.$$

Thus Π is a generalized contraction operator on the conic interval $\langle u, v\rangle$.

From Theorem 10.1 and inequalities (10.20) and (10.21) we have

Theorem 10.3. *Let a conical interval $\langle u, v\rangle$ $(u,\ v \in K_{\text{in}})$ be invariant with respect to an operator Π which is uniformly concave on $\langle u, v\rangle$.*

Then equation (10.14) has a unique solution x^* on $\langle u, v\rangle$. The successive approximations (10.15) converge to x^* for any initial approximation $x_0 \in \langle u, v\rangle$.

A nonlinear operator Π is said to be *strongly positive* (as in the linear case) if, for any nonzero element $x \in K$, Πx is an interior point of K. If Π is uniformly concave on every conic interval $\langle u, v\rangle$ $(u,\ v \in K_{\text{in}})$, we say that Π is *uniformly concave on* K.

Theorem 10.4. *Let Π be strongly positive and uniformly concave on K. Let u, v be nonzero elements of K such that*

$$\Pi u \geqslant u, \quad \Pi v \leqslant v. \tag{10.23}$$

Then equation (10.14) has a unique solution x^* on K and the successive approximations (10.15) converge to x^* for any initial approximation x_0 in K.

To prove the theorem, first note (see Krasnosel'skii /1/, p. 211) that (10.23) implies $u \leqslant v$, and then that Π leaves any interval $\langle \alpha^{-1}u, \alpha v\rangle$ $(\alpha > 1)$ invariant. Finally since Π is strongly positive, we may assume that u and v are interior points of K. It remains to apply Theorem 10.3.

The first theorems on concave operators were obtained in Krasnosel'skii and Ladyzhenskaya /1/. Uniformly concave operators were first examined in Bakhtin /1/. Our proof follows mainly the article of Krasnosel'skii and Stetsenko /1/.

10.5. Equations with monotone nonlinearities

We return to the investigation of equation (10.5).

Let the ap-operator L be regular and its Green's function $G(t, s)$ of constant sign (see Chapter 2) with respect to a solid cone-valued ap-function $K(t)$. Let m be the order of the differential operator L. It is convenient to write equation (10.5) as

$$\alpha L x = g(t, x), \tag{10.24}$$

where $\alpha = 1$ if $G(t, s)$ is nonnegative, $\alpha = -1$ if $G(t, s)$ is nonpositive.

We assume that the function $g(t, x)$ satisfies the condition: for every t,

$$0 \leqslant g(t, y) - g(t, x) \leqslant A(t)(y - x) \quad (u(t) \leqslant x \leqslant y \leqslant v(t)), \tag{10.25}$$

where $u(t)$, $v(t)$ are fixed ap-functions with continuous derivatives up to order m bounded on $(-\infty, \infty)$, $A(t)$ is an ap-matrix, nonnegative with respect to the cone $K(t)$ $(A(t)K(t) \subset K(t))$ for each t.

If $g(t, x)$ is a scalar function, inequality (10.25) means that $g(t, x)$ is nondecreasing in x on the interval $u(t) \leqslant x \leqslant v(t)$ and satisfies a Lipschitz condition in x with coefficient $A(t)$. If R^n is many-dimensional and for all t the cone $K(t)$ is the cone of vectors with nonnegative components, then (10.25) means that $g(t, x)$ satisfies a Lipschitz condition in the space variables. In the general case condition (10.25) is less intuitive, but in concrete cases it is not difficult to check.

Theorem 10.5. Let the right-hand side of equation (10.24) satisfy (10.25) and suppose that the spectral radius of the linear operator

$$Sx(t) = \alpha \int_{-\infty}^{\infty} G(t, s) A(s) x(s)\, ds \tag{10.26}$$

is less than 1. Let $u(t)$ and $v(t)$ satisfy the differential inequalities

$$\alpha L u(t) \leqslant g[t, u(t)] \tag{10.27}$$

and

$$\alpha L v(t) \geqslant g[t, v(t)]. \tag{10.28}$$

Then equation (10.24) has a unique ap-solution on the conic interval $\langle u, v \rangle$.

To prove this, we consider in $B(R^n)$ the integral operator

$$\Pi x(t) = \alpha \int_{-\infty}^{\infty} G(t, s)\, g[s, x(s)]\, ds. \tag{10.29}$$

The differential inequalities (10.27), (10.28) and the monotonicity of the operator Π on $\langle u, v \rangle$ imply that the latter maps $\langle u, v \rangle$ into itself. From (10.25) it follows that Π satisfies (10.16). It remains to refer to Theorem 10.2.

To apply Theorem 10.5 we need, on the one hand, functions $u(t)$ and $v(t)$ satisfying the differential inequalities (10.27), (10.28) and, on the other, estimates for the spectral radius of the operator (10.26).

The first problem will be considered below. To estimate the spectral radius one can use the method of test functions (see Theorem 5.1) — the spectral radius of operator (10.26) is less than 1 if there exists an ap-function $x_0(t) \in B^m(R^n)$ such that

$$x_0(t) \gg 0, \qquad \alpha L x_0(t) \gg A(t) x_0(t). \tag{10.30}$$

10.6. Equations with uniformly concave nonlinearities

We continue our investigation of equation (10.24).

We shall say that the function $g(t,x)$ is *uniformly concave* on $K(t)$ if it is

1. monotone for each t: $g(t,y) - g(t,x) \in K(t)$ for $x, y, y-x \in K(t)$;
2. uniformly concave with respect to t: for any $\tau \in (0,1)$, there exists a positive $\eta = \eta(\tau;\, r,\, r_1)$ $(r,\, r_1 > 0)$ such that

$$g(t,\ \tau x) - (1+\eta)\tau g(t,x) \in K(t) \tag{10.31}$$

for all points x contained in $K(t)$ together with their spherical neighborhoods of radius r and such that $\|x\| \leqslant r_1$. A simple computation shows that for all τ in any fixed interval $[a,\ b] \subset (0,\ 1)$ one can select a common $\eta = \eta(a,\ b;\ r,\ r_1)$. Condition (10.31) implies that $g(t,0) \in K(t)$. Thus a uniformly concave function $g(t,x)$ is positive on $K(t)$.

We shall say that a uniformly concave function $g(t,x)$ is *regularly concave on the cone* $K(t)$ if $g(t,x) \neq 0$ for $x \neq 0$ $(x \in K(t))$ for a set of t dense on $(-\infty,\infty)$.

Theorem 10.6. Let the Green's function $G(t,s)$ of the operator L be strongly positive or strongly negative with respect to a cone-valued function $K(t)$. Let $g(t,x)$ be regularly concave on $K(t)$. Suppose that there exist nonzero ap-functions $u(t),\ v(t) \in K(t)$ in $B^m(R^n)$ which satisfy the differential inequalities (10.27)—(10.28).

Then, equation (10.24) has a unique nontrivial ap-solution $x^*(t)$ in $\hat{K}(t)$ such that

$$u(t) \leqslant x^*(t) \leqslant v(t) \qquad (-\infty < t < \infty). \tag{10.32}$$

For the proof, construct the operator (10.29), note that it is strongly positive and uniformly concave on $\hat{K}(t)$, and use Theorem 10.4.

Let conditions 1—2 hold for every t only for x satisfying the inequalities $0 \leqslant x \leqslant v(t)$. Suppose moreover that $g(t,x) \neq 0$ for $x \neq 0$ $(0 \leqslant x \leqslant v(t))$ for a set of t dense on $(-\infty,\infty)$. We shall then say that $g(t,x)$ is *regularly concave on the variable conic interval* $\langle 0,\ v(t) \rangle$.

Theorem 10.3 implies

Theorem 10.7. Let the Green's function $G(t, s)$ *of the operator* L *be strongly positive or strongly negative with respect to the cone-valued function* $K(t)$. *Assume that there exist nonzero ap-functions* $u(t)$, $v(t) \in \hat{K}(t)$, $v(t) - u(t) \in \hat{K}(t)$ *in* $B^m(R^n)$ *satisfying the differential inequalities (10.27), (10.28). Let* $g(t, x)$ *be regularly concave on the variable conic interval* $\langle 0, v(t) \rangle$.

Then equation 10.24 has a unique solution $x^*(t) \in B^m(R^n)$ *on the interval* $\langle u(t), v(t) \rangle$.

It is clear that Theorems 10.6 and 10.7 remain valid if instead of strong positivity or strong negativity of the Green's function one assumes strong positivity of the linear operator

$$T_\alpha f(t) = \alpha \int_{-\infty}^{\infty} G(t, s) f(s)\, ds.$$

Theorem 10.7 is useful in applications concerning equations of the type

$$Lx = f(t, x) \tag{10.33}$$

with nonlinearities convex in x. By analogy with the notion of concavity, we define the function $f(t, x)$ to be uniformly convex if the following two conditions are satisfied:

1. $f(t, x)$ is monotone in x for each t;
2. for any interval $[a, b] \subset (0, 1)$, there exists

$$\xi = \xi(a, b; r, r_1) \qquad (r, r_1 > 0),$$

such that for $a \leqslant \tau \leqslant b$

$$(1 - \xi)\tau f(t, x) - f(t, \tau x) \in K(t) \tag{10.34}$$

for all x contained in $K(t)$ together with their spherical neighborhoods of radius r and such that $|x| \leqslant r_1$.

If $f(t, x)$ is uniformly convex, it is sometimes convenient to replace equation (10.33) by the equation

$$L_1 x = g(t, x), \tag{10.35}$$

where

$$L_1 x = A(t)x - Lx, \tag{10.36}$$

and

$$g(t, x) = A(t)x - f(t, x). \tag{10.37}$$

The function (10.37) is, generally speaking, regularly concave on some variable conic interval. If the operator (10.36) has a strongly positive inverse, one can try to use Theorem 10.7.

10.7. Use of majorants and minorants

Application of Theorem 10.6 does not require actual construction of functions $u(t)$, $v(t) \in K(t)$ satisfying the differential inequalities (10.27), (10.28), but only the fact that they exist. In this connection, the method of linear majorants and minorants may prove useful; see, for example, Krasnosel'skii /1, 2/, where this method is used to investigate periodic solutions. However, additional difficulties arise here, because the linear operator (10.25) generally has no eigenfunctions. An example is the integral operator

$$Sx(t) = \int_{-\infty}^{t} e^{-(t-s)+\int_{s}^{t} a(\tau)d\tau} x(s)\, ds, \tag{10.38}$$

where $a(t)$ is an ap-function with zero mean and unbounded indefinite integral.

Suppose first that $g(t, 0) \not\equiv 0$. Then the nonzero ap-function

$$u(t) = \alpha \int_{-\infty}^{\infty} G(t, s)\, g(s, 0)\, ds \in K(t)$$

satisfies (10.27). Assume moreover that for all $t \in (-\infty, \infty)$ and $x \in K(t)$

$$g(t, x) \leqslant A(t)x + a(t), \tag{10.39}$$

where $a(t)$ is an ap-function and $A(t)$ an ap-matrix such that the spectral radius of the integral operator

$$Sx = \alpha \int_{-\infty}^{\infty} G(t, s)\, A(s)\, x(s)\, ds \tag{10.40}$$

is less than 1. Then the ap-operator

$$L_1 x = \alpha Lx - A(t)x \tag{10.41}$$

is positively invertible. Hence there exists an ap-function $v(t) \gg 0$ $(v(t) \in B^m(R^n))$ satisfying the differential inequality

$$\alpha Lv(t) - A(t)v(t) \gg 0 \tag{10.42}$$

($v(t)$ may be an ap-solution of any linear equation $\alpha Lx - A(t)x = f(t)$, where $f(t) \gg 0$). Since inequality (10.42) is homogeneous, we may assume without loss of generality that

$$\alpha Lv(t) - A(t)v(t) \geqslant a(t). \tag{10.43}$$

From (10.39) and (10.43) it follows that the ap-function $v(t)$ satisfies (10.28).

Note that an ap-operator is positively invertible if and only if the ap-operator $\alpha Lx - \mu A(t)x = L_\mu x$ is regular for each $\mu \in [0, 1]$.

Now let $g(t, 0) \equiv 0$. Suppose there exists $r_0 > 0$ such that, for $x \in K(t)$ and $|x| \leqslant r_0$,

$$g(t, x) \geqslant B(t)x \qquad (-\infty < t < \infty), \tag{10.44}$$

where $B(t)$ is a nonnegative ap-matrix on $K(t)$. Suppose moreover that there is a nonzero ap-function $u(t) \in \hat{K}(t)$ $(u(t) \in B^m(R^n);$ $|u(t)| \leqslant r_0;\ -\infty < t < \infty)$ such that

$$\alpha Lu(t) - B(t)u(t) \leqslant 0. \tag{10.45}$$

By (10.44), $u(t)$ satisfies inequality (10.27).

For minihedral cones, condition (10.45) can be substantially weakened.

Let the values of the ap-function $K(t)$ be minihedral cones. Then it follows from Lemma 5.4 that $\hat{K}(t)$ is minihedral.

Lemma 10.1. Suppose there exists an ap-function $x_0(t) \in B^m(R^n)$, *such that* $x_0(t) \overline{\in} -\hat{K}(t)$ and

$$\alpha Lx_0(t) - B(t)x_0(t) \leqslant 0. \tag{10.46}$$

Then there exists a nonzero ap-function $u(t) \in B^m(R^n)$, $u(t) \in \hat{K}(t)$, *satisfying the differential inequality (10.45).*

Proof. Consider the integral operator

$$S_1x(t) = \alpha \int_{-\infty}^{\infty} G(t, s)\, B(s)\, x(s)\, ds. \tag{10.47}$$

It follows from (10.46) that

$$S_1x_0(t) \geqslant x_0(t). \tag{10.48}$$

Now set $u_0(t) = [x_0(t)]_+$. It follows from the assumptions of the lemma that $u_0(t)$ is a nonzero ap-function. We have

$$S_1u_0(t) \geqslant [S_1x_0(t)]_+ \geqslant u_0(t).$$

It remains to set $u(t) = S_1 u_0(t)$. Q. E. D.

In the one-dimensional case the differential inequalities (10.27), (10.28) are automatically valid if $g(t,x)$ is a function of the type $\sqrt{x}$, i. e., $g(t,x)$ increases "slowly" for large x and "rapidly" for small x. Similar properties of $g(t,x)$ in the many-dimensional case are easily determined.

Lemma 10.2. Let $\varkappa_i \to \infty$, $\delta_i \to 0$, $r_i \to 0$ *be sequences of positive numbers such that*

$$g(t, x) \geqslant \varkappa_i B(t) x \qquad (x \in K(t),\ |x| \leqslant r_i) \tag{10.49}$$

and

$$g(t, x) \leqslant \delta_i A(t) x + a_i(t) \qquad (x \in K(t)), \tag{10.50}$$

where $a_i(t)$ *is a sequence of ap-functions,* $A(t)$ *and* $B(t)$ *nonzero, nonnegative ap-matrices.*

Then there exist nonzero ap-functions $u(t)$, $v(t) \in \widehat{K}(t)$ *satisfying the differential inequalities (10.27) and (10.28).*

Proof. Since the matrix $B(t)$ is nonzero and the operator αL^{-1} strongly positive, there exist a nonzero element $u_1(t) \in \widehat{K}(t)$ and a number $\beta > 0$ such that

$$S_1 u_1(t) \geqslant \beta u_1(t). \tag{10.51}$$

Select i_0 so that $\beta \varkappa_{i_0} > 1$. Then it follows from (10.51) that

$$\varkappa_{i_0} S_1 u_1(t) \geqslant u_1(t).$$

Thus the function $u(t) = \varkappa_{i_0} S_1 u_1(t)$, multiplied by a sufficiently small positive number, will satisfy (10.27).

Now let $v_1(t)$ be an interior element of the cone $\widehat{K}(t)$. Then, for some $\gamma > 0$,

$$\alpha \int_{-\infty}^{\infty} G(t, s)[A(s) + I] v_1(s)\, ds \ll \gamma v_1(t). \tag{10.52}$$

Denote the left-hand side of (10.52) by $v_2(t)$. Obviously,

$$[A(t) + I] v_2(t) \ll \gamma \cdot \alpha L v_2(t) \tag{10.53}$$

and $v_2(t) \gg 0$. We first choose i_0 such that $\gamma \delta_{i_0} < 1$, and then k such that $k v_2(t) \gg \gamma a_{i_0}(t)$. Then it follows from (10.53) that

$$\alpha L[k v_2(t)] \gg \delta_{i_0} A(t)[k v_2(t)] + a_{i_0}(t),$$

which implies that the function $v(t) = kv_2(t)$ satisfies (10.27). Q. E. D.

10.8. Examples

a) Consider the problem of ap-solutions of the first-order system of differential equations

$$\frac{dx_i}{dt} = \sum_{j=1}^{n} a_{ij}(t)\, x_j + f_i(t, x_1, \ldots, x_n) \qquad (i = 1, \ldots, n). \quad (10.54)$$

Let $A(t)$ denote the ap-matrix whose elements are the ap-functions $a_{ij}(t)$. It is assumed below that for all $t \in (-\infty, \infty)$ the off-diagonal elements of $A(t)$ are positive, and for some $t_0 \in (-\infty, \infty)$ it has a nondegeneracy path (see Section 6.4). The functions $f_i(t, x_1, \ldots, x_n)$ $(i = 1, \ldots, n)$ are assumed to be monotone in the space variables for nonnegative $x_1, \ldots, x_n$, almost periodic in t, and

$$f_i(t, 0, \ldots, 0) \geqslant 0 \qquad (i = 1, \ldots, n). \quad (10.55)$$

In addition, we assume that for any $\tau \in (0, 1)$ there exists $\eta = \eta(\tau; r, r_1) > 0$ $(r, r_1 > 0)$ such that

$$f_i(t, \tau x_1, \ldots, \tau x_n) \geqslant (1 + \eta)\, \tau f_i(t, x_1, \ldots, x_n) \quad (i = 1, \ldots, n), \quad (10.56)$$

where $r \leqslant x_1, \ldots, x_n \leqslant r_1$. Then the vector-function

$$f(t, x) = \{f_1(t, x_1, \ldots, x_n), \ldots, f_n(t, x_1, \ldots, x_n)\}$$

is uniformly concave on the cone K of vectors with nonnegative coordinates

Theorem 10.8. Suppose that at least one of inequalities (10.55) is not an identity, and that for nonnegative $x_1, \ldots, x_n$,

$$f_i(t, x_1, \ldots, x_n) \leqslant \sum_{j=1}^{n} b_{ij}(t)\, x_j + b_i(t) \qquad (i = 1, \ldots, n), \quad (10.57)$$

where $b_{ij}(t)$ *and* $b_i(t)$ *are ap-functions. Let* a *be a vector with positive components* $a_1, \ldots, a_n$ *such that*

$$\sup_{-\infty < t < \infty} \sum_{j=1}^{n} [a_{ij}(t) + b_{ij}(t)]\, a_j < 0 \qquad (i = 1, \ldots, n). \quad (10.58)$$

Then system (10.54) has a unique ap-solution in the cone $\hat{K}$.

Proof. It follows from (10.58) that $A(t)a \ll 0$ and thus, by Theorem 6.3, the ap-operator $Lx = dx/dt - A(t)x$ is regular and its Green's function is given by (6.5), where $U(t, s) = U(t)U^{-1}(s)$ is the translation operator of the differential equation

$$\frac{dx}{dt} - A(t)\,x = 0.$$

In addition (see Section 7.6), the linear integral operator

$$Sx(t) = \int_{-\infty}^{t} U(t, s)\, x(s)\, ds$$

is strongly positive with respect to the cone $\hat{K}$.

Now consider the nonlinear integral operator

$$\Pi x(t) = \int_{-\infty}^{t} U(t, s)\, f[s, x(s)]\, ds.$$

Let $x(t) \in \hat{K}$. Then it follows from the inequality

$$\Pi x(t) \geqslant \int_{-\infty}^{t} U(t, s)\, f(s, 0)\, ds \gg 0$$

that Π is strongly positive.

Inequalities (10.56) imply that Π is uniformly concave on $\hat{K}$.

The theorem now follows from Theorem 10.4, since by (10.58) the spectral radius of the operator

$$S_1 x(t) = \int_{-\infty}^{t} U(t, s)\, B(s)\, x(s)\, ds$$

is less than 1 (see Section 10.7). Here $B(t)$ is the ap-matrix with elements $b_{ij}(t)$. Q. E. D.

We now consider the case in which system (10.54) has a trivial solution, i. e.,

$$f_i(t, 0, \ldots, 0) \equiv 0 \qquad (i = 1, \ldots, n). \tag{10.59}$$

We assume moreover that $f(t, x) \neq 0$ for nonzero $x \in K$ and for t in a set dense on $(-\infty, \infty)$.

Theorem 10.9. Let (10.57) and (10.58) hold. Suppose that for $x \in K$ *and* $|x| \leqslant r_0$

$$f_i(t, x_1, \ldots, x_n) \geqslant \sum_{j=1}^{n} c_{ij}(t) x_j \qquad (i = 1, \ldots, n), \quad (10.60)$$

where $c_{ij}(t)$ *are nonnegative ap-functions. Let* $d = \{d_1, \ldots, d_n\}$ *be a vector such that*

$$\inf_{-\infty < t < \infty} \sum_{j=1}^{n} [a_{ij}(t) + c_{ij}(t)] d_j > 0 \qquad (i = 1, \ldots, n), \quad (10.61)$$

and suppose that at least one of the coordinates of d *is positive.*

Then system (10.54) has a unique nontrivial ap-solution in the cone $\hat{K}$.

This proposition follows from Theorem 10.6 and Lemma 10.1.

b) We now consider the problem of ap-solutions of a second-order system of differential equations

$$\frac{d^2 x_i}{dt^2} + \sum_{j=1}^{n} a_{ij}(t) x_j + f_i(t, x_1, \ldots, x_n) = 0 \qquad (i = 1, \ldots, n). \quad (10.62)$$

Theorem 10.10. Let the functions $a_{ij}(t)$, $f_i(t, x_1, \ldots, x_n)$ *satisfy the assumptions of Theorem 10.8.*

Then system (10.62) has a unique ap-solution in the cone $\hat{K}$.

Theorem 10.11. Let the functions $a_{ij}(t)$, $f_i(t, x_1, \ldots, x_n)$ *satisfy the assumptions of Theorem 10.9.*

Then system (10.62) has a unique nontrivial ap-solution in the cone $\hat{K}$.

The proofs of these theorems are similar to those of Theorems 10.8 and 10.9, except that the strong positivity of the relevant nonlinear integral operator is established with the help of Theorem 7.5.

c) Theorems 10.8 – 10.11 are of course only examples — they are concerned with narrow and rather special types of systems. Some more interesting examples will be cited in the following sections of this and following chapters.

Here we present a simple example of a system of two differential equations whose ap-solutions are easily investigated using variable cones.

Consider the system

$$\left.\begin{aligned} x_1' &= [a(t) + 2\sin t]\, x_1 - b_0 x_2 + f_1(t, x_1, x_2), \\ x_2' &= x_1 + a(t)\, x_2 + f_2(t, x_1, x_2). \end{aligned}\right\} \quad (10.63)$$

Here $a(t)$ is an arbitrary ap-function, b_0 a positive number.

As before, we assume that the functions $f_1(t, x_1, x_2)$ and $f_2(t, x_1, x_2)$ are nonnegative for nonnegative x_1, x_2, monotone and uniformly concave.

The results of Section 6.4 imply that the translation operator $U(t, s)$ of the system of differential equations

$$\left.\begin{aligned} x_1' &= [a(t) + 2 \sin t]\, x_1 - b_0 x_2, \\ x_2' &= x_1 + a(t)\, x_2 \end{aligned}\right\} \qquad (10.64)$$

cannot be positive with respect to the cone of vectors with nonnegative components. Suppose that the translation operator is positive with respect to a cone-valued ap-function $K(t)$. It is easy to see that it is then positive with respect to the cone-valued function $K_\alpha(t)$ determined by the inequalities

$$x_1 + \alpha(t)\, x_2 \geqslant 0, \qquad x_2 \geqslant 0, \qquad (10.65)$$

where $\alpha(t)$ is an ap-function. We claim that for $b_0 < 0.48$ such a cone indeed exists.

It is easy to verify that the translation operator of system (10.64) is positive with respect to the cone (10.65) if and only if the function $\alpha(t)$ satisfies the differential inequality

$$D^*\alpha(t) \geqslant \alpha^2(t) + 2\alpha(t) \sin t + b_0 \qquad (-\infty < t < \infty), \qquad (10.66)$$

where $D^*\alpha(t)$ is the upper right derivative of $\alpha(t)$.

By the results of Section 9.8, for any $b_0 < 0.48$ the second-order linear differential equation

$$\frac{d^2x}{dt^2} - 2 \sin t \frac{dx}{dt} + b_0 x = 0$$

is coarsely nonoscillatory on $(-\infty, \infty)$. Thus there exists an ap-function $\alpha(t)$ satisfying (10.66); $\alpha(t)$ may be, for example, a 2π-periodic solution of the Riccati equation

$$\frac{dz}{dt} = z^2 + 2z \sin t + b_0. \qquad (10.67)$$

Note that if we let $\alpha(t)$ be the upper extremal solution of the Riccati equation (10.67) then, as is easily checked, $\alpha(t) \geqslant 0$ for all $t \in (-\infty, \infty)$.

Since no constant can satisfy (10.66), the translation operator $U(t, s)$ of system (10.64) cannot be positive with respect to a constant cone.

Using the results of Sections 10.5—10.7, one can indicate conditions for the system (10.63) to have a unique nontrivial ap-solution in the variable cone (10.65).

d) As a last example we consider the problem of ap-solutions of a third-order scalar differential equation

$$\frac{d^3x}{dt^3} + a_1(t)\frac{d^2x}{dt^2} + a_2(t)\frac{dx}{dt} + a_3(t)x = f(t, x). \tag{10.68}$$

Consider the functions

$$q_1(t) = \frac{-a_1(t) - \sqrt{a_1^2(t) - 3a_2(t)}}{3}$$

and

$$q_2(t) = \frac{-a_1(t) + \sqrt{a_1^2(t) - 3a_2(t)}}{3}.$$

We shall assume that

$$a_1^2(t) \geqslant 3a_2(t), \tag{10.69}$$

so that the functions $q_1(t)$ and $q_2(t)$ are real.

Set

$$q_1^0 = \sup_{-\infty < t < \infty} q_1(t) \quad \text{and} \quad q_2^0 = \inf_{-\infty < t < \infty} q_2(t).$$

We shall assume that $q_1^0 \leqslant q_2^0$ and

$$(q_2^0)^3 + a_1(t)(q_2^0)^2 + a_2(t)q_2^0 + a_3(t) \leqslant 0 \qquad (-\infty < t < \infty).$$

Our last assumption is

$$a_3^0 = \inf_{-\infty < t < \infty} a_3(t) > 0.$$

By Theorem 8.9, the ap-operator

$$Lx = \frac{d^3x}{dt^3} + a_1(t)\frac{d^2x}{dt^2} + a_2(t)\frac{dx}{dt} + a_3(t)x$$

is regular and its Green's function $G(t, s)$ is nonnegative.

Theorem 10.12. Suppose there exist constants α *and* β $(\alpha < \beta)$ *such that*

$$\alpha a_3(t) \leqslant f(t, \alpha) \qquad (-\infty < t < \infty)$$

and

$$\beta a_3(t) \geqslant f(t, \beta) \qquad (-\infty < t < \infty).$$

Let $f(t, x)$ *be continuously differentiable with respect to* x *for* $x \in [\alpha, \beta]$ *and assume that its derivative* $f'_x(t, x)$ *satisfies the inequalities*

$$0 \leqslant f'_x(t, x) \leqslant a(t) \qquad (\alpha \leqslant x \leqslant \beta; \ -\infty < t < \infty),$$

where $a(t)$ *is an ap-function such that*

$$\inf_{-\infty < t < \infty} [a_3(t) - a(t)] > 0.$$

Then equation (10.68) has a unique ap-solution $x^*(t)$ *such that*

$$\alpha \leqslant x^*(t) \leqslant \beta \qquad (-\infty < t < \infty).$$

This proposition follows from Theorem 10.5.

We shall return to scalar equations of higher order in Section 12.

§ 11. STABILITY OF AP-SOLUTIONS OF EQUATIONS WITH MONOTONE AND CONCAVE NONLINEARITIES

11.1. Stability of the trivial solution

In this subsection we need a well-known theorem on integral inequalities.

Lemma 11.1. Let $v(t)$ *be a nonnegative continuous scalar function satisfying the integral inequality*

$$v(t) \leqslant e^{-\gamma t} v_0 + \beta \int_0^t e^{-\gamma(t-s)} v(s)\, ds \qquad (t \geqslant 0), \tag{11.1}$$

where $v_0, \beta \geqslant 0$.

Then

$$v(t) \leqslant e^{-(\gamma-\beta)t} v_0 \qquad (t \geqslant 0). \tag{11.2}$$

Consider the equation

$$Lx = \omega(t, x), \tag{11.3}$$

where L is a regular ap-operator of order m, $\omega(t,x)$ is jointly continuous in the variables $x \in R^n$ and t, $-\infty < t < \infty$, almost periodic in t uniformly with respect to x in some ball $|x| \leqslant r_0$ of sufficiently small radius r_0.

We assume that all first-order terms of $\omega(t,x)$ with respect to the space variables vanish, that is to say, $\omega(t, 0) \equiv 0$ and

$$|\omega(t, x)| \leqslant q(r)|x| \qquad (|x| \leqslant r), \tag{11.4}$$

where $q(r)$ is monotone and

$$\lim_{r \to 0} q(r) = 0. \tag{11.5}$$

Equation (11.3) has a trivial solution. We are interested in its Lyapunov stability.

We recall a few definitions.

Equation (11.3) can be rewritten as a first-order equation

$$\tilde{L}u \equiv \frac{du}{dt} + Q(t)u = \tilde{\omega}(t, u) \tag{11.6}$$

in R^{mn} (see p. 8), where

$$u = \{x, x', \ldots, x^{(m-1)}\},$$
$$\tilde{\omega}(t, u) = \{0, 0, \ldots, \omega(t, x)\},$$

and $Q(t)$ is the matrix (2.6).

The trivial solution of equation (11.3) is said to be *stable in Lyapunov's sense* if for any $\varepsilon > 0$ there exists $\delta > 0$ such that any solution $u(t)$ of equation (11.6) with $|u(0)| < \delta$ satisfies the inequality $|u(t)| < \varepsilon$ $(0 \leqslant t < \infty)$. If there exist δ_0, γ_0, $M_0 > 0$ such that if $|u(0)| < \delta_0$ then

$$|u(t)| \leqslant M_0 e^{-\gamma_0 t}|u(0)| \qquad (t \geqslant 0), \tag{11.7}$$

the trivial solution of equation (11.3) is said to be *exponentially stable*.

If the trivial solution is not stable, it is said to be *unstable*.

Recall that regularity of an ap-operator L (see Section 3.2) implies an exponential dichotomy of the solutions of the equation $\tilde{L}u = 0$, i. e., there exist projection operators P_+ and P_- $(P_+ + P_- = I)$

onto the subspaces of initial values E_+ and E_- for which the solutions of the equation $\tilde{L}u = 0$ decrease exponentially as $t \to \infty$ and as $t \to -\infty$, respectively. If $P_- = 0$, the trivial solution of the equation $Lx = 0$ is exponentially stable; if $P_- \neq 0$, it is unstable. Accordingly, we shall say that the operator L is stable if $P_- = 0$, unstable if $P_- \neq 0$.

Theorem 11.1. Let L be a stable regular ap-operator. Then the trivial solution of equation (11.3) is exponentially stable.

This theorem, as well as theorem 11.2 (in a more general case: equations with arbitrary bounded coefficients) is in effect due to Lyapunov, and in its final form to Perron /1/. The proof follows.

Any solution $u(t)$ of equation (11.6) is at the same time a solution of the integral equation

$$u(t) = U(t)\,u(0) + \int_0^t U(t, s)\,\tilde{\omega}[s, u(s)]\,ds, \qquad (11.8)$$

where $U(t, s) = U(t)U^{-1}(s)$, with $U(t)$ $(U(0) = I)$ the fundamental matrix of solutions of the homogeneous equation $\tilde{L}u=0$. Regularity and stability of the ap-operator L imply the estimate

$$|U(t, s)| \leqslant Me^{-\gamma(t-s)} \qquad (t \geqslant s), \qquad (11.9)$$

where M and γ are positive (see (4.10)). Thus, for $t \geqslant 0$,

$$|u(t)| \leqslant Me^{-\gamma t}|u(0)| + \int_0^t Me^{-\gamma(t-s)}|\tilde{\omega}[s, u(s)]|\,ds. \qquad (11.10)$$

Let $\varepsilon > 0$ be a given number, so small that $\beta = Mq(\varepsilon) \leqslant \gamma/2$.

Set $\delta = \varepsilon/(1 + M)$. We first show that if $|u(0)| < \delta$ then $|u(t)| < \varepsilon$ for $t > 0$. In fact, if t_0 is the first instant of time at which $|u(t)| = \varepsilon$, then by (11.10), for $0 \leqslant t \leqslant t_0$,

$$|u(t)| \leqslant Me^{-\gamma t}|u(0)| + \beta \int_0^t e^{-\gamma(t-s)}|u(s)|\,ds, \qquad (11.11)$$

and by Lemma 11.1,

$$\varepsilon = |u(t_0)| \leqslant Me^{-(\gamma-\beta)t_0}|u(0)|,$$

so that

$$\varepsilon < Me^{-\gamma t_0/2} \cdot \frac{\varepsilon}{M} < \varepsilon.$$

We have arrived at a contradiction.

Consequently, inequality (11.11) holds for all $t \geqslant 0$ and, by the same Lemma 11.1, it follows that

$$|u(t)| \leqslant Me^{-\gamma t/2}|u(0)| \qquad (t \geqslant 0,\ |u(0)| \leqslant \delta).$$

Q. E. D.

11.2. Instability theorem

We continue our examination of equation (11.3), assuming now that $\omega(t, x)$ satisfies a Lipschitz condition

$$|\omega(t, x) - \omega(t, y)| \leqslant p(r)|x - y| \quad (|x|,\ |y| \leqslant r), \qquad (11.12)$$

where $p(r) \to 0$ as $r \to 0$.

Theorem 11.2. Let L be an unstable operator.

Then the trivial solution of equation (11.3) is unstable.

To prove this, it is sufficient to establish the existence of a number $r_0 > 0$ such that any neighborhood of the origin contains the initial value $u(0)$ of some solution $u(t)$ $(0 \leqslant t < \infty)$ of equation (11.6) which does not remain within the ball $|u| \leqslant r_0$.

A direct check shows that the bounded solutions $u(t)$ of equation (11.6) on $[0, \infty)$ are exactly the solutions of the nonlinear integral equation

$$u(t) = Tu(t), \qquad (11.13)$$

where

$$Tu(t) = U(t)P_+u(0) + \int_0^t U(t)P_+U^{-1}(s)\tilde{\omega}[s, u(s)]\,ds - \\ - \int_t^\infty U(t)P_-U^{-1}(s)\tilde{\omega}[s, u(s)]\,ds. \qquad (11.14)$$

Let $u_1(t)$ and $u_2(t)$ be two solutions of (11.13), bounded on $[0, \infty)$, such that $P_+u_1(0) = P_+u_2(0)$ and $|u_1(t)|,\ |u_2(t)| \leqslant r_0$ for $t \geqslant 0$. Then

$$|u_1(t) - u_2(t)| = \\ = |Tu_1(t) - Tu_2(t)| \leqslant Mp(r_0)\int_0^t e^{-\gamma(t-s)}|u_1(s) - u_2(s)|\,ds + \\ + Mp(r_0)\int_t^\infty e^{\gamma(t-s)}|u_1(s) - u_2(s)|\,ds$$

and so

$$|u_1(t)-u_2(t)|\leqslant Mp(r_0)\cdot\frac{1}{\gamma}(2-e^{-\gamma t})\sup_{0\leqslant s<\infty}|u_1(s)-u_2(s)|. \quad (11.15)$$

Let r_0 be such that $2Mp(r_0)<\gamma$. Then it follows from (11.15) that $u_1(t)\equiv u_2(t)$.

We have proved that the intersection of the ball $|u(0)|\leqslant r_0$ with any hyperplane $u=P_+u(0)+v$ $(v\in E_-)$ contains at most one point which is the initial point of a solution that remains within the ball. In other words, "almost all" initial values in each neighborhood determine solutions which leave the ball $|u|\leqslant r_0$ for some $t>0$. This means that the trivial solution of equation (11.3) is unstable. Q. E. D.

It is well known that Theorem 11.2 remains valid without the assumption (11.12), but the simplification gained by adopting this not too restrictive condition is considerable.

11.3. Stability of the operator L and the sign of its Green's function

The examples considered in Section 8 show that the Green's function $G(t,s)$ of a regular operator L may be nonnegative with respect to a solid almost periodic cone-valued function $K(t)$, whether L is stable or unstable. It is interesting to note that if L is stable and its Green's function $G(t,s)$ is of constant sign, then it is necessarily nonnegative.

To prove this, recall that if L is stable, then

$$G(t,s)=\begin{cases}K(t,s), & \text{if } t\geqslant s,\\ 0, & \text{if } t<s,\end{cases} \quad (11.16)$$

where $K(t,s)$ is the Cauchy function of the operator L. The Cauchy function is defined as the solution of the matrix equation $LX=0$ with initial values

$$K(s,s)=K'_t(s,s)=\ldots=K^{(m-2)}_{t^{m-2}}(s,s)=0$$

and

$$K^{(m-1)}_{t^{m-1}}(s,s)=I.$$

11.4. Stability of ap-solutions of equations with monotone nonlinearities

We now proceed to investigate the stability of nontrivial ap-solutions of the equation

$$\alpha L x = g(t, x). \tag{11.17}$$

By definition, a solution $x^*(t)$ of equation (11.17) is *stable* if the trivial solution of the perturbed equation

$$\alpha L x = g[t, x^*(t) + x] - g[t, x^*(t)] \tag{11.18}$$

is stable. The *exponential stability (instability) of the solution* $x^*(t)$ is defined in a natural way.

Let the assumptions of Theorem 10.5 hold.

In this subsection we shall also assume that $g(t, x)$ is differentiable with respect to x in a neighborhood Ω of the ap-solution $x^*(t)$, which exists by virtue of Theorem 10.5, with $g'_x(t, x)$ uniformly continuous in t, x and almost periodic in t.

Theorem 11.3. Under the assumptions of Theorem 10.5, let the operator L *be stable.*

Then the ap-solution $x^*(t) \in \langle u, v \rangle$ *of equation (11.17) is exponentially stable.*

Proof. Set

$$\omega(t, x, h) = g(t, x+h) - g(t, x) - g'_x(t, x)\, h. \tag{11.19}$$

Obviously,

$$g(t, x+h) - g(t, x) = \int_0^1 g'_x(t, x + \theta h)\, h\, d\theta,$$

and so

$$\omega(t, x, h) = \int_0^1 \left[g'_x(t, x + \theta h) - g'_x(t, x)\right] h\, d\theta.$$

Hence

$$|\omega(t, x, h)| \leqslant q(r)\,|h| \qquad (|h| \leqslant r),$$

where

$$q(r) = \sup_{-\infty < t < \infty}\ \sup_{|x-y| \leqslant r;\ x, y \in \Omega} \left|g'_x(t, x) - g'_x(t, y)\right|, \tag{11.20}$$

and, by the uniform continuity of $g'_x(t, x)$ with respect to x,

$$\lim_{r\to 0} q(r) = 0.$$

By Theorem 11.1, it will suffice to show that the ap-operator

$$L_* x = Lx - \alpha g'_x[t, x^*(t)]\, x \tag{11.21}$$

is regular and stable.

We define an auxiliary family of the ap-operators

$$L(\mu)\, x = \alpha L x - \mu g'_x[t, x^*(t)]\, x, \tag{11.22}$$

where $0 \leqslant \mu \leqslant 1$.

It follows from (10.25) that

$$0 \leqslant g'_x[t, x^*(t)]\, x \leqslant A(t)\, x \quad (x \in K(t),\ -\infty < t < \infty). \tag{11.23}$$

Furthermore, since the spectral radius of the operator (10.26) is less than 1, there exists an ap-function $x_0(t) \in B^m(R^n)$ such that

$$x_0(t) \gg 0,\ \ \alpha L x_0(t) - A(t)\, x_0(t) \gg 0. \tag{11.24}$$

Comparing (11.23) and (11.24), we obtain

$$L(\mu)\, x_0(t) \gg 0 \qquad (0 \leqslant \mu \leqslant 1). \tag{11.25}$$

From (11.25) and Lemma 7.1 it follows that the operator $L(\mu)$ is positively invertible for all $\mu \in [0, 1]$. The regularity and stability of the ap-operator (11.21) now follow from the results of Section 4.3. Q. E. D.

T h e o r e m 11.4. *Under the assumptions of Theorem 10.5, let the operator* L *be unstable.*

Then the ap-solution $x^*(t) \in \langle u, v\rangle$ *of equation (11.17) is unstable.*

P r o o f. Let $\omega(t, x, h)$ be the function (11.19). Obviously,

$$\omega(t, x, h_1) - \omega(t, x, h_2) = \\ = \int_0^1 \{g'_x[t, x + h_2 + \theta(h_1 - h_2)] - g'_x(t, x)\}(h_1 - h_2)\, d\theta.$$

Thus

$$|\,\omega(t, x, h_1) - \omega(t, x, h_2)\,| \leqslant q(r)\,|\,h_1 - h_2\,| \quad (|\,h_1\,|,\ |\,h_2\,| \leqslant r),$$

where $q(r)$ is the function (11.20). This inequality is essentially condition (11.12) of Theorem 11.2.

Now we need only repeat the reasoning in the proof of Theorem 11.3. Q. E. D.

By the arguments of Section 11.3, one always has $\alpha = 1$ in Theorem 11.3; in Theorem 11.4 the factor α may be either 1 or -1.

11.5. Lemma on concave operators

Let K be a solid cone in a Banach space E. Let W be a continuous operator defined on K and differentiable at interior elements of K.

The operator W is said to be concave if

$$W(\tau x) \geqslant \tau W x \tag{11.26}$$

for all $x \in K$ and $0 \leqslant \tau \leqslant 1$.

Lemma 11.2. The operator W is concave if and only if

$$Wx \geqslant W'(x)\, x \tag{11.27}$$

for all $x \gg 0$.

Proof. Sufficiency. Since W is a continuous operator, it is sufficient to prove (11.26) for interior elements x of the cone K and $0 < \tau < 1$. Suppose that inequality (11.26) fails for some interior element x_0 of the cone K and $\tau_0 \in (0, 1)$. It follows that $y_0 = W(\tau_0 x_0) - \tau_0 W(x_0)$ is not a point of K. Hence (see, e. g., Dunford and Schwartz /1/) there exists a continuous linear functional l such that $l(y_0) < 0$ and $l(x) \geqslant 0$ for any $x \in K$. Consider the scalar function

$$\gamma(\tau) = l\left[\frac{1}{\tau} W(\tau x_0) - W(x_0)\right] \qquad (0 < \tau \leqslant 1).$$

This function is differentiable,

$$\gamma'(\tau) = -\frac{l\,[W(\tau x_0) - W'(\tau x_0)\,\tau x_0]}{\tau^2}.$$

It follows from inequality (11.27) that $\gamma'(\tau) \leqslant 0$ $(0 < \tau \leqslant 1)$. Thus,

$$\gamma(\tau_0) \geqslant \gamma(1) = 0.$$

On the other hand, $\gamma(\tau_0) = \frac{1}{\tau_0} l(y_0) < 0$ and we have derived a contradiction.

Necessity. Suppose that for some $x_0 \gg 0$ the element

$$z_0 = Wx_0 - W'(x_0)x_0$$

is not in K, but nevertheless

$$W(\tau x_0) \geqslant \tau W x_0 \qquad (0 < \tau \leqslant 1). \tag{11.28}$$

As before, we define a linear functional l_1, negative at z_0 and nonnegative on K. The inequality $l_1(z_0) < 0$ means that the derivative of the function

$$\gamma_1(\tau) = l_1\left[\frac{1}{\tau} W(\tau x_0) - W x_0\right]$$

is positive at $\tau = 1$. Thus, for values of τ less than but near unity, $\gamma_1(\tau)$ assumes values smaller than $\gamma_1(1)$, i.e., negative values. This contradicts (11.28). Q.E.D.

11.6. Stability of ap-solutions of equations with concave nonlinearities

We now consider the stability of positive ap-solutions $x^*(t)$ of equation (11.17), under the assumptions of Theorems 10.6 and 10.7. As in Section 11.4, we assume that $g(t, x)$ is continuously differentiable with respect to x in a neighborhood Ω of the ap-solution $x^*(t)$ and that its derivative $g'_x(t, x)$ is uniformly continuous in $x \in \Omega$ with respect to $t \in (-\infty, \infty)$. It then follows from Lemma 11.2 that

$$g(t, x) - g'_x(t, x)x \in K(t) \quad (x \in \Omega, \ -\infty < t < \infty). \tag{11.29}$$

We shall also assume that for some t_0 and for all $x \in \Omega$

$$g(t_0, x) - g'_x(t_0, x)x \neq 0. \tag{11.30}$$

Theorem 11.5. Under the assumptions of one of Theorems 10.6 or 10.7, let the operator L *be stable.*

Then the positive ap-solution $x^*(t)$ *of equation (11.17) is exponentially stable in Lyapunov's sense.*

Proof. Since the ap-operator L is stable and its Green's function $G(t, s)$ is strongly positive with respect to a solid cone-valued ap-function $K(t)$, it follows (see Section 11.3) that $\alpha = 1$. Thus, we must show (see Theorem 11.1) that the operator

$$L_* x = Lx - g'_x[t, x^*(t)]\, x \tag{11.31}$$

is regular and stable.

Let $f(t)$ be an arbitrary function in $B(R^n)$. Let us consider the ap-solutions of the differential equation

$$Lx - g'_x[t, x^*(t)]\, x = f(t). \tag{11.32}$$

Clearly, such solutions exist if and only if the operator equation

$$x - S_0 x = f_0 \tag{11.33}$$

is solvable in the space $B(R^n)$, where

$$S_0 x(t) = \int_{-\infty}^{\infty} G(t, s)\, g'_x[s, x^*(s)]\, x(s)\, ds$$

and

$$f_0(t) = \int_{-\infty}^{\infty} G(t, s) f(s)\, ds.$$

Set

$$f^*(t) = Lx^*(t) - g'_x[t, x^*(t)]\, x^*(t) = \\ = g[t, x^*(t)] - g'_x[t, x^*(t)]\, x^*(t).$$

From (11.29) and (11.30) it follows that $f^*(t)$ is a nonzero ap-function in the cone $\hat{K}(t)$. Thus

$$x^*(t) - S_0 x^*(t) = \int_{-\infty}^{\infty} G(t, s) f^*(s)\, ds \gg 0. \tag{11.34}$$

Note moreover that $x^*(t) \gg 0$ and S_0 is positive on $\hat{K}(t)$. Hence (see Theorem 5.1) the spectral radius of S_0 is less than 1.

Thus, the operator equation (11.33) is uniquely solvable and so the ap-operator (11.31) is regular.

The foregoing argument clearly proves that all the ap-operators

$$L(\mu)\, x = Lx - \mu g'_x[t, x^*(t)]\, x \qquad (0 \leqslant \mu \leqslant 1)$$

are regular. Together with the results of Section 4.3, this implies the stability of the operator (11.31). Q. E. D.

Similarly, one can prove

Theorem 11.6. Under the assumptions of one of Theorems 10.6 or 10.7, let the operator L be unstable.

Then the positive ap-solution $x^*(t)$ of equation (11.17) is unstable.

11.7. Stability in the cone of first-order systems with concave nonlinearities

In stability theory of differential equations (see, e.g., Lur'e /1/, Krasovskii /1/, Barbashin /1/, Pliss /1/, Lefschetz /1/, Aizerman and Gantmakher /1/ and others), it is of major importance to estimate the domain of initial values in the phase space for which the corresponding solutions asymptotically approach a given solution $x^*(t)$. If this domain exhausts the whole phase space, the solution $x^*(t)$ is said to be **stable in the large.**

In this subsection we prove that under the assumptions of Theorem 10.6 all solutions $x(t)$ of the first-order system

$$\frac{dx}{dt} + A(t)x = g(t, x) \tag{11.35}$$

with initial values in $K_{\text{in}}(0)$ asymptotically approach a positive ap-solution $x^*(t)$ as $t \to \infty$. In this case the solution $x^*(t)$ is said to be **stable in the cone** $K(0)$.

First we prove a general lemma for an equation

$$\frac{dx}{dt} + A(t)x = f(t, x), \tag{11.36}$$

whose right-hand side is an arbitrary function, jointly continuous in all variables and positive on $K(t)$: $f(t,x) \in K(t)$ for $x \in K(t)$ $(-\infty < t < \infty)$.

Lemma 11.3. Let the translation operator $U(t, s)$ of the equation

$$\frac{dx}{dt} + A(t)x = 0 \tag{11.37}$$

be positive with respect to a solid cone-valued ap-function $K(t)$.

Then for any initial value $x(s) = x \in K(s)$ equation (11.36) has a solution $x(t)$ such that

$$x(t) \in K(t) \tag{11.38}$$

for all $t \geqslant s$ in its domain of definition.

Proof. Set

$$P(t)\,x = \begin{cases} x, & \text{if} \quad x \in K(t), \\ q_0(t, x)\,x_0(t) + x, & \text{if} \quad x \overline{\in} K(t), \end{cases}$$

where $q_0(t, x)$ is the functional (5.32):

$$q_0(t, x) = \min_{x + r x_0(t) \in K(t)} r$$

Here $x_0(t)$ is an ap-function in the interior of the cone $\hat{K}(t)$. Recall (see Section 5.5), that the functional $q_0(t, x)$ satisfies a Lipschitz condition in x with a constant independent of t, and for any fixed x the functional is almost periodic in t.

Consider the equation

$$\frac{dy}{dt} + A(t)\,y - f[t, P(t)\,y]. \tag{11.39}$$

Let x be an element of $K(s)$. Let $y(t)$ be a solution of (11.39) with initial value $y(s) = x$. Then

$$y(t) = U(t, s)\,x + \int_s^t U(t, \sigma)\,f[\sigma, P(\sigma)\,y(\sigma)]\,d\sigma. \tag{11.40}$$

Since $P(t)y(t) \in K(t)$, it follows from (11.40) that $y(t) \in K(t)$ for $t \geqslant s$. But then $y(t)$ is a solution of (11.36) with initial value $y(s) = x$. Q. E. D.

We return to our investigation of equation (11.35). Let the assumptions of Theorem 10.6 hold.

To simplify matters, we shall assume that there is a unique solution (on some interval) of equation (11.35) for each initial value in $K_{in}(s)$.

Lemma 11.4. Let $x(t)$ be a solution of (11.35) and let $x(s) \in K_{in}(s)$.

Then, for all $t > s$ in the domain of definition of $x(t)$, we have $x(t) \in K_{in}(t)$.

Proof. For the values of t near s the relation $x(t) \in K_{in}(t)$ is obvious.

If the assertion of lemma is false, there exists a first $t_0 > s$ such that $x(t_0) \in K(t_0)$ but $x(t_0) \overline{\in} K_{in}(t_0)$. Since

$$x(t) = U(t, s)\,x(s) + \int_s^t U(t, \sigma)\,g[\sigma, x(\sigma)]\,d\sigma$$

it follows that

$$x(t) \geqslant U(t, s)\,x(s) \qquad (s \leqslant t \leqslant t_0).$$

But the operator $U(t,s)$, which is positive for $t>s$, maps the interior points of $K(s)$ onto interior points of $K(t)$. Hence $x(t_0)\in K_{\text{in}}(t_0)$. This is a contradiction. Q. E. D.

By Lemma 11.4, each inital value in $K_{\text{in}}(s)$ determines a solution of (11.35), which is unique for all $t\geqslant s$ in its domain of definition.

Lemma 11.5. *Let $x_1(t)$ and $x_2(t)$ be two solutions satisfying the conditions*

$$x_1(s),\ x_2(s)\in K_{\text{in}}(s),\quad x_2(s)-x_1(s)\in K(s).$$

Then, for all $t>s$ in the domain of definition of both solutions, $x_1(t)\leqslant x_2(t)$, i. e.,

$$x_2(t)-x_1(t)\in K(t)\qquad (t\geqslant s).$$

Proof. Consider the function

$$y(t)=x_2(t)-x_1(t)\qquad (t\geqslant s).$$

It is a solution of the differential equation

$$\frac{dy}{dt}+A(t)\,y=g\,[t,\,x_1(t)+y]-g\,[t,\,x_1(t)],\tag{11.41}$$

which satisfies the initial condition

$$y(s)=x_2(s)-x_1(s).$$

By Lemma (11.3), equation (11.41) has a solution $y_0(t)$ satisfying the same initial condition

$$y_0(s)=x_2(s)-x_1(s)$$

and such that

$$y_0(t)\in K(t)\qquad (t\geqslant s).$$

Then the function

$$x_0(t)=y_0(t)+x_1(t)$$

is a solution of (11.35) with initial value $x_0(s)=x_2(s)$. The uniqueness of solutions of (11.35) with initial values in $K_{\text{in}}(s)$ implies that $x_0(t)\equiv x_2(t)$, so that $y_0(t)\equiv y(t)$. Hence

$$y(t)\in K(t)\qquad (t\geqslant s).$$

Q. E. D.

Lemma 11.6. Let $x_1(t)$ *be a solution of equation (11.35) with initial value* $x_1(s) \in K_{in}(s)$, *and* $x_2(t)$ *a solution of (11.35) with initial value* $x_2(s) = \tau x_1(s)$; *where* $\tau \in (0, 1)$.

Then, for all $t \geqslant s$ *in the domain of definition of both solutions,* $x_2(t) \geqslant \tau x_1(t)$, *i.e.,*

$$x_2(t) - \tau x_1(t) \in K(t) \qquad (t \geqslant s).$$

To prove this, we consider the function

$$y(t) = x_2(t) - \tau x_1(t) \qquad (t \geqslant s)$$

and note that it is a solution of the equation

$$\frac{dy}{dt} + A(t)y = \{g[t, \tau x_1(t) + y] - g[t, \tau x_1(t)]\} + \\ + \{g[t, \tau x_1(t)] - \tau g[t, x_1(t)]\}, \qquad (11.42)$$

with zero initial value for $t = s$. The right-hand side of (11.42) is a positive function, since $g(t, x)$ is monotone and concave with respect to x. Thus the assertion follows from Lemma 11.3.

Let $x^*(t)$ be a strictly positive ap-solution of equation (11.35) (which exists by virtue of Theorem 10.6). For any point $x \in K_{in}(s)$ there exist positive constants α and β such that

$$\alpha x^*(s) \leqslant x \leqslant \beta x^*(s). \qquad (11.43)$$

Without loss of generality, we may assume that $\alpha < 1$ and $\beta > 1$. Denote by $x(t)$ the solution of (11.35) with initial value $x(s) = x$.

By Lemma 11.5,

$$x_1(t) \leqslant x(t) \leqslant x_2(t) \qquad (t \geqslant s),$$

where $x_1(t)$ and $x_2(t)$ are solutions of (11.35) satisfying initial conditions

$$x_1(s) = \alpha x^*(s), \qquad x_2(s) = \beta x^*(s).$$

It follows from Lemma 11.6 that

$$\alpha x^*(t) \leqslant x_1(t), \qquad x_2(t) \leqslant \beta x^*(t) \qquad (t \geqslant s),$$

and therefore

$$\alpha x^*(t) \leqslant x(t) \leqslant \beta x^*(t) \qquad (t \geqslant s). \qquad (11.44)$$

We shall need the family of norms $|x|_t$ $(-\infty < t < \infty)$ in R^n defined by (6.80). Recall that $|x|_t$ is defined by

$$|x|_t = \inf\{\mu: -\mu x^*(t) \leqslant x \leqslant \mu x^*(t)\} \quad (-\infty < t < \infty). \tag{11.45}$$

Since the ap-function $x^*(t)$ is strictly positive, it follows (see (6.81) that there exist positive constants m_1 and m_2 such that

$$m_1|x| \leqslant |x|_t \leqslant m_2|x| \qquad (x \in R^n, \; -\infty < t < \infty). \tag{11.46}$$

From (11.44) and (11.46) it follows that the solution $x(t)$ is uniformly bounded in norm for $t \geqslant s$. Thus (see, e. g., Krasnosel'skii /2/) we may assume that $x(t)$ is defined for all $t \in [s, \infty)$.

The operator $V(t,s)$ which maps each point x onto the point $x(t)$, where $x(t)$ is the solution of equation (11.35) with initial value $x(s) = x$:

$$V(t, s)\,x = x(t) \qquad (t \geqslant s), \tag{11.47}$$

is called *the translation operator* (*along the trajectories*) *of the nonlinear equation* (11.35). The translation operator is defined for all $t \geqslant s$ at points x through which there is a unique trajectory of equation (11.35), defined for all $t \geqslant s$.

It follows from the foregoing discussion that under our assumptions the translation operator $V(t,s)$ $(t \geqslant s)$ is defined on $K_{\text{in}}(s)$ Moreover, the translation operator is positive on $K_{\text{in}}(s)$ $(V(t,s)K_{\text{in}}(s) \subset K_{\text{in}}(t)$ for $t > s$); by Lemma 11.5 it is monotone (if $x_1, x_2 \in K_{\text{in}}(s)$ and $x_2 - x_1 \in K(s)$ then $V(t,s)x_2 - V(t,s)x_1 \in K(t)$ for $t \geqslant s$); by Lemma 11.6 it is concave (if $x \in K_{\text{in}}(s)$ and $0 < \tau < 1$, then $V(t,s)(\tau x) - \tau V(t,s)x \in K(t)$ for $t \geqslant s$). By general theorems on the continuous dependence of solutions of differential equations on initial values and parameters, $V(t,s)x$ $(-\infty < s \leqslant t < \infty, \; x \in K_{\text{in}}(s))$ is a continuous function of all its variables.

Lemma 11.7. The operator $V(t,s)$ *is uniformly concave in the following sense: there is a positive number* Δ *such that, for any* α, β $(0<\alpha<\beta)$ *and any* a, b $(0 < a < b < 1)$, *if*

$$\alpha x^*(s) \leqslant x \leqslant \beta x^*(s) \tag{11.48}$$

and

$$a \leqslant \tau \leqslant b \tag{11.49}$$

then for $t - s \geqslant \Delta$,

$$V(t, s)(\tau x) \geqslant (1+\xi)\tau V(t, s)x, \tag{11.50}$$

where

$$\xi = \xi(\alpha, \beta;\ a, b) > 0. \tag{11.51}$$

Proof. Let x and τ satisfy (11.48) and (11.49). Set

$$h(t,s) = V(t,s)(\tau x) - \tau V(t,s)x.$$

The function $h(t,s)$ (see proof of Lemma 11.6) is a solution of the integral equation

$$h(t, s) = \int_s^t U(t, \sigma)\{g[\sigma, \tau V(\sigma, s)x + h(\sigma, s)] - g[\sigma, \tau V(\sigma, s)x]\}\,d\sigma + \int_s^t U(t, \sigma)\{g[\sigma, \tau V(\sigma, s)x] - \tau g[\sigma, V(\sigma, s)x]\}\,d\sigma.$$

The monotonicity of $g(t,x)$ implies that for $t \geqslant s$

$$h(t, s) \geqslant \int_s^t U(t, \sigma)\{g[\sigma, \tau V(\sigma, s)x] - \tau g[\sigma, V(\sigma, s)x]\}\,d\sigma. \tag{11.52}$$

Without loss of generality, we may assume that $\alpha < 1$ and $\beta > 1$. Then it follows from Lemmas 11.5 and 11.6 that

$$\alpha x^*(t) \leqslant V(t, s)x \leqslant \beta x^*(t). \tag{11.53}$$

Thus, since $g(t,x)$ is uniformly concave in x with respect to t, there exists a positive number $\eta = \eta(\alpha, \beta;\ a, b)$ such that

$$g[t, \tau V(t, s)x] \geqslant (1+\eta)\tau g[t, V(t, s)x] \qquad (t \geqslant s).$$

Consequently, it follows from (11.52) that

$$h(t, s) \geqslant \eta\tau \int_s^t U(t, \sigma)\, g[\sigma, V(\sigma, s)x]\,d\sigma$$

and, by (11.53),

$$h(t, s) \geqslant \alpha\eta\tau \int_s^t U(t, \sigma)\, g[\sigma, x^*(\sigma)]\,d\sigma. \tag{11.54}$$

Since the assumptions of the lemma imply that the Green's function $G(t, s)$ of the operator

$$Lx = \frac{dx}{dt} + A(t)\,x$$

is nonnegative, it follows, by Lemma 6.1, that

$$G(t, s) = \begin{cases} U(t, s) & \text{for } t \geqslant s, \\ 0 & \text{for } t < s. \end{cases}$$

Thus

$$x^*(t) = \int_{-\infty}^{t} U(t, \sigma)\, g[\sigma, x^*(\sigma)]\, d\sigma$$

and it follows from (11.54) that for $t \geqslant s$

$$h(t, s) \geqslant \alpha\eta\tau \left\{ x^*(t) - \int_{-\infty}^{s} U(t, \sigma)\, g[\sigma, x^*(\sigma)]\, d\sigma \right\}. \qquad (11.55)$$

Since

$$|U(t, \sigma)| \leqslant Me^{-\gamma(t-\sigma)} \qquad (t \geqslant \sigma),$$

where $M, \gamma > 0$, it follows that for $t - s \geqslant \Delta$

$$\left| \int_{-\infty}^{s} U(t, \sigma)\, g[\sigma, x^*(\sigma)]\, d\sigma \right| \leqslant M_1 e^{-\gamma\Delta}, \qquad (11.56)$$

where

$$M_1 = \frac{M}{\gamma} \sup_{-\infty < t < \infty} |g[t, x^*(t)]|.$$

From (11.56) it follows that for sufficiently large Δ

$$\int_{-\infty}^{s} U(t, \sigma)\, g[\sigma, x^*(\sigma)]\, d\sigma \leqslant \delta x^*(t) \qquad (t \geqslant s + \Delta),$$

where $\delta < 1$. It then follows from (11.55) that for $t - s \geqslant \Delta$

$$h(t, s) \geqslant \alpha\eta\tau(1 - \delta)\, x^*(t).$$

Together with (11.53), this gives (11.50) with

$$\xi = \frac{\alpha\eta(1+\delta)}{\beta}.$$

Q. E. D.

We can now prove the main result of this subsection.

Theorem 11.7. Let the assumptions of Theorem 10.6 hold for equation (11.35).

Then the positive ap-solution $x^*(t)$ of equation (11.35) that exists under these assumptions is stable in the cone $K(0)$.

Proof. Denote by $x_1(t)$ the solution of (11.35) with initial value

$$x_1(0) = \alpha_0 x^*(0),$$

where $0 < \alpha_0 < 1$. By Lemmas 11.5 and 11.6,

$$\alpha_0 x^*(t) \leqslant x_1(t) \leqslant x^*(t) \qquad (t \geqslant 0).$$

Denote by $\alpha(t)$ the largest α for which the inequality $\alpha x^*(t) \leqslant x_1(t)$ holds. Then

$$\alpha(t)\, x^*(t) \leqslant x_1(t) \leqslant x^*(t) \qquad (t \geqslant 0). \tag{11.57}$$

$\alpha(t)$ is a nondecreasing function (it can be shown that it is continuous and strictly increasing, but this is unnecessary).

We set

$$\alpha_1 = \lim_{t\to\infty} \alpha(t)$$

and show that $\alpha_1 = 1$. Assuming the contrary, we conclude from Lemma 11.7 that for all $k = 1, 2, \ldots$

$$x^*(k\Delta) \geqslant x_1(k\Delta) \geqslant (1+\xi_0)^k \cdot \alpha_0 x^*(k\Delta),$$

where $\xi_0 = \xi(\alpha_0, 1;\ \alpha_0, \alpha_1) > 0$, and so, for all $k = 1, 2, \ldots,$

$$(1+\xi_0)^k \alpha_0 \leqslant \alpha_1.$$

This is a contradiction.

Thus $\alpha_1 = 1$. Then it follows from (11.57) that

$$\lim_{t\to\infty} |x_1(t) - x^*(t)| = 0. \tag{11.58}$$

Similarly, one proves that

$$\lim_{t\to\infty} |x_2(t) - x^*(t)| = 0, \tag{11.59}$$

where $x_2(t)$ is the solution of (11.35) with initial value

$$x_2(0) = \beta_0 x^*(0),$$

where $\beta_0 > 1$.

Now let $x(t)$ be an arbitrary solution of equation (11.35) with initial value in $K_{\text{in}}(0)$. Choose α_0 and β_0 $(0<\alpha_0<1<\beta_0)$ such that

$$\alpha_0 x^*(0) \leqslant x(0) \leqslant \beta_0 x^*(0). \tag{11.60}$$

Then it follows from the monotonicity of $V(t, s)$ that

$$x_1(t) \leqslant x(t) \leqslant x_2(t) \qquad (t \geqslant 0),$$

where $x_1(t)$ and $x_2(t)$ are solutions of (11.35) satisfying inequalities (11.58) and (11.59) respectively. Therefore,

$$\lim_{t\to\infty} |x(t) - x^*(t)| = 0. \tag{11.61}$$

Q. E. D.

Note that the limit in (11.61) is uniform with respect to all initial values $x(0)$ satisfying (11.60).

11.8. Stability in the cone of solutions of higher-order systems with concave nonlinearities

Consider the equation

$$Lx \equiv \frac{d^m x}{dt^m} + A_1(t)\frac{d^{m-1}x}{dt^{m-1}} + \ldots + A_m(t)x = g(t, x). \tag{11.62}$$

We shall assume that the conditions of Theorem 10.6 hold and that the operator L is stable. Recall that the Green's function of the operator L is given by (11.16) and is positive with respect to the solid cone-valued ap-function $K(t)$ considered in Theorem 10.6.

As usual, we shall write equation (11.62) as a system

$$\tilde{L}u \equiv \frac{du}{dt} + Q(t)u = \tilde{g}(t, u), \tag{11.63}$$

where

$$u = \{x,\ x',\ \ldots,\ x^{(m-1)}\},$$
$$\tilde{g}(t,\ u) = \{0,\ 0,\ \ldots,\ 0,\ g(t,\ x)\},$$

and

$$Q(t) = \begin{Vmatrix} 0 & -I & 0 & \ldots & 0 \\ 0 & 0 & -I & \ldots & 0 \\ \cdot & \cdot & \cdot & \cdot & \cdot \\ 0 & 0 & 0 & \ldots & -I \\ A_m(t) & A_{m-1}(t) & A_{m-2}(t) & \ldots & A_1(t) \end{Vmatrix}.$$

System (11.63) is a first-order equation in the phase space R^{mn}.

Denote by $U(t, s)$ the translation operator, acting in R^{mn}, of the homogeneous equation

$$\frac{du}{dt} + Q(t)\,u = 0.$$

It will be convenient to consider cone-valued functions $\tilde{K}(t)$ whose values are solid cones in R^{mn} satisfying following three conditions:

1. $U(t, s)\,\tilde{K}(s) \subset \tilde{K}(t)\ (-\infty < s \leqslant t < \infty)$.
2. If $x \in K(t)$, then $\{0, 0, \ldots, 0, x\} \in \tilde{K}(t)$.
3. If $\{u_1, u_2, \ldots, u_m\} \in \tilde{K}(t)$, then $u_1 \in K(t)$.

Note that we are not assuming the function $\tilde{K}(t)$ to be almost periodic.

Cone-valued functions with these properties can be constructed in various ways. We shall describe two of the principal methods.

Let $\tilde{K}_{\max}(t)$ denote the set

$$\tilde{K}_{\max}(t) = \{u\colon\ [U(\sigma,\ t)\,u]_1 \in K(\sigma)\quad \text{for}\quad \sigma \geqslant t\}, \qquad (11.64)$$

where $[v]_1$ denotes the vector of R^n whose components are the first n components of $v \in R^{mn}$.

We denote by $\tilde{K}_{\min}(t)$* the closure of the convex hull of the cones $U(t, \sigma)\,K_0(\sigma)\ (-\infty < \sigma \leqslant t)$, where $K_0(\sigma)$ denotes the cone in R^{mn} consisting of all points $u = \{0, 0, \ldots, x\}\ (x \in K(\sigma))$.

It is easy to verify that $\tilde{K}_{\max}(t)$ and $\tilde{K}_{\min}(t)$ are solid cones with properties 1—3, and

$$\tilde{K}_{\min}(t) \subset \tilde{K}_{\max}(t) \qquad (-\infty < t < \infty). \qquad (11.65)$$

* Analogous cones were introduced by Kolesov /2/ in the periodic case.

Moreover, if $\tilde{K}(t)$ is any other cone with properties 1—3, then

$$\tilde{K}_{\min}(t) \subset \tilde{K}(t) \subset \tilde{K}_{\max}(t) \qquad (-\infty < t < \infty). \qquad (11.66)$$

Below we shall use the fact that the cones $\tilde{K}_{\max}(t)$ are *uniformly normal*, in the sense that, for each t, if

$$u, \ v, \ v-u \in \tilde{K}_{\max}(t)$$

then $|u| \leqslant N|v|$, where N does not depend on t.

To prove this, suppose the contrary: there exist sequences t_j and $u_j, \ v_j \in \tilde{K}_{\max}(t_j)$, such that $|u_j| = 1$, $|v_j| \to 0$, $v_j - u_j \in \tilde{K}_{\max}(t_j)$. By the definition of the cone (11.64), it follows that for all $t \geqslant 0$

$$[U(t+t_j, \ t_j)\,u_j]_1 \in K(t+t_j) \qquad (j=1, \ 2, \ \ldots) \qquad (11.67)$$

and

$$[U(t+t_j, \ t_j)(v_j - u_j)]_1 \in K(t+t_j), \ (j=1, \ 2, \ \ldots). \qquad (11.68)$$

As usual, we may assume that the u_j converge to a point u^* ($|u^*| = 1$), the sequences $A_i(t+t_j)$ ($i = 1, \ldots, m$) converge uniformly to ap-matrices $A_i^*(t)$, and the sequence of cone-valued ap-functions $K(t+t_j)$ converges to a cone-valued ap-function $K^*(t)$. It follows from Theorem 4.4 that the sequence $U(t+t_j, \ t_j)$ converges for $t \geqslant 0$ to $U^*(t, 0)$, where $U^*(t, s)$ is the translation operator of the system

$$\frac{du}{dt} + Q^*(t)\,u = 0$$

corresponding to the equation

$$L^*x \equiv \frac{d^m x}{dt^m} + A_1^*(t)\,\frac{d^{m-1}x}{dt^{m-1}} + \ \ldots \ + A_m^*(t)\,x = 0. \qquad (11.69)$$

Letting $j \to \infty$ in (11.67) and (11.68), we obtain the identity

$$[U^*(t, 0)\,u^*]_1 = 0 \qquad (t \geqslant 0).$$

But the function

$$x(t) = [U^*(t, 0)\,w]_1$$

is a solution of the homogeneous equation (11.69) for any $w \in R^{mn}$. By the uniqueness theorem, $u^* = 0$. We have arrived at a contradiction.

We are interested in the stability of the positive ap-solution $x^*(t)$ of equation (11.62) whose existence follows from Theorem 10.6. In other words, we wish to study the stability of the ap-solution

$$u^*(t) = \{x^*(t), [x^*(t)]', \ldots, [x^*(t)]^{(m-1)}\}$$

of equation (11.63). Note that $u^*(t) \in K_{\min}(t)$ $(-\infty < t < \infty)$.

Theorem 11.8. Under the assumptions of Theorem 10.6, let the operator L *be stable. Let the function* $g(t, x)$ *satisfy the following additional condition: there exists* t_0 *such that*

$$g(t_0, x) \in K_{\text{in}}(t_0) \qquad (x \in K_{\text{in}}(t_0)). \tag{11.70}$$

Then the solution $u^*(t)$ *of equation (11.63) is stable in the cone* $\tilde{K}_{\max}(0)$.

The proof follows the same method as that of Theorem 11.7. We need only note that this method does not use the almost periodicity of the relevant cone-valued function. The properties of the nonlinear function $\tilde{g}(t, u)$ (positivity, monotonicity and uniform concavity) and the general properties of the solutions of equation (11.63) are established as in Section 10. To complete the proof we need only use the uniform normality of the cones $\tilde{K}_{\max}(t)$ and the fact that for every t the solution $u^*(t)$ lies in $\tilde{K}_{\max}(t)$ together with a spherical neighborhood of fixed radius $r_0 > 0$.

We shall prove a more general proposition: for every t, $u^*(t)$ lies in $\tilde{K}_{\min}(t)$ together with a spherical neighborhood of some fixed radius.

Otherwise, there exist sequences t_j and u_j such that

$$|u_j| \to 0, \quad u^*(t_j) + u_j \,\overline{\in}\, \tilde{K}_{\min}(t_j). \tag{11.71}$$

Hence there exist vectors c_j, $|c_j| = 1$ such that

$$(c_j, u^*(t_j) + u_j) < 0 \tag{11.72}$$

and

$$(c_j, u) \geqslant 0 \qquad (u \in \tilde{K}_{\min}(t_j)). \tag{11.73}$$

As usual, we may assume that the sequence c_j is convergent:

$$\lim_{j \to \infty} c_j = c,$$

the sequence of ap-matrices $Q(t + t_j)$ uniformly convergent:

$$\lim_{j \to \infty} Q(t + t_j) = Q^*(t), \tag{11.74}$$

the sequence of ap-functions $g(t + t_j; u)$ uniformly convergent in $t \in (-\infty, \infty)$ and in all u from a fixed ball:

$$\lim_{j\to\infty} \tilde{g}(t + t_j, u) = \tilde{g}^*(t, u), \tag{11.75}$$

and the sequence of solid cone-valued ap-functions $K(t + t_j)$ uniformly convergent to a solid cone-valued ap-function $K^*(t)$.

Let $U^*(t, s)$ denote the translation operator of the equation

$$\frac{du}{dt} + Q^*(t)u = 0.$$

Obviously, for any $\sigma \leqslant 0$ the sequence of cones $U(t_j, t_j+\sigma)K_0(t_j+\sigma)$ converges to the cone $U^*(0, \sigma)K_0^*(\sigma)$. It therefore follows from (11.73) that

$$(c, u) \geqslant 0 \qquad (u \in U^*(0, \sigma)K_0^*(\sigma)) \tag{11.76}$$

and so

$$(c, u) \geqslant 0 \qquad (u \in K_{\min}^*(0)). \tag{11.77}$$

From (11.74) and (11.75) it follows that the sequence of ap-functions $u^*(t + t_j)$ converges uniformly to an ap-solution $u^{**}(t)$ of the equation

$$\frac{du}{dt} + Q^*(t)u = \tilde{g}^*(t, u). \tag{11.78}$$

Hence it follows from the equality

$$u^{**}(t) = \int_{-\infty}^{t} U^*(t, \sigma)\tilde{g}(\sigma)\,d\sigma, \tag{11.79}$$

where

$$\tilde{g}(t) = \{0, \ldots, 0, g^*[t, x^{**}(t)]\}$$

and

$$x^{**}(t) = \lim_{j\to\infty} x^*(t + t_j) \qquad (-\infty < t < \infty),$$

that

$$u^{**}(t) \in \tilde{K}_{\min}^*(t) \qquad (-\infty < t < \infty).$$

But then, by (11.77),

$$(c,\ u^{**}(0)) \geqslant 0.$$

On the other hand, it follows from (11.72) that

$$(c,\ u^{**}(0)) \leqslant 0.$$

Hence

$$(c,\ u^{**}(0)) = 0,$$

whence, by (11.76) and (11.79),

$$(c,\ U^{*}(0,\ \sigma)\,\tilde{g}(\sigma)) = 0 \qquad (\sigma \leqslant 0). \tag{11.80}$$

By assumption

$$g[t_0,\ x^{*}(t_0)] \in K_{\text{in}}(t_0).$$

Since the functions $K(t)$ and $g[t,x^{*}(t)]$ are almost periodic, there exists $t^{*}<0$ such that

$$g^{*}[t^{*},\ x^{**}(t^{*})] \in K_{\text{in}}^{*}(t^{*}).$$

Let Δ be a sufficiently small neighborhood of t^{*}. Then there exists δ such that for all $\sigma \in \Delta$

$$g^{*}[\sigma,\ x^{**}(\sigma)] + \delta x \in K_{\text{in}}^{*}(\sigma) \qquad (x \in R^{n},\ |x| \leqslant 1).$$

By (11.77) and (11.80),

$$(c,\ U^{*}(0,\ \sigma)\,\tilde{x}) = 0 \qquad (\sigma \in \Delta), \tag{11.81}$$

where

$$\tilde{x} = \{0,\ \ldots,\ 0,\ x\} \qquad (x \in R^{n}).$$

We can rewrite (11.81) as

$$(b,\ U^{*}(t^{*},\ \sigma)\,\tilde{x}) \equiv 0 \qquad (\sigma \in \Delta,\ x \in R^{n}), \tag{11.82}$$

where

$$b = [U^{*}(0,\ t^{*})]^{*}\,c = \{b_1,\ \ldots,\ b_m\}.$$

Setting $\sigma = t^*$ in (11.82), we obtain the equality

$$(b_m, x) \equiv 0 \qquad (x \in R^n), \tag{11.83}$$

whence it follows that $b_m = 0$. Differentiating (11.82) with respect to σ and letting $\sigma = t^*$, we obtain (using (11.83) the equality

$$(b_{m-1}, x) \equiv 0 \qquad (x \in R^n),$$

which implies that $b_{m-1} = 0$. Continuing this process, we finally obtain $b = 0$. Thus $c = 0$, and this is a contradiction. Q. E. D.

In many cases one can prove that the cone-valued functions $\tilde{K}_{\min}(t)$ and $\tilde{K}_{\max}(t)$ are almost periodic. It would be interesting to know if this holds in the general case.

11.9. Stability of ap-solution under small perturbations of the right-hand side of the equation

We end this section with a brief discussion of the stability of the ap-solution $x^*(t)$ of an equation with monotone or concave nonlinearity

$$\alpha L x = g(t, x)$$

under small perturbations of the right-hand side. We shall assume that the conditions of the theorems of Section 11.4 or 11.6 hold. Recall that in this situation the ap-operator

$$L_1 x = Lx - \alpha g'_x [t, x^*(t)] x \tag{11.84}$$

is regular.

Consider the equation

$$\alpha L x = g(t, x) + \mu f(t, x, x', \ldots, x^{(m-1)}), \tag{11.85}$$

where the function $f(t, x, y_1, \ldots, y_{m-1})$ is almost periodic in t for any fixed $x, y_1, \ldots, y_{m-1}$ and satisfies a Lipschitz condition (with constant independent of t) with respect to $x, y_1, \ldots, y_{m-1}$, in a domain of R^{mn} which contains the range of the vector-function $u^*(t) = \{x^*(t), [x^*(t)]', \ldots, [x^*(t)]^{(m-1)}\}$. Obviously, for all sufficiently small μ, equation (11.85) has an ap-solution $x^*(t; \mu)$ which is continuous with respect to μ uniformly in $t \in (-\infty, \infty)$, $x^*(t, 0) = x^*(t)$. To prove this one first replaces equation (11.85) by a first-order equation in the phase space R^{mn}, and then, using the regularity of the first-order ap-operator corresponding to (11.84), considers a suitable integral equation and uses the contracting-mapping principle.

11.10. Examples

In this section we have been studying the stability of the ap-solutions whose existence was established in Section 10. The results yield a more exhaustive analysis of the examples of Section 10.8.

In Theorems 10.8 and 10.9, one can state without additional assumptions (by Theorem 11.7) that a positive ap-solution is stable in the cone K. If we assume that the right-hand sides of (10.54) are smooth, then (by Theorem 11.5) a positive ap-solution is exponentially stable.

By Theorem 11.6, the positive ap-solution whose existence is ensured by Theorems 10.10 and 10.11 is unstable for both increasing and decreasing time (though Theorem 11.6 is not applicable unless the nonlinearities in (10.68) are sufficiently smooth).

Finally, the ap-solution x^* of equation (10.68) which exists by Theorem 10.12 is exponentially stable (use Theorem 11.3).

§ 12. ALMOST PERIODIC OSCILLATIONS IN AUTOMATIC CONTROL SYSTEMS

12.1. Introductory remarks

There is a voluminous literature dedicated to oscillations in automatic control systems. Greatest consideration has been given to the problem of stability in the large of a system with one equilibrium state. The classical results of Lur'e, Pliss, Kalman, Popov, Yakubovich and others (see, e. g., the monographs of Aizerman and Gantmakher /1/, Lefschetz /1/, and the review article of Pyatnitskii /1/) are well known. Considerably less attention has been paid to forced periodic and almost periodic oscillations of such systems. We refer to the important papers of Naumov and Tsypkin /1/, Yakubovich /2, 3/, Aizerman and Yakubovich /1/.

The investigation of automatic control systems with several equilibrium states is only in its infancy; here we mention the papers of Brayton and Moser /1/, Moser /3/.

In this section we study forced ap-oscillations of certain automatic control systems with two or three equilibrium states; we indicate conditions for the existence of such oscillations, estimate their number, and study their stability properties. The method is based on the general theory elaborated above and on special constructions involving various "cones."

12.2. Undisturbed automatic control system

Consider the system of differential equations

$$\left.\begin{aligned}\frac{d\xi}{dt} &= \lambda_0\xi + d_1\varphi(\sigma),\\ \frac{dx}{dt} &= Bx + d\varphi(\sigma),\end{aligned}\right\} \tag{12.1}$$

where

$$\sigma = c_1\xi + (c, x), \tag{12.2}$$

with ξ a scalar and x a vector variable. In equations (12.1), the coefficients λ_0, d_1, c_1 are scalars, B a constant square matrix of order m, d and c vectors. The scalar function $\varphi(\sigma)$ is known as the *characteristic function* [see Lefschetz /1/] *of the control system.* As a rule, $\varphi(\sigma)$ satisfies the inequality

$$\sigma\varphi(\sigma) \geqslant 0 \qquad (-\infty < \sigma < \infty). \tag{12.3}$$

We assume $\varphi(\sigma)$ to be continuous, so that $\varphi(0) = 0$.

Throughout the sequel we shall assume that $\lambda_0 < 0$ and the eigenvalues λ of the matrix B satisfy the inequality

$$\operatorname{Re}\lambda < \lambda_0. \tag{12.4}$$

Furthermore, we shall assume that

$$d_1 > 0, \qquad c_1 > 0, \tag{12.5}$$

$$(d, c) < 0, \tag{12.6}$$

$$(d, c) + d_1c_1 \geqslant 0, \tag{12.7}$$

$$(Bd, c) - \lambda_0(d, c) > 0 \tag{12.8}$$

and finally

$$\operatorname{Re}(B^{-1}_{\lambda_0+i\omega}d, c) < 0 \qquad (-\infty < \omega < \infty), \tag{12.9}$$

where

$$B_\mu = \mu I - B.$$

The assumptions (12.4)—(12.9) define a special (but fairly large) class of automatic control systems; other classes will also be examined.

By virtue of (12.6), (12.8) and (12.9), Theorem 8.4 implies the existence of a positive-definite matrix H such that

$$H(B - \lambda_0 I) + (B^* - \lambda_0 I) H < 0 \tag{12.10}$$

and

$$Hd + c = 0. \tag{12.11}$$

Set

$$\rho = \sqrt{\frac{c_1}{d_1}} \tag{12.12}$$

and denote by K the solid cone selected in the space R^{m+1} by the functional

$$q(\xi, x) = -\rho\xi + \sqrt{(Hx, x)}. \tag{12.13}$$

The dual cone K^* consists by definition, of all vectors $\{\eta, y\}$ such that

$$\xi\eta + (x, y) \geqslant 0$$

for all $\{\xi, x\}$ satisfying the inequality

$$\sqrt{(Hx, x)} \leqslant \rho\xi.$$

In other words, $\{\eta, y\} \in K^*$ if and only if

$$\xi\eta + \min_{\sqrt{(Hx, x)} \leqslant \rho\xi} (x, y) \geqslant 0.$$

Since

$$\min_{\sqrt{(Hx, x)} \leqslant \rho\xi} (x, y) = \rho\xi \sqrt{(H^{-1}y, y)},$$

it follows that $\{\eta, y\} \in K^*$ if and only if

$$\xi\eta - \rho\xi \sqrt{(H^{-1}y, y)} \geqslant 0.$$

Thus the dual cone K^* is selected by the functional

$$q^*(\eta, y) = -\frac{1}{\rho}\eta + \sqrt{(H^{-1}y, y)}. \tag{12.14}$$

To simplify the notation, we set

$$\bar{B} = \left\| \begin{matrix} \lambda_0 & 0 \\ 0 & B \end{matrix} \right\|, \qquad \bar{c} = \{c_1, c\}, \qquad \bar{d} = \{d_1, d\}. \tag{12.15}$$

We claim that $\bar{d} \in K$ and $\bar{c} \in K^*$. To prove this, we must prove the inequalities

$$\rho d_1 \geqslant \sqrt{(Hd, d)}, \qquad \frac{1}{\rho} c_1 \geqslant \sqrt{(H^{-1}c, c)},$$

which, by (12.12), can be rewritten

$$d_1 c_1 \geqslant (Hd, d), \qquad d_1 c_1 \geqslant (H^{-1}c, c).$$

By virtue of (12.11) both these inequalities are equivalent to (12.7).

We shall assume that each initial value

$$z(s) = z = \{\xi(s), x(s)\}$$

uniquely determines a solution

$$z(t) = \{\xi(t), x(t)\} = W(t, s)z \qquad (t \geqslant s)$$

of system (12.1). Since the nonlinear characteristic function $\varphi(\sigma)$ may be rapidly increasing, we do not assume that all the solutions of system (12.1) are globally continuable. Thus the translation operator $W(t, s)$ of this system need not be defined for all $t > s$.

The following lemma is convenient in examination of translations along the trajectories of system (12.1).

Lemma 12.1. Let $\{\xi(t), x(t)\}$ be a solution of system (12.1). Then for all t such that $v(t) \leqslant 0$ the function

$$v(t) = \rho^2\xi^2(t) - (Hx(t), x(t)) \tag{12.16}$$

satisfies the differential inequality

$$\frac{dv(t)}{dt} \geqslant 2\sigma(t)\varphi[\sigma(t)] + \varepsilon(x(t), x(t)), \tag{12.17}$$

where ε *is a positive number.*

P r o o f. It is obvious that

$$\frac{dv(t)}{dt} = 2\lambda_0\rho^2\xi^2(t) - ((B^*H + HB)x(t),\ x(t)) + 2\sigma(t)\varphi[\sigma(t)].$$

If $v(t) \leqslant 0$, then

$$\lambda_0\rho^2\xi^2(t) \geqslant \lambda_0(Hx(t),\ x(t)).$$

Consequently, for such t,

$$\frac{dv(t)}{dt} \geqslant -([H(B - \lambda_0 I) + (B^* - \lambda_0 I)H]\,x(t),\ x(t)) + \\ + 2\sigma(t)\varphi[\sigma(t)].$$

It remains to use property (12.10) of the matrix H. Q. E. D.

Lemma 12.1 implies that $W(t, s)z$ is an interior point of the cone K for $t > s$, if $z \in K$ and $z \neq 0$. Similarly, the operators $W(t, s)$ are also strongly positive for $t > s$ with respect to the cone $-K$, which is selected by the functional

$$q_-(\xi, x) = \rho\xi + \sqrt{(Hx, x)}. \tag{12.18}$$

Obviously, $W(t, s) = e^{\overline{B}(t-s)}$ if $\varphi(\sigma) \equiv 0$; thus the linear operators $e^{\overline{B}t}\ (t > 0)$ are strongly positive with respect to both K and $-K$.

Since $\varphi(0) = 0$, system (12.1) has an equilibrium state at the origin. To examine its stability, one can use the ordinary Lyapunov method, i. e., analysis of stability in the first approximation (see, e. g., Krasovskii /1/). The equations of the first approximation are also systems of type (12.1), with

$$\varphi(\sigma) = h\sigma; \tag{12.19}$$

it follows from (12.3) that $h \geqslant 0$. If $\varphi(\sigma)$ is of the form (12.19), then system (12.1) has the form

$$\left.\begin{aligned} \frac{d\xi}{dt} &= \lambda_0\xi + d_1 h\sigma, \\ \frac{dx}{dt} &= Bx + dh\sigma \end{aligned}\right\} \tag{12.20}$$

or, what amounts to the same,

$$\frac{dz}{dt} = \overline{B}z + \bar{d}h(\bar{c}, z). \tag{12.21}$$

Denote by $D(h)$ $(h \geqslant 0)$ the matrix of system (12.21). As we know, for $t>0$ the operators $e^{D(h)t}$ are strongly positive with respect to the cone K. Well-known theorems on strongly positive linear operators (see, for example, Krasnosel'skii /1/) yield the following proposition:

Each matrix $D(h)$ *has a simple real eigenvalue* $\lambda_0(h)$, *which is strictly greater than the real parts of all the other eigenvalues.*

We claim that

$$\lambda_0(h_1) < \lambda_0(h_2) \tag{12.22}$$

for $0 \leqslant h_1 < h_2$. To prove this, we first show that for $t>0$

$$e^{D(h_1)t}z \ll e^{D(h_2)t}z \qquad (h_1 < h_2;\ z \in K,\ z \neq 0). \tag{12.23}$$

Set

$$z_1(t) = e^{D(h_1)t}z, \quad z_2(t) = e^{D(h_2)t}z, \quad u(t) = z_2(t) - z_1(t).$$

Obviously,

$$\frac{du(t)}{dt} = D(h_2)u(t) + \bar{d}(h_2 - h_1)(\bar{c}, z_1(t)).$$

Since $u(0) = 0$,

$$u(t) = \int_0^t e^{D(h_2)(t-s)}\bar{d}(h_2 - h_1)(\bar{c}, z_1(s))\,ds,$$

which implies (12.23).

Denote by $g_0(h)$ the eigenvector of $D(h)$ corresponding to the eigenvalue $\lambda_0(h)$. From (12.23) it follows that

$$e^{D(h_1)t}g_0(h_2) \ll e^{\lambda_0(h_2)t}g_0(h_2) \qquad (t>0).$$

Thus, by Theorem 5.1, the spectral radius of the operator $e^{D(h_1)t}$, which is equal to $e^{\lambda_0(h_1)t}$, is strictly less than $e^{\lambda_0(h_2)t}$. This proves (12.22).

Set

$$\alpha_0 = -(\bar{B}^{-1}\bar{d}, \bar{c}) = -(B^{-1}d, c) - \frac{d_1c_1}{\lambda_0}. \tag{12.24}$$

Since

$$-\bar{B}^{-1} = \int_0^\infty e^{\bar{B}s}\,ds$$

and the operators $e^{\bar{B}s}$ $(s>0)$ are strongly positive, the operator $-\bar{B}^{-1}$ is strongly positive with respect to the cone K; hence $\alpha_0>0$.

It is easy to verify that

$$\lambda_0(\alpha_0^{-1})=0. \tag{12.25}$$

The strict monotonicity of $\lambda_0(h)$ implies that for $0\leqslant h<\alpha_0^{-1}$ all the eigenvalues of $D(h)$ are situated in the left half-plane, while for $h>\alpha_0^{-1}$ the matrix $D(h)$ has a unique positive eigenvalue, all other eigenvalues lying in the left half-plane.

By the general Lyapunov theorems (see, e. g., Malkin /1/) these arguments entail

Theorem 12.1. Let the nonlinear characteristic function $\varphi(\sigma)$ be twice continuously differentiable in a neighborhood of the origin.

If

$$\alpha_0\varphi'(0)<1, \tag{12.26}$$

then the trivial solution of system (12.1) is exponentially stable.

If

$$\alpha_0\varphi'(0)>1 \tag{12.27}$$

or

$$\alpha_0\varphi'(0)=1, \qquad \varphi''(0)\neq 0, \tag{12.28}$$

then the trivial solution of (12.1) is unstable.

System (12.1) may have equilibrium states distinct from the origin. To find them, one must solve the system

$$\begin{gathered}\lambda_0\xi+d_1\varphi(\sigma)=0,\\ Bx+d\varphi(\sigma)=0,\\ \sigma=c_1\xi+(c,x),\end{gathered}$$

which can be written

$$\begin{gathered}\bar{B}z+d\varphi(\sigma)=0,\\ \sigma=(\bar{c},z).\end{gathered}$$

To determine σ we have a scalar equation.

$$\sigma=\alpha_0\varphi(\sigma), \tag{12.29}$$

where α_0 is given by (12.24).

If equation (12.29) has a nontrivial solution σ_0, system (12.1) has a nonzero equilibrium state:

$$z_0 = -\varphi(\sigma_0)\bar{B}^{-1}\bar{d} \qquad (12.30)$$

or, what is the same, an equilibrium state

$$\xi_0 = -\frac{d_1}{\lambda_0}\varphi(\sigma_0), \qquad x_0 = -\varphi(\sigma_0)B^{-1}d. \qquad (12.31)$$

As remarked above, the operator $-\bar{B}^{-1}$ is strongly positive with respect to the cone K and $\bar{d} \in K$. Thus, any nonzero equilibrium state lies either in K or in $-K$.

Investigation of the nonzero equilibrium state (12.31) is not difficult if $\varphi(\sigma)$ is monotone. In this case we can apply Theorem 12.1.

We shall consider system (12.1) more closely for the case of a convex function $\varphi(\sigma)$. To simplify matters, we assume that $\varphi(\sigma)$ is twice continuously differentiable,

$$\varphi''(\sigma) > 0 \qquad (-\infty < \sigma < \infty), \qquad (12.32)$$

and

$$\lim_{\sigma\to\infty}\frac{\varphi(\sigma)}{\sigma} = \infty, \qquad \lim_{\sigma\to-\infty}\frac{\varphi(\sigma)}{\sigma} = 0 \qquad (12.33)$$

(Figure 12.1). For example, the functions $\varphi(\sigma) = e^{\gamma\sigma} - 1$ $(\gamma > 0)$ have these properties.

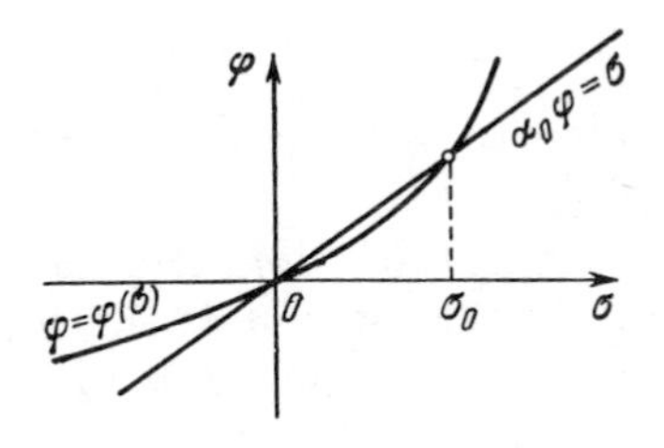

FIGURE 12.1

We shall assume that the origin is a stable equilibrium state of system (12.1). Then, by Theorem 12.1, we have inequality (12.26). Thus, it follows from (12.33) that equation (12.29) has a positive solution σ_0. The nontrivial solution is unique by virtue of (12.32).

Since $\varphi(\sigma)$ is convex,

$$\alpha_0\varphi'(\sigma_0) > 1.$$

Hence, the equilibrium state (12.31) is unstable.

One can give a general description of the configuration (Figure 12.2) of the integral curves of system (12.1) in the phase space R^{m+1}. The integral curves trace out an m-dimensional surface Γ partitioning R^{m+1} into two regions R_1 and R_2. All solutions of (12.1) with initial values in R_1 approach the origin as t increases; all solutions with initial values in R_2 "escape to infinity" as t increases. Solutions with initial values on Γ lie entirely on Γ and approach the nonzero equilibrium state z_0 asymptotically as $t\to\infty$.

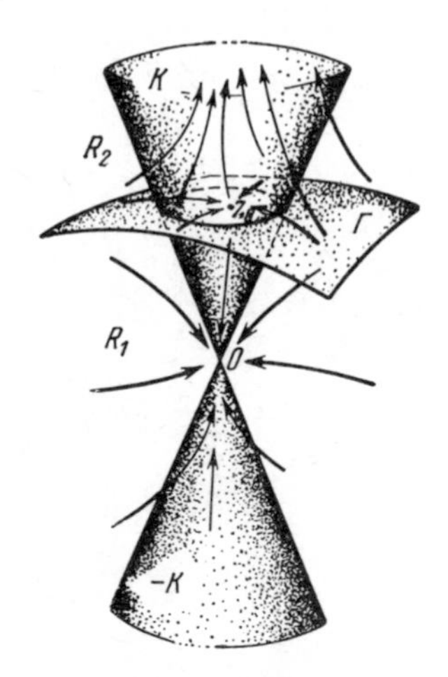

FIGURE 12.2.

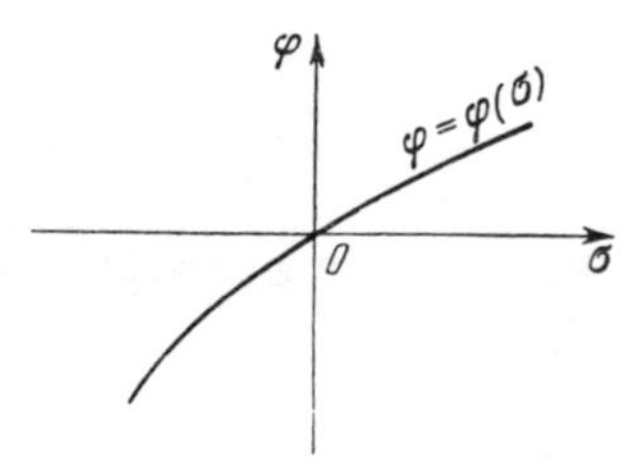

FIGURE 12.3.

The proofs are omitted, since they are implicit in the constructions of the following subsections.

If the trivial solution is unstable and inequality (12.27) holds, system (12.1) has a stable equilibrium state z_0 in $-K$.

The substitution $u=z-z_0$ brings us back to the case just considered.

If the nonlinear characteristic function $\varphi(\sigma)$ is concave (see Figure 12.3), i. e., $\varphi'(\sigma) > 0$ and $\varphi''(\sigma) < 0$, the substitution $u = -z$ achieves the same end.

To end the subsection, we note that systems with characteristic functions of the type illustrated in Figure 12.4 are treated in a similar manner.

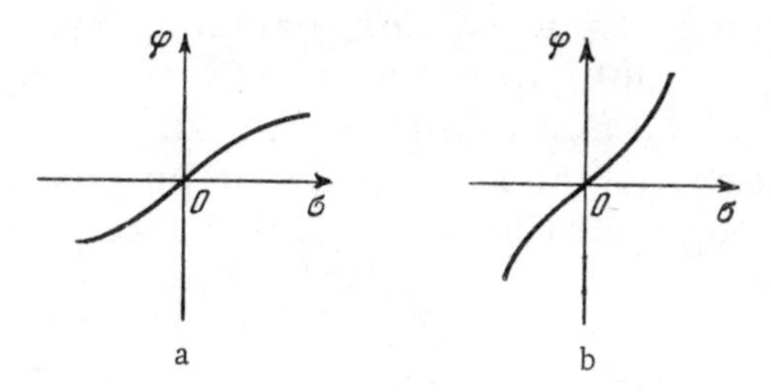

FIGURE 12.4.

12.3. Existence of an ap-solution of the disturbed system

Consider the disturbed system

$$\left.\begin{aligned} \frac{d\xi}{dt} &= \lambda_0\xi + d_1\varphi(\sigma) + f_1(t), \\ \frac{dx}{dt} &= Bx + d\varphi(\sigma) + f(t), \end{aligned}\right\} \tag{12.34}$$

where $f_1(t)$ and $f(t)$ are ap-functions (the first being scalar, the second having values in R^m). We write system (12.34) in the more convenient vector form

$$\frac{dz}{dt} = \bar{B}z + \bar{d}\varphi(\sigma) + \bar{f}(t), \tag{12.35}$$

where

$$\bar{f}(t) = \{f_1(t), f(t)\}. \tag{12.36}$$

In this subsection, we are interested in the existence of a stable ap-solution of system (12.34). The special case in which the undisturbed system is stable in the large was examined in detail by Yakubovich /2, 3/.

We examine system (12.39) on the assumption that the undisturbed system satisfies the conditions of subsection 12.2. In other words, we assume that the inequalities (12.4)—(12.9) are valid and that the nonlinear characteristic function $\varphi(\sigma)$ is twice continuously differentiable and satisfies (12.3), (12.32), (12.33). Moreover, we suppose that the origin is a stable equilibrium state of system (12.1), i. e., we suppose that (12.26) holds.

It is clear that under small almost periodic disturbances (12.36), system (12.34) has a stable ap-solution. On the other hand, examples are easily given of disturbances (12.36) under which system (12.34) has no ap-solutions. Below we give bounds for (12.36) such that (12.34) will have a stable ap-solution.

Consider the function

$$\psi(\sigma) = \sigma - \alpha_0\varphi(\sigma) \quad (-\infty < \sigma < \infty).$$

By (12.32) this function is concave. By (12.26), it is positive for small positive σ, and by (12.33) we have

$$\lim_{\sigma\to\infty} \psi(\sigma) = -\infty.$$

Hence $\psi(\sigma)$ has a unique maximum point σ^* (Figure 12.5). Set

$$\beta^* = \psi(\sigma^*) = \sigma^* - \alpha_0\varphi(\sigma^*).$$

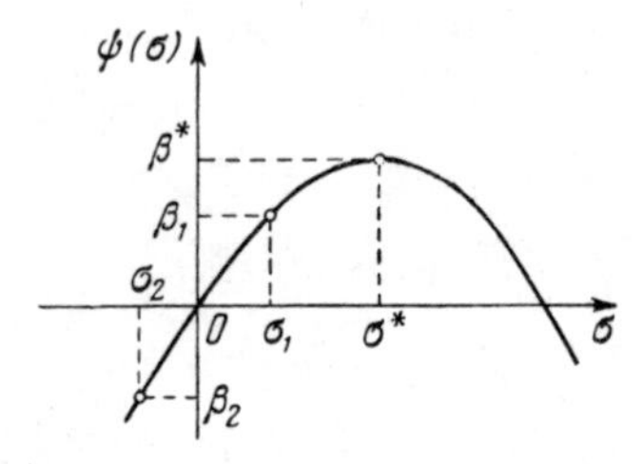

FIGURE 12.5.

We set

$$g(t) = \int_{-\infty}^{t} (e^{\bar{B}(t-s)}\bar{f}(s), \bar{c})\, ds. \tag{12.37}$$

Theorem 12.2. Let the function (12.37) satisfy

$$\sup_{-\infty < t < \infty} g(t) < \beta^*. \tag{12.38}$$

Then the disturbed system (12.34) has a unique ap-solution

$$z^*(t) = \{\xi^*(t), x^*(t)\}, \tag{12.39}$$

satisfying

$$\sup_{-\infty < t < \infty} \sigma^*(t) = \sup_{-\infty < t < \infty} (z^*(t), \bar{c}) < \sigma^*. \tag{12.40}$$

Moreover, this ap-solution is exponentially stable.

Proof. Consider the nonlinear integral operator

$$\Pi\sigma(t) = \int_{-\infty}^{t} (e^{\bar{B}(t-s)}\bar{d}, \bar{c})\, \varphi[\sigma(s)]\, ds + g(t) \tag{12.41}$$

in the space $B(R^1)$ of scalar ap-functions.

Let $\sigma_1 < \sigma^*$ be sufficiently close to σ^* (see Figure 12.5), so that

$$\sup_{-\infty < t < \infty} g(t) < \beta_1, \tag{12.42}$$

where

$$\beta_1 = \sigma_1 - \alpha_0\varphi(\sigma_1).$$

Then

$$\Pi\sigma_1 = \alpha_0\varphi(\sigma_1) + g(t) < \alpha_0\varphi(\sigma_1) + \beta_1 = \sigma_1 \qquad (-\infty < t < \infty). \tag{12.43}$$

Let σ_2 be a negative number, so large in absolute value that

$$g(t) \geqslant \beta_2 \qquad (-\infty < t < \infty), \tag{12.44}$$

where

$$\beta_2 = \sigma_2 - \alpha_0\varphi(\sigma_2).$$

Clearly,

$$\Pi\sigma_2 = \alpha_0\varphi(\sigma_2) + g(t) \geqslant \alpha_0\varphi(\sigma_2) + \beta_2 = \sigma_2 \qquad (-\infty < t < \infty). \tag{12.45}$$

The kernel $(e^{\bar{B}(t-s)}\bar{d}, \bar{c})$ of the integral operator (12.41) is positive for $t > s$. Thus, as $\varphi(\sigma)$ is monotone, the operator (12.41) is monotone in $B(R^1)$ with respect to the semi-order defined by the cone of nonnegative functions. By (12.43) and (12.45), this operator maps the conic interval $\langle \sigma_2, \sigma_1 \rangle$ into itself.

We now show that (12.41) satisfies the hypotheses of Theorem 10.2. Let $\eta_1(t), \eta_2(t) \in B(R^1)$ and $\eta_1(t) \leqslant \eta_2(t) \leqslant \sigma_1$.

Then from the convexity of $\varphi(\sigma)$ we have

$$0 \leqslant \Pi\eta_2(t) - \Pi\eta_1(t) \leqslant S[\eta_2(t) - \eta_1(t)],$$

where

$$S\eta(t) = \int_{-\infty}^{t} (e^{\bar{B}(t-s)}\bar{d}, \bar{c})\, \varphi'(\sigma_1)\, \eta(s)\, ds.$$

The linear integral operator S is positive and

$$\| S \|_{B(R^1)} = \alpha_0 \varphi'(\sigma_1).$$

Since $\sigma_1 < \sigma^*$ and $\alpha_0\varphi'(\sigma^*) = 1$, it follows that

$$\alpha_0\varphi'(\sigma_1) < 1.$$

Consequently, Π indeed satisfies the hypotheses of Theorem 10.2 and therefore has a unique fixed point $\sigma^*(t)$ in the conic interval $\langle \sigma_2, \sigma_1 \rangle$.

It follows that $\sigma^*(t)$ is the unique fixed point of Π in the set of ap-functions $\sigma(t)$ satisfying the inequality

$$\sup_{-\infty < t < \infty} \sigma(t) < \sigma^*.$$

Given $\sigma^*(t)$, let us construct the ap-solution of equation (12.35). Consider the inhomogeneous linear equation

$$\frac{dz}{dt} = \bar{B}z + \bar{d}\varphi[\sigma^*(t)] + \bar{f}(t).$$

It has a unique ap-solution

$$z^*(t) = \int_{-\infty}^{t} e^{\bar{B}(t-s)} \{\bar{d}\varphi[\sigma^*(s)] + \bar{f}(s)\}\, ds.$$

Clearly, $(z^*(t), \bar{c}) = \Pi\sigma^*(t)$, so that

$$(z^*(t), \bar{c}) = \sigma^*(t).$$

Hence $z^*(t)$ is an ap-solution of equation (12.35).

It remains only to show that $z^*(t)$ is exponentially stable. By Theorem 11.1, it suffices to show that the ap-operator

$$Lz = \frac{dz}{dt} - \bar{B}z - \bar{d}\varphi'[\sigma^*(t)](\bar{c}, z)$$

is regular and stable. To this end (see subsection 4.3) we need only verify the regularity of all ap-operators

$$L(\mu)z = \frac{dz}{dt} - \bar{B}z - \mu\, \bar{d}\varphi'[\sigma^*(t)](\bar{c}, z) \quad (0 \leqslant \mu \leqslant 1), \tag{12.46}$$

since the operator $L(0)$ is obviously stable.

An operator (12.46) is regular if the equation

$$\frac{dz}{dt} - \bar{B}z = \mu \bar{d}\varphi' [\sigma^* (t)] (\bar{c}, z) + \bar{f} (t) \tag{12.47}$$

has an ap-solution for any ap-function $\bar{f}(t)$. To construct it, we examine the solution of the integral equation

$$z(t) = \mu \int_{-\infty}^{t} e^{\bar{B}(t-s)} \bar{d}\varphi' [\sigma^* (s)] (\bar{c}, z(s)) \, ds + \int_{-\infty}^{t} e^{\bar{B}(t-s)} \bar{f}(s) \, ds. \tag{12.48}$$

It is convenient then to find the function $\sigma(t) = (z(t), \bar{c})$ from the equation

$$\sigma(t) = \mu \int_{-\infty}^{t} (e^{\bar{B}(t-s)} \bar{d}, \bar{c}) \varphi' [\sigma^* (s)] \sigma (s) \, ds + \\ + \int_{-\infty}^{t} (e^{\bar{B}(t-s)} \bar{f}(s), \bar{c}) \, ds, \tag{12.49}$$

and first to determine $z(t)$ using the right-hand side of equation (12.48).

We now note that all equations (12.49) will have solutions if the norm of the linear integral operator

$$S_0 \sigma (t) = \int_{-\infty}^{t} (e^{\bar{B}(t-s)} \bar{d}, \bar{c}) \varphi' [\sigma^* (s)] \sigma (s) \, ds$$

is less than 1.

But S_0 is positive with respect to the cone of nonnegative functions, so indeed

$$\| S_0 \|_{B(R^1)} \leqslant \alpha_0 \varphi' (\sigma_1) < 1.$$

12.4. Existence of an unstable ap-solution of the disturbed system

We continue our examination of the disturbed system (12.35).

From Theorem 12.2 we deduce that, under disturbances satisfying (12.38), a stable equilibrium state of system (12.1) becomes a stable ap-solution of system (12.35). It turns out that an unstable equilibrium state of system (12.1), under the same disturbances, becomes an unstable ap-solution of the disturbed system. Moreover, the disturbed system has no other ap-solutions. We prove these propositions in this subsection.

Theorem 12.3. Under the hypotheses of Theorem 12.2, the disturbed system has only one ap-solution other than the stable ap-solution of Theorem 12.2. This second ap-solution is unstable.

We give the proof in several steps.

Step I. By setting

$$u = z - z^*(t), \tag{12.50}$$

where $z^*(t)$ is the stable ap-solution of the disturbed system existing by Theorem 12.2, we go from equation (12.35) to a new equation

$$\frac{du}{dt} = \overline{B}u + \bar{d}\chi(t, \sigma), \tag{12.51}$$

where

$$\chi(t, \sigma) = \varphi[\sigma + \sigma^*(t)] - \varphi[\sigma^*(t)] \tag{12.52}$$

and

$$\sigma = (u, \bar{c}). \tag{12.53}$$

To establish Theorem 12.3, it is clearly sufficient to show that system (12.51) has one and only one nontrivial ap-solution and that this solution is unstable.

Step II. We show that equation (12.51) has a bounded solution $u^*(t)$ on $(-\infty, \infty)$ lying entirely in the cone K and satisfying the inequalities

$$r_0 \leqslant |u^*(t)| \leqslant r_1 \qquad (-\infty < t < \infty), \tag{12.54}$$

where r_0 and r_1 are positive.

Set

$$\sigma_{\max} = \sup_{-\infty < t < \infty} \sigma^*(t).$$

By (12.40), $\sigma_{\max} < \sigma^*$, so that

$$\alpha_0\varphi'(\sigma_{\max}) < 1.$$

Choose $\delta_0 > 0$ such that $\delta_0 + \sigma_{\max} < \sigma^*$. Then for $-\infty < t < \infty$, $0 \leqslant \sigma \leqslant \delta_0$, we have by the mean-value theorem that

$$\chi(t, \sigma) = \varphi'[\theta(t)\sigma + \sigma^*(t)]\sigma \leqslant \varphi'[\sigma + \sigma_{\max}]\sigma \leqslant$$
$$\leqslant \varphi'(\delta_0 + \sigma_{\max})\sigma,$$

whence it follows, in turn, that

$$\chi(t, \sigma) \leqslant (\alpha_0^{-1} - \varepsilon_0)\sigma \qquad (-\infty < t < \infty, \; 0 \leqslant \sigma \leqslant \delta_0), \tag{12.55}$$

where ε_0 is positive.

We have by (12.33) that for $\varepsilon_1 > 0$ there exists $\eta_1 > 0$ such that

$$\chi(t, \sigma) \geqslant (\alpha_0^{-1} + \varepsilon_1)\sigma - \eta_1 \qquad (-\infty < t < \infty, \; \sigma \geqslant 0). \tag{12.56}$$

Now consider two linear equations of the form (12.21):

$$\frac{du}{dt} = \overline{B}u + \bar{d}(\alpha_0^{-1} - \varepsilon_0)\sigma \tag{12.57}$$

and

$$\frac{du}{dt} = \overline{B}u + \bar{d}(\alpha_0^{-1} + \varepsilon_1)\sigma. \tag{12.58}$$

The trivial solution of (12.57) is exponentially stable, while the second equation has an exponentially increasing solution.

By (12.55) for $u \in K$ and $|\bar{c}| \cdot |u| \leqslant \delta_0$ we have the conic inequality

$$\overline{B}u + \bar{d}\chi(t, \sigma) \leqslant \overline{B}u + \bar{d}(\alpha_0^{-1} - \varepsilon_0)\sigma \qquad (-\infty < t < \infty),$$

while (12.56) implies that for all $u \in K$

$$\overline{B}u + \bar{d}\chi(t, \sigma) \geqslant \overline{B}u + \bar{d}(\alpha_0^{-1} + \varepsilon_1)\sigma - \bar{d}\eta_1 \qquad (-\infty < t < \infty).$$

The last two inequalities — by a general existence principle due to Krasnosel'skii /2/ (p. 191, Theorem 8. 5) guarantee the existence of a positive solution $u^*(t)$, separated from zero and bounded on the whole real line. Though all Krasnosel'skii's constructions are for the case in which K is the cone of vectors with nonnegative components, the proof makes no use of this fact. This general principle implies the existence of a bounded solution satisfying (12.54).

Step III. Let

$$u_1(t) = \{\xi_1(t), x_1(t)\}, \qquad u_2(t) = \{\xi_2(t), x_2(t)\}$$

be two solutions of system (12.51), bounded on the real line. We claim that the points $u_1(0)$ and $u_2(0)$ are comparable with respect to the cone K, i. e., either $u_2(0) - u_1(0) \in K$, or $u_1(0) - u_2(0) \in K$.

Let us suppose the contrary.

Set

$$\eta(t)=\xi_1(t)-\xi_2(t), \quad y(t)=x_1(t)-x_2(t),$$
$$\sigma_1(t)=(u_1(t), \bar{c}), \quad \sigma_2(t)=(u_2(t), \bar{c}), \quad \sigma(t)=\sigma_1(t)-\sigma_2(t),$$
$$\chi_0(t, \sigma)=\chi[t, \sigma_2(t)+\sigma]-\chi[t, \sigma_2(t)].$$

Then

$$\frac{d\eta}{dt}=\lambda_0\eta+d_1\chi_0(t, \sigma),$$
$$\frac{dy}{dt}=By+d\chi_0(t, \sigma).$$

Consider the scalar function

$$w(t)=\rho^2\eta^2(-t)-(Hy(-t), y(-t)) \qquad (0\leqslant t<\infty),$$

where the constant ρ is determined by (12.12), and the symmetric positive-definite matrix H satisfies (12.10) and (12.11).

Obviously,

$$\frac{dw}{dt}=-2\rho^2\lambda_0\eta^2(-t)+((HB+B^*H)y(-t), y(-t))-$$
$$-2\{\rho^2\eta(-t)d_1+(Hd, y(-t))\}\chi_0[-t, \sigma(-t)],$$

so that, by (12.12) and (12.11),

$$\frac{dw}{dt}=-2\rho^2\lambda_0\eta^2(-t)+((HB+B^*H)y(-t), y(-t))-$$
$$-2\sigma(-t)\chi_0[-t, \sigma(-t)].$$

Suppose that $w(t)<0$ for some $t\geqslant 0$. Then, for this t, we have by (12.59) that

$$\frac{dw}{dt}<([H(B+\lambda_0 I)+(B^*+\lambda_0 I)H]y(-t), y(-t))-$$
$$-2\sigma(-t)\chi_0[-t, \sigma(-t)] \qquad (12.60)$$

and, by (12.10), we further have that

$$\frac{dw}{dt}<0.$$

Since $w(0)<0$ by assumption, it follows that $w(t)<0$ for all $t\geqslant 0$ and that (12.60) also holds for all $t\geqslant 0$. Hence, $w'(t)<0$ for all $t\geqslant 0$, so that $w(t)$ is decreasing.

In particular,

$$w(t)\leqslant w(0)<0 \qquad (t\geqslant 0).$$

This last inequality yields

$$(Hy(-t),\ y(-t)) \geqslant -w(0) > 0,$$

so that

$$|y(-t)| \geqslant r > 0 \qquad (t \geqslant 0).$$

Therefore, by (12.60) and (12.10), there exists $k > 0$ such that

$$\frac{dw}{dt} \leqslant -k \qquad (0 \leqslant t < \infty).$$

This contradicts the boundedness of $w(t)$.

Step IV. This is an auxiliary step. We consider the translation operator $W(t, s)$ of system (12.51) and note some of its properties. Recall that at certain points $u \in R^{m+1}$ the translation operator may not be defined for all $t > s$.

The translation operator is positive for $t > s$ with respect to both the cone K and the cone $-K$. The translation operator is *strongly monotone* on R^{m+1}, i. e., whenever $u_1 \leqslant u_2$ and $u_1 \neq u_2$,

$$W(t, s)u_2 - W(t, s)u_1 \gg 0 \qquad (t > s). \tag{12.61}$$

We shall not go into the proofs of these properties here. It suffices to repeat the arguments used in subsection 12.2 to prove the strong positivity of the translation operator of system (12.1).

We shall have occasion to use one additional property of the translation operator. Namely, it turns out that this operator is convex in the following sense: for any $u \gg 0$ and $\tau \in (0, 1)$

$$\tau W(t, s)u \gg W(t, s)(\tau u) \qquad (t > s). \tag{12.62}$$

Since $\chi(t, 0) \equiv 0$, it follows that

$$\tau\chi(t, \sigma) - \chi(t, \tau\sigma) =$$
$$= \int_0^1 \int_0^1 \varphi''[\sigma^*(t) + s\tau\sigma + ss_1(1-\tau)\sigma]\, s\tau(1-\tau)\sigma^2\, ds\, ds_1$$

and by (12.32), for every $\sigma > 0$

$$\tau\chi(t, \sigma) - \chi(t, \tau\sigma) > 0 \quad (0 < \tau < 1, \quad -\infty < t < \infty). \tag{12.63}$$

Let $u \gg 0$, $\tau \in (0, 1)$. Set

$$p(t) = \tau W(t, s)u - W(t, s)(\tau u) \qquad (t \geqslant s)$$

and

$$\sigma(t) = (W(t, s)u, \bar{c}), \qquad \sigma_\tau(t) = (W(t, s)(\tau u), \bar{c}).$$

$p(t)$ is the solution of the differential equation

$$\frac{dp}{dt} = \bar{B}p + \bar{d}\{\chi[t, (\bar{c}, p) + \sigma_\tau(t)] - \chi[t, \sigma_\tau(t)]\} + \\ + \bar{d}\{\tau\chi[t, \sigma(t)] - \chi[t, \tau\sigma(t)]\}, \qquad (12.64)$$

with initial value $p(s) = 0$. From the monotonicity of $\chi(t, \sigma)$ in the second variable and from (12.63) we have that equation (12.64) satisfies the hypotheses of Lemma 11.3. Hence

$$p(t) \geqslant 0 \qquad (t \geqslant s). \qquad (12.65)$$

By (12.64),

$$p(t) = \int_s^t e^{\bar{B}(t-\theta)}\bar{d}\{\chi[\theta, (\bar{c}, p(\theta)) + \sigma_\tau(\theta)] - \chi[\theta, \sigma_\tau(\theta)]\}\, d\theta + \\ + \int_s^t e^{\bar{B}(t-\theta)}\bar{d}\{\tau\chi[\theta, \sigma(\theta)] - \chi[\theta, \tau\sigma(\theta)]\}\, d\theta,$$

and thus, by (12.65),

$$p(t) \geqslant \int_s^t e^{\bar{B}(t-\theta)}\bar{d}\{\tau\chi[\theta, \sigma(\theta)] - \chi[\theta, \tau\sigma(\theta)]\}\, d\theta. \qquad (12.66)$$

Since $W(t, s)$ is strongly positive for $t > s$, we have $\sigma(t) > 0$ for $t > s$. Thus (12.66) and (12.63) imply (12.62).

Let $\mathfrak{M}$ denote the set of scalar ap-functions which are uniform limits of sequences of the form $\sigma^*(t + t_j)$ $(j = 1, 2, \ldots)$. For every function $\sigma^{**} \in \mathfrak{M}$ we consider the equation

$$\frac{du}{dt} = \bar{B}u + \bar{d}\chi^*(t, \sigma), \qquad (12.67)$$

where

$$\chi^*(t, \sigma) = \varphi[\sigma^{**}(t) + \sigma] - \varphi[\sigma^{**}(t)].$$

Let $W^*(t, s)$ denote the translation operator of equation (12.67).

Clearly, every equation (12.67) satisfies all properties mentioned regarding $W(t, s)$. Thus every operator $W^*(t, s)$, too, is strongly positive with respect to both K and $-K$, strongly monotone on all of R^{m+1}, and convex.

Step V. Let $u^*(t)$ be a solution of equation (12.51) which is bounded in $(-\infty, \infty)$ and takes values in K; its existence was established in the Step II. Recall that $u^*(t)$ satisfies (12.54). The strong positivity of $W(t, s)$ $(t > s)$ implies

$$u^*(t) \gg 0 \qquad (-\infty < t < \infty). \tag{12.68}$$

We now prove that every nontrivial solution $u^*(t)$ of (12.51), distinct from $u(t)$, approaches the origin as $t \to \infty$ if

$$0 \leqslant u(0) \leqslant u^*(0).$$

Note that since the operators $W(t, s)$ $(t > s)$ are strongly monotone,

$$u^*(t) \gg u(t) \qquad (t > 0). \tag{12.69}$$

We define a scalar function α by

$$\alpha(t) = \inf\{\alpha\colon\ \alpha u^*(t) \geqslant u(t)\}.$$

By (12.69),

$$0 < \alpha(t) < 1 \qquad (t > 0). \tag{12.70}$$

Furthermore, $\alpha(t)$ is strictly decreasing, i. e., $0 < t_1 < t_2$ implies $\alpha(t_1) > \alpha(t_2)$. In fact, by the monotonicity of the operator $W(t_2, t_1)$,

$$W(t_2, t_1)[\alpha(t_1) u^*(t_1)] \geqslant W(t_2, t_1) u(t_1).$$

Thus, by (12.70) and (12.62),

$$\alpha(t_1) W(t_2, t_1) u^*(t_1) \gg W(t_2, t_1) u(t_1),$$

i. e.,

$$\alpha(t_1) u^*(t_2) \gg u(t_2).$$

But then $\alpha(t_2) < \alpha(t_1)$.

Since $\alpha(t)$ is monotone, the limit

$$\alpha_0 = \lim_{t\to\infty} \alpha(t)$$

exists.

Our proposition will be proved if we show that $\alpha_0 = 0$.

Suppose that $\alpha_0 > 0$.

Choose a sequence $h_j \to \infty$ such that the sequence of ap-functions $\sigma^*(t + h_j)$ converges uniformly to an ap-function $\sigma^{**}(t) \in \mathfrak{M}$ and the sequences $u^*(t + h_j)$ and $u(t + h_j)$ converge uniformly on every finite interval to functions $v^*(t)$ and $v(t)$, respectively. The functions $v^*(t)$ and $v(t)$ are solutions of equation (12.67) bounded on $(-\infty, \infty)$. Moreover, by the strong positivity of $W^*(t, s)$, $v^*(t)$ takes values in K_{in}.

Since the inequality

$$\alpha(t+h_j)\,u^*(t+h_j) \geqslant u(t+h_j),$$

holds for any $t \in (-\infty, \infty)$ and sufficiently large j, it follows that

$$\alpha_0 v^*(t) \geqslant v(t) \qquad (-\infty < t < \infty)$$

and the convexity and monotonicity of $W^*(t, s)$ yield

$$\alpha_0 v^*(t) \gg v(t) \qquad (-\infty < t < \infty). \tag{12.71}$$

Let

$$\beta_0 = \inf\{\beta : \beta v^*(0) - v(0) \in K\}.$$

By (12.71), we have $\beta_0 < \alpha_0$.

The inclusion $\beta_0 v^*(0) - v(0) \in K$ may be written

$$\beta_0 u^*(h_j) - u(h_j) + \omega_j \in K \qquad (j = 1, 2, \ldots), \tag{12.72}$$

where

$$\omega_j = \beta_0 [v^*(0) - u^*(h_j)] - v(0) + u(h_j).$$

By construction,

$$\lim_{j \to \infty} \omega_j = 0. \tag{12.73}$$

Since $v^*(0) \gg 0$ and $u(h_j) \to v^*(0)$, we have for sufficiently large j that elements $u^*(h_j)$ are in K together with spherical neighborhoods of fixed radius. Thus, for sufficiently large j, (12.73) yields

$$\frac{\alpha_0 - \beta_0}{2} u^*(h_j) - \omega_j \in K. \tag{12.74}$$

It follows from (12.72) that

$$\frac{\alpha_0 + \beta_0}{2} u^*(h_j) - u(h_j) \in K, \tag{12.75}$$

whence

$$\alpha(h_j) \leqslant \frac{\alpha_0 + \beta_0}{2},$$

and $\alpha(h_j) < \alpha_0$. This is a contradiction.

Step VI. Completion of proof. From the results of Steps III, IV and V, it follows that every equation (12.67) has a unique solution bounded on $(-\infty, \infty)$, taking values in K, and satisfying inequalities (12.54). It follows (see, for example, Demidovich /1/, p. 441) that these bounded solutions are almost periodic.

The results of Step III imply that any two ap-solutions of equation (12.51) are comparable with respect to the cone K. All values of the nontrivial ap-solutions, then, lie either in K or in $-K$. By Step V, the ap-solution taking values in K is unique and unstable.

By the first part of Theorem 12.2, there are no nontrivial ap-solutions with values in $-K$, completing the proof of Theorem 12.2.

12.5. Applications to scalar equations of higher order

Completely controllable systems (Kalman /1/) form an important class of undisturbed automatic control systems (12.1). The condition for complete controllability is that the vectors

$$\bar{d},\ \bar{B}\bar{d},\ \ldots,\ \bar{B}^m\bar{d}$$

form a basis for R^{m+1}. Every completely controllable system may be described (see, e. g., Lefschetz /1/, p. 113) by a scalar differential equation

$$\frac{d^{m+1}x}{dt^{m+1}} + a_1\frac{d^m x}{dt^m} + \ldots + a_{m+1}x = \varphi(\sigma), \qquad (12.76)$$

where

$$\sigma - b_{m+1}x + b_m x' + \ldots + b_1 x^{(m)}, \qquad (12.77)$$

$a_1, \ldots, a_{m+1}, b_1, \ldots, b_{m+1}$ being constants and $\varphi(\sigma)$ the nonlinear characteristic function in equations (12.1).

In subsection 12.1 we assumed that system (12.1) satisfies conditions (12.4)—(12.9). Let us examine these restrictions in terms of the coefficients $a_1, \ldots, a_{m+1}, b_1, \ldots, b_{m+1}$ of equation (12.76).

We first rewrite (12.76) as

$$\frac{du}{dt} + Qu = e_{m+1}\varphi[(b, u)], \tag{12.78}$$

where Q is the matrix (8.13):

$$Q = \begin{Vmatrix} 0 & -1 & 0 & \dots & 0 \\ 0 & 0 & -1 & \dots & 0 \\ \dots & \dots & \dots & \dots & \dots \\ 0 & 0 & 0 & \dots & -1 \\ a_{m+1} & a_m & a_{m-1} & \dots & a_1 \end{Vmatrix} \tag{12.79}$$

and

$$e_{m+1} = \{0, \dots, 0, 1\}, \qquad b = \{b_{m+1}, \dots, b_1\}.$$

By condition (12.4), the characteristic polynomial

$$\chi(\lambda) = \lambda^{m+1} + a_1\lambda^m + \dots + a_{m+1} \tag{12.80}$$

of the differential equation

$$\frac{d^{m+1}x}{dt^{m+1}} + a_1\frac{d^m x}{dt^m} + \dots + a_{m+1}x = 0 \tag{12.81}$$

is a Hurwitz polynomial having λ_0 as leading root. In other words, the equation $\chi(\lambda) = 0$ has a simple negative root λ_0 such that all other roots λ satisfy Re $\lambda < \lambda_0$.

The existence of a leading root λ_0 of the characteristic polynomial enables us (see, e. g., Gantmakher /1/) to choose a basis in R^{m+1} so that the matrix $-Q$ is of the form of the matrix $\bar{B}$ in (12.15). With respect to this basis, the system (12.78) is of the form (12.1).

The actual construction of this basis is unimportant. We need only note that its first vector e_0 must be an eigenvector of $-Q$ corresponding to the eigenvalue λ_0, while the remaining elements are m arbitrary linearly independent vectors, orthogonal to an eigenvector g_0 of the adjoint matrix $-Q^*$ which corresponds to the same eigenvalue λ_0. The vectors e_0 and g_0 are normalized so that

$$(e_0, g_0) = 1. \tag{12.82}$$

Let e_0 be the vector

$$e_0 = \{1, \lambda_0, \dots, \lambda_0^m\}.$$

It turns out that condition (12.82) will be satisfied if the eigenvector

$$g_0 = \{\eta_1, \eta_2, \ldots, \eta_{m+1}\}$$

of the matrix $-Q^*$ is normalized by the equality

$$\eta_{m+1} = \frac{1}{\chi'(\lambda_0)} \tag{12.83}$$

(here the denominator is positive, since λ_0 is the largest simple real root of the polynomial (12.80). By definition, in fact, the components η_i of the vector g_0 satisfy the system

$$\begin{aligned} \eta_1 - a_m\eta_{m+1} &= \lambda_0\eta_2, \\ \eta_2 - a_{m-1}\eta_{m+1} &= \lambda_0\eta_3, \\ &\ldots\ldots\ldots \\ \eta_m - a_1\eta_{m+1} &= \lambda_0\eta_{m+1}. \end{aligned}$$

Multiplying the second equation by $2\lambda_0$, the third by $3\lambda_0^2$, etc., the last by $m\lambda_0^{m-1}$, and then adding, we obtain

$$\begin{aligned} \eta_1 + 2\lambda_0\eta_2 + \ldots + m\lambda_0^{m-1}\eta_m - \left[\chi'(\lambda_0) - (m+1)\lambda_0^m\right]\eta_{m+1} = \\ = \lambda_0\eta_2 + 2\lambda_0^2\eta_3 + \ldots + m\lambda_0^m\eta_{m+1}, \end{aligned}$$

so that

$$(e_0, g_0) = \eta_1 + \lambda_0\eta_2 + \ldots + \lambda_0^m\eta_{m+1} = \chi'(\lambda_0)\,\eta_{m+1} = 1.$$

Clearly

$$d_1 = (e_{m+1}, g_0)$$

and by (12.83),

$$d_1 = \frac{1}{\chi'(\lambda_0)}.$$

Thus, the first of inequalities (12.5) always holds.

Set

$$\varkappa(\lambda) = b_1\lambda^m + \ldots + b_m\lambda + b_{m+1}. \tag{12.84}$$

Since $c_1 = (e_0, b)$, we have $c_1 = \varkappa(\lambda_0)$. Thus the second of inequalities (12.5) may now be written

$$\varkappa(\lambda_0) > 0. \tag{12.85}$$

From inequality (12.7), we get $(e_{m+1}, b) \geqslant 0$, i.e.,

$$b_1 \geqslant 0. \tag{12.86}$$

To examine condition (12.8), we first observe that

$$(Bd, c) - \lambda_0(d, c) = (\bar{B}\bar{d}, \bar{c}) - \lambda_0(\bar{d}, \bar{c}),$$

whence it follows that

$$(Bd, c) - \lambda_0(d, c) = -(Qe_{m+1}, b) - \lambda_0(e_{m+1}, b) = \\ = b_2 - (a_1 + \lambda_0) b_1.$$

In our case, then, (12.8) is of the form

$$b_2 > (a_1 + \lambda_0) b_1. \tag{12.87}$$

We now turn to the more complicated condition (12.9). Clearly, for every real nonzero ω,

$$\operatorname{Re}(B^{-1}_{\lambda_0+i\omega}d, c) \equiv \operatorname{Re}(\bar{B}^{-1}_{\lambda_0+i\omega}\bar{d}, \bar{c}),$$

and so (in our case)

$$\operatorname{Re}(B^{-1}_{\lambda_0+i\omega}d, c) = \operatorname{Re}([(\lambda_0 + i\omega) I + Q]^{-1} e_{m+1}, b). \tag{12.88}$$

We need a simple and well-known equality (see, for example, Lefschetz /1/, p. 114):

$$([\mu I + Q]^{-1} e_{m+1}, b) = \frac{\varkappa(\mu)}{\chi(\mu)}.$$

Together with (12.88) this equality implies that we may write condition (12.9) for nonzero ω as

$$\operatorname{Re} \frac{\varkappa(\lambda_0 + i\omega)}{\chi(\lambda_0 + i\omega)} < 0. \tag{12.89}$$

It is easy to see that the function $\varkappa(\lambda_0 + i\omega)/\chi(\lambda_0 + i\omega)$ is continuous at $\omega = 0$, and

$$\lim_{\omega \to 0} \operatorname{Re} \frac{\varkappa(\lambda_0 + i\omega)}{\chi(\lambda_0 + i\omega)} = \frac{\varkappa'(\lambda_0)}{\chi'(\lambda_0)} - \frac{1}{2} \frac{\chi''(\lambda_0)\varkappa(\lambda_0)}{[\chi'(\lambda_0)]^2}. \tag{12.90}$$

Hence (12.9) is replaced here by two conditions: (12.88) and the inequality

$$2\varkappa'(\lambda_0)\chi'(\lambda_0) < \varkappa(\lambda_0)\chi''(\lambda_0). \qquad (12.91)$$

In the general case, (12.6) follows from (12.9) (as may be seen by integrating (12.9) over $[0, \infty)$).

We have thus proved

Lemma 12.2. Conditions (12.4) – (12.9) for equation (12.76) mean that:

1. *The characteristic polynomial (12.80) has leading negative root* λ_0.

2. $\varkappa(\lambda_0) > 0, \quad b_1 \geqslant 0, \quad b_2 > (a_1 + \lambda_0) b_1$.

3. *Inequalities (12.89) and (12.91) hold.*

We leave it to the reader to verify that for the second-order equation

$$\frac{d^2x}{dt^2} + a_1 \frac{dx}{dt} + a_2 x = \varphi(\sigma),$$

where

$$\sigma = b_1 \frac{dx}{dt} + b_2 x,$$

conditions 1 — 3 mean that

$$0 < a_2 < \frac{1}{4} a_1^2; \quad a_1 > 0; \quad b_2 > \frac{1}{2}\left(a_1 + \sqrt{a_1^2 - 4a_2}\right) b_1 \geqslant 0.$$

As to the third-order equation

$$\frac{d^3x}{dt^3} + a_1 \frac{d^2x}{dt^2} + a_2 \frac{dx}{dt} + a_3 x = \varphi(\sigma), \qquad (12.92)$$

where

$$\sigma = b_1 \frac{d^2x}{dt^2} + b_2 \frac{dx}{dt} + b_3 x, \qquad (12.93)$$

conditions 1 — 3 are naturally a little more complicated. Condition 1 must be verified directly. If it holds and λ_0 is the leading root of the polynomial

$$\chi(\lambda) = \lambda^3 + a_1\lambda^2 + a_2\lambda + a_3,$$

then condition 2 may be written

$$\varkappa(\lambda_0) > 0, \ b_1 \geqslant 0, \ b_2 > (a_1 + \lambda_0) b_1.$$

Condition 3 then follows from condition 2.

We now turn from equation (12.76) to the disturbed equation

$$\frac{d^{m+1}x}{dt^{m+1}}+a_1\frac{d^m x}{dt^m}+\ldots+a_{m+1}x=\varphi(\sigma)+h(t), \tag{12.94}$$

where $h(t)$ is a scalar ap-function. From equation (12.94) we get the disturbed version

$$\frac{du}{dt}+Qu=e_{m+1}\varphi(\sigma)+e_{m+1}h(t), \qquad \sigma=(b,\,u) \tag{12.95}$$

of system (12.78).

Assume that the nonlinear characteristic function $\varphi(\sigma)$ satisfies the hypotheses of Theorems 12.1 – 12.3. Then we may use Theorem 12.3 to examine the ap-solutions of equation (12.95).

Before we formulate the corresponding assertions, note that in the case of equation (12.95), the number (12.24) is defined by

$$\alpha_0=(Q^{-1}e_{m+1},\,b)=\frac{b_{m+1}}{a_{m+1}}.$$

Since $a_{m+1}>0$ and α_0 is positive for all systems (12.1) considered here, it follows that $b_{m+1}>0$.

The main result of this subsection is

Theorem 12.4. Suppose that conditions 1–3 of Lemma 12.2 hold. Let

$$b_{m+1}\varphi'(0)<a_{m+1}.$$

Let $h(t)$ satisfy the inequality

$$\sup_{-\infty<t<\infty} h(t)<\frac{a_{m+1}}{b_{m+1}}\sigma^*-\varphi(\sigma^*),$$

where σ^* is a root of the equation

$$b_{m+1}\varphi'(\sigma)=a_{m+1}.$$

Then equation (12.94) has two ap-solutions, one stable and the other unstable.

12.6. Existence of three ap-solutions

We now examine a special case of equation (12.94), namely the equation

$$Lx=\varphi(x)+h(t), \tag{12.96}$$

where L is a different operator

$$Lx = \frac{d^m x}{dt^m} + a_1 \frac{d^{m-1} x}{dt^{m-1}} + \ldots + a_m x \qquad (12.97)$$

with constant coefficients.

Let the nonlinear characteristic function $\varphi(x)$ be of type x^3, i. e., $\varphi(x)$ is monotone increasing, $\varphi(0) = 0$, $x\varphi''(x) > 0$ for $x \neq 0$, and

$$\lim_{|x| \to \infty} \frac{\varphi(x)}{x} = \infty \qquad (12.98)$$

(see Figure 12.4b). If

$$\varphi'(0) < a_m, \qquad (12.99)$$

the the undisturbed equation

$$Lx = \varphi(x) \qquad (12.100)$$

has, by (12.98), exactly three equilibrium states. This implies, in general, that for small $h(t)$ the disturbed equation (12.96) has at least three distinct ap-solutions.

We shall now use the results of §§ 10 and 11 to derive a global existence theorem for three solutions and to examine them for stability.

Throughout this subsection we assume that the characteristic polynomial $\chi(\lambda)$ of the differential operator (12.97) is a Hurwitz polynomial and that the Green's function $G(t, s)$ of this operator is nonnegative (see § 8). We further assume that for some $\gamma > 0$ the differential ap-operator

$$L(\gamma)x = Lx - \gamma x \qquad (12.101)$$

is regular and that its Green's function $G(t, s; \gamma)$ is nonpositive.

Note that for second-order operators (12.97) the Green's function of operator (12.101) will be nonpositive for all sufficiently large γ. To see this, observe that $L(\gamma)u_0(t) = a_m - \gamma$, where $u_0(t) \equiv 1$, and then use Theorem 9.9.

For higher-order operators (12.97) the existence of such numbers γ is a restrictive assumption.

We first consider third-order ap-operators. Since the characteristic polynomial $\chi(\lambda)$ is a Hurwitz polynomial and the Green's function $G(t, s)$ is nonnegative (see subsection 8.4), there exists a negative root λ_0 of $\chi(\lambda)$ such that the other two roots λ satisfy the inequality $\mathrm{Re}\,\lambda \leqslant \lambda_0$. Since the Green's function $G(t, s; \gamma)$ is assumed nonpositive, it follows that all three roots of $\chi(\lambda)$ are real and that $\chi(\lambda) - \gamma = 0$ has two negative roots and one positive root. In other words, the graph of $\chi(\lambda)$ is of the form shown in Figure 12.6. The above conditions are satisfied by all γ in the interval $(\gamma_1, \gamma_0]$ shown in Figure 12.6.

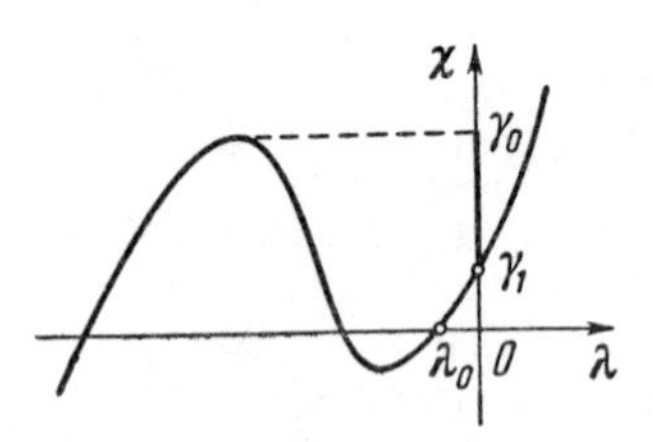

FIGURE 12.6.

It follows from our argument that the Green's function $G(t, s; \gamma)$ of the third-order operator cannot be nonpositive for any γ if the graph of the characteristic polynomial is as in Figure 12.7a or b.

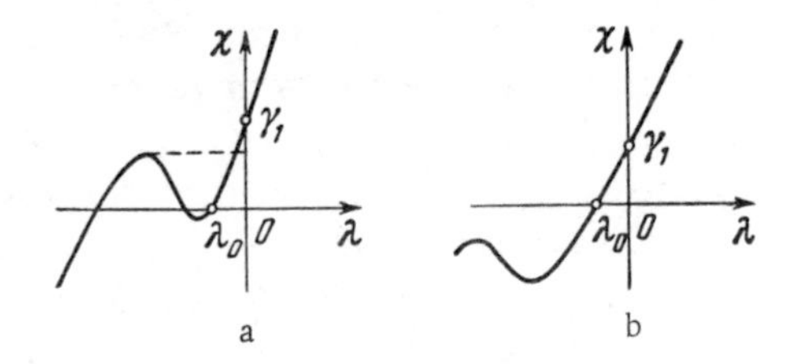

FIGURE 12.7.

Analysis of fourth-order ap-operators also reduces (see subsection 8.4) to examination of the graph of the characteristic polynomial. For example, if $\chi(\lambda)$ has the form illustrated in Figure 12.8, then it is a Hurwitz polynomial, the Green's function $G(t, s)$ is nonnegative and the Green's functions $G(t, s; \gamma)$ are nonpositive for γ in the interval $(\gamma_1, \gamma_0]$ shown in Figure 12.8.

The fact that the corresponding γ for third- and fourth-order operators form a finite interval $(\gamma_1, \gamma_0]$ is not accidental. It can be shown that such is the case for all higher-order operators (provided, of course, that such γ exist!).

We introduce new notation. Let x_1 and x_2 be the solutions of the equation

$$\varphi'(x) = a_m$$

(Figure 12.9). Let $\gamma > a_m$ and $y_1(\gamma)$, $y_2(\gamma)$ be the solutions of the equation

$$\varphi'(x) = \gamma$$

(see Figure 12.9). Set

$$m(\gamma) = \max\{a_m x_1 - \varphi(x_1),\ a_m y_2(\gamma) - \varphi[y_2(\gamma)]\}$$

and

$$M(\gamma) = \min\{a_m x_2 - \varphi(x_2),\ a_m y_1(\gamma) - \varphi[y_1(\gamma)]\}.$$

We shall consider γ such that $m(\gamma) < M(\gamma)$.

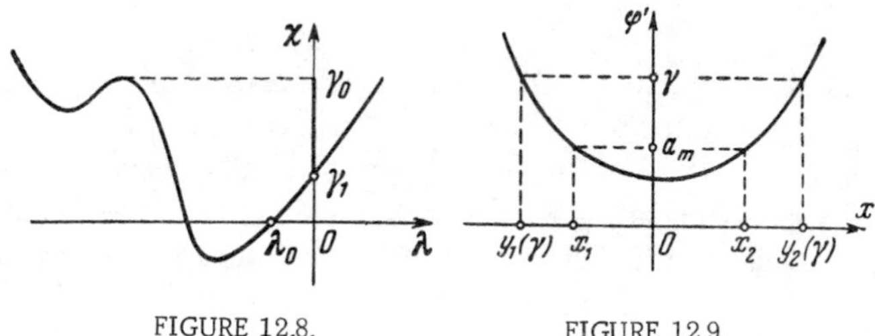

FIGURE 12.8. FIGURE 12.9.

Theorem 12.5. Let $h(t)$ be an ap-function satisfying

$$m(\gamma) < h(t) < M(\gamma) \qquad (-\infty < t < \infty), \tag{12.102}$$

where γ is a number such that $G(t, s; \gamma)$ is nonpositive.

Then equation (12.96) has at least one stable ap-solution and at least two unstable ones.

We only outline the proof. In studying the ap-solutions of equation (12.96), we may turn to the equivalent integral equation

$$x(t) = \int_{-\infty}^{\infty} G(t, s)\{\varphi[x(s)] + h(s)\}\, ds.$$

The operator Π defined by the right-hand side of this equation leaves the conic interval $\langle x_1, x_2\rangle$ invariant, is monotone there, and satisfies the hypotheses of Theorem 10.2 on the conic interval $\langle \Pi x_1, \Pi x_2\rangle$. Hence equation (12.96) has a unique ap-solution $x^*(t)$ satisfying the inequalities

$$x_1 < x^*(t) < x_2 \qquad (-\infty < t < \infty). \tag{12.103}$$

This solution is stable by Theorem 11.3.

As to the existence of unstable ap-solutions, it is convenient to replace equation (12.96) by the integral equation

$$x(t) = -\int_{-\infty}^{\infty} G(t, s; \gamma)\{\gamma x(s) - \varphi[x(s)] - h(s)\}\, ds.$$

The integral operator $\Pi(\gamma)$ defined by the right-hand side leaves each of the conic intervals $\langle y_1(\gamma), x_1\rangle$ and $\langle x_2, y_2(\gamma)\rangle$ invariant and satisfies the hypotheses of Theorem 10.2 on the conic intervals $\langle \Pi(\gamma) y_1(\gamma), \Pi(\gamma) x_1\rangle$ and $\langle \Pi(\gamma) x_2, \Pi(\gamma) y_2(\gamma)\rangle$. By this theorem, there exist ap-solutions $x^{**}(t)$ and $x^{***}(t)$ satisfying the inequalities

$$y_1(\gamma) \leqslant x^{**}(t) \leqslant x_1, \; x_2 \leqslant x^{***}(t) \leqslant y_2(\gamma) \quad (-\infty < t < \infty), \tag{12.104}$$

each solution being unique on the corresponding conic interval. Theorem 11.4 implies that $x^{**}(t)$ and $x^{***}(t)$ are unstable solutions. Q. E. D.

Clearly, the greater the γ for which $G(t, s; \gamma)$ is nonpositive, the less restrictive are conditions (12.102). We have seen that γ may be taken arbitrarily large for second-order equations, in which case (12.102) may be replaced by

$$a_m x_1 - \varphi(x_1) < h(t) < a_m x_2 - \varphi(x_2) \quad (-\infty < t < \infty). \tag{12.105}$$

It is useful to note that the constraints (12.105) do not depend on γ.

As an example, consider the well-known Duffing equation

$$\frac{d^2x}{dt^2} + c\frac{dx}{dt} + x = \beta x^3 + h(t), \tag{12.106}$$

where $\beta > 0$ and $h(t)$ is a scalar function. Suppose that

$$c \geqslant 2; \tag{12.107}$$

then the ap-operator

$$Lx = \frac{d^2x}{dt^2} + c\frac{dx}{dt} + x$$

satisfies the conditions of this subsection. For the case of Duffing's equation, inequalities (12.105) assume the form

$$|h(t)| < \frac{2}{3\sqrt{3\beta}} \qquad (-\infty < t < \infty). \tag{12.108}$$

Assuming (12.104) and (12.108), we infer from Theorem 12.5 that *Duffing's equation has three ap-solutions,* $x_1(t)$, $x_2(t)$, $x_3(t)$, *satisfying*

$$x_1(t) < -\frac{1}{\sqrt{3\beta}} < x_2(t) < \frac{1}{\sqrt{3\beta}} < x_3(t) \qquad (-\infty < t < \infty). \quad (12.109)$$

The solutions $x_1(t)$ *and* $x_3(t)$ *are unstable, while* $x_2(t)$ *is stable. Moreover, Duffing's equation has no ap-solutions other than these three (if (12.107) and (12.108) hold).*

To prove the last assertion, suppose that equation (12.106) has an ap-solution $x_0(t)$ distinct from $x_1(t)$, $x_2(t)$ and $x_3(t)$. By the proof of Theorem 12.5, at least one of the functions

$$g_1(t) = \frac{1}{\sqrt{3\beta}} - x_0(t) \qquad (12.110)$$

or

$$g_2(t) = -\frac{1}{\sqrt{3\beta}} - x_0(t) \qquad (12.111)$$

changes sign. To fix ideas, let this be $g_1(t)$. Let t_1 and t_2 be points at which

$$g_1(t) > 0 \qquad (t_1 < t < t_2) \qquad (12.112)$$

and

$$g_1(t_1) = g_1(t_2) = 0. \qquad (12.113)$$

Set

$$h_0(t) = \frac{1}{\sqrt{3\beta}} - \beta x_0^3(t) - h(t) \qquad (t_1 \leqslant t \leqslant t_2).$$

From (12.108) and (12.112),

$$h_0(t) > 0 \qquad (t_1 \leqslant t \leqslant t_2). \qquad (12.114)$$

Clearly, $Lg_1(t) = h_0(t)$. Thus

$$g_1(t) = K(t, t_1)\frac{dg_1(t_1)}{dt} + \int_{t_1}^{t} K(t, s)\, h_0(s)\, ds,$$

where $K(t,s)$ is the Cauchy function of the equation $Lx = 0$. Hence

$$\int_{t_1}^{t_2} K(t_2, s)\, h_0(s)\, ds = -K(t_2, t_1)\frac{dg_1(t_1)}{dt}.$$

By (12.11) the left-hand side of this equality is strictly positive, since $K(t, s)$ is positive for $t > s$ (recall that by assumption the solutions of the equation $Lx = 0$ are nonoscillatory on the whole real line). On the other hand, the right-hand side of the last equality cannot be positive, since (12.112) and (12.113) imply $g_1'(t_1) \geqslant 0$ We have thus arrived at a contradiction.

§ 13. POSITIVE ALMOST-PERIODIC SOLUTIONS OF SECOND-ORDER EQUATIONS

13.1. Statement of results

In § 12 we established several theorems on ap-solutions of scalar second-order equations, examining ap-oscillations produced in autonomous systems by an almost-periodic applied force.

In this section we study differential equations

$$\frac{d^2x}{dt^2} + p(t)\frac{dx}{dt} + q(t)x = f(t, x) \qquad (13.1)$$

and

$$\frac{d^2x}{dt^2} + p(t)\frac{dx}{dt} + q(t)x = f(t, \sigma), \qquad (13.2)$$

where

$$\sigma = \alpha(t)x + \frac{dx}{dt}, \qquad (13.3)$$

and $\alpha(t)$ is a function in $B^1(R^1)$. The main difference between equations (13.1), (13.2) and the equations examined in § 12 is that the coefficients $p(t)$ and $q(t)$ of the ap-operator

$$Lx = \frac{d^2x}{dt^2} + p(t)\frac{dx}{dt} + q(t)x \qquad (13.4)$$

need not be constant.

In this section we assume that the functions $f(t, u)$ $(-\infty < t < \infty, \quad u \geqslant 0)$ appearing on the right of (13.1) and (13.2) satisfy the following conditions:

1. $f(t, u)$ is jointly continuous in $-\infty < t < \infty, \quad u \geqslant 0$, and almost periodic in t uniformly with respect to u on every interval $[a, b] \subset [0, \infty)$.
2. For $u > 0$, f has a continuous derivative $f_u'(t, u)$, which is almost periodic in t uniformly with respect to u on every interval $[a, b] \subset (0, \infty)$.
3. $f(t, 0) \equiv 0$ and $f(t, u)$ is nondecreasing in u for nonnegative u.

In stating the results of this section, it is convenient to introduce two more restricted classes of functions, $\mathfrak{F}_1$ and $\mathfrak{F}_2$, satisfying conditions 1—3.

We say that $f(t, u) \in \mathfrak{F}_1$ if:

4'. $f(t, u)$ is concave uniformly in t, i. e., for any $\tau \in (0, 1)$ and any $u > 0$:

$$f(t, \tau u) - \tau(1+\eta) f(t, u) \geqslant 0 \qquad (-\infty < t < \infty), \qquad (13.5)$$

where $\eta = \eta(\tau, u) > 0$.

5'. There exists $t_0 \in (-\infty, \infty)$ such that for all $u > 0$

$$u f'_u(t_0, u) - f(t_0, u) < 0. \qquad (13.6)$$

6'. The following equalities hold:

$$\lim_{u \to +0} \inf_{-\infty < t < \infty} \frac{f(t, u)}{u} = \infty \qquad (13.7)$$

$$\lim_{u \to \infty} \sup_{-\infty < t < \infty} \frac{f(t, u)}{u} = 0. \qquad (13.8)$$

A typical element of $\mathfrak{F}_1$ is the function u^ν, where $0 < \nu < 1$.

We say that $f(t, u) \in \mathfrak{F}_2$ if:

4''. $f(t, u)$ is convex uniformly in t, i. e., for every $\tau \in (0, 1)$ and $u > 0$

$$f(t, \tau u) - \tau(1-\eta) f(t, u) \leqslant 0 \qquad (-\infty < t < \infty), \qquad (13.9)$$

where $\eta = \eta(\tau, u) > 0$.

5''. There exists $t_0 \in (-\infty, \infty)$ such that for all $u > 0$

$$u f'_u(t_0, u) - f(t_0, u) > 0. \qquad (13.10)$$

6''. For $u > 0$ we have $\inf_{-\infty < t < \infty} f(t, u) > 0$,

$$\lim_{u \to +0} \sup_{-\infty < t < \infty} \frac{f(t, u)}{u} = 0 \qquad (13.11)$$

and

$$\lim_{u \to \infty} \inf_{-\infty < t < \infty} \frac{f(t, u)}{u} = \infty. \qquad (13.12)$$

Typical elements of $\mathfrak{F}_2$ are the functions u^ν $(\nu > 1)$, $e^u - u - 1$, $u \ln(1+u)$, etc.

We note that conditions 5' and 5'' are inessential additions to conditions 4' and 4''; by Lemma 11.2 the latter conditions imply, respectively, that for all $u > 0$, $-\infty < t < \infty$

$$uf'_u(t, u) - f(t, u) \leqslant 0 \tag{13.13}$$

and

$$uf'_u(t, u) - f(t, u) \geqslant 0. \tag{13.14}$$

Throughout this section, we assume that

$$M(p) \geqslant 0. \tag{13.15}$$

Recall (see § 9) that (13.15) is not a serious restriction.

We first state the theorems concerning equation (13.1).

Theorem 13.1. Let the ap-operator (13.4) be regular, with nonnegative Green's function. Let $f(t, u) \in \mathfrak{F}_1$.

Then equation (13.1) has a unique nontrivial nonnegative ap-solution. This solution is strictly positive and exponentially stable in Lyapunov's sense.

Theorem 13.2. Assume that the coefficients $p(t)$, $q(t)$ *of the ap-operator (13.4) satisfy the inequality*

$$M^2(p) - M(p^2) + 4M(q) > 0. \tag{13.16}$$

Let $f(t, u) \in \mathfrak{F}_2$.

Then equation (13.1) has a unique nontrivial nonnegative ap-solution. This solution is strictly positive and bilaterally unstable.

The results for equation (13.2) are as follows.

Theorem 13.3. Let the ap-operator (13.4) be regular, with nonnegative Green's function. Let $\alpha(t)$ *in (13.3) satisfy the inequality*

$$M(\alpha) \geqslant \frac{1}{2} M(p) \tag{13.17}$$

and the differential inequality

$$\frac{d\alpha(t)}{dt} \leqslant \alpha^2(t) - p(t)\alpha(t) + q(t) \qquad (-\infty < t < \infty). \tag{13.18}$$

Moreover, assume that for at least one $t_0 \in (-\infty, \infty)$

$$\frac{d\alpha(t_0)}{dt} < \alpha^2(t_0) - p(t_0)\alpha(t_0) + q(t_0). \tag{13.19}$$

Let $f(t, u) \in \mathfrak{F}_1$.

Then equation (13.2) has a unique solution in the class of nonzero ap-functions in $B^1(R^1)$ *satisfying*

$$x(t) \geqslant 0, \quad \alpha(t)\,x(t) + \frac{dx(t)}{dt} \geqslant 0 \quad (-\infty < t < \infty). \qquad (13.20)$$

This ap-solution is strictly positive and exponentially stable in Lyapunov's sense.

Theorem 13.4. *Assume that the inequality (13.16) holds and let the coefficient* $\alpha(t)$ *satisfy* $M(\alpha) > 0$ *and conditions (13.18)–(13.19). Let* $f(t, u) \in \mathfrak{F}_2$.

Then equation (13.2) has a unique solution in the class of nonzero ap-functions in $B^1(R^1)$ *satisfying (13.20). This ap-solution is strictly positive and bilaterally unstable.*

13.2. Proof of Theorem 13.1

Let $G(t,s)$ denote the Green's function of the ap-operator (13.3). The ap-solutions of (13.3) coincide with the fixed points of the nonlinear integral operator

$$\Pi x(t) = \int_{-\infty}^{\infty} G(t, s) f[s, x(s)]\,ds \qquad (13.21)$$

in the space $B(R^1)$.

From the assumption that $G(t,s)$ is nonnegative and from the properties of the function $f(t,x)$t is immediate that the operator (13.21) is uniformly concave (see p.) on the cone $\hat{K}$ of nonnegative functions in $B(R^1)$.

From (13.7) and (13.8) it follows that for sufficiently small $\varepsilon > 0$,

$$\Pi\varepsilon \geqslant \varepsilon, \quad \Pi\left(\frac{1}{\varepsilon}\right) \leqslant \frac{1}{\varepsilon} \qquad (-\infty < t < \infty).$$

Thus, by Theorem 10.6, there exists a unique nontrivial nonnegative ap-solution $x^*(t)$ of equation (13.1). This solution is strictly positive, since it satisfies $x^*(t) \geqslant \varepsilon$.

It follows from our hypotheses and from Theorems 9.6 and 9.8 that the ap-operator (13.4) is stable. Thus, by Theorem 11.5, the ap-solution $x^*(t)$ is exponentially stable. Q. E. D.

We complete Theorem 13.1 by estimating the domain of initial values for which the corresponding solutions approach $x^*(t)$ as $t \to \infty$.

Under the hypotheses of Theorem 13.1, the solutions of the homogeneous equation

$$Lx = 0$$

are nonoscillatory on $(-\infty, \infty)$ (see subsection 9.1). Thus, the Riccati equation

$$\frac{dz}{dt} = z^2 - p(t)z + q(t) \qquad (13.22)$$

has bounded solutions on the real line. As usual (see subsection 9.2), we denote its extremal solutions by $z_1(t), z_2(t)$. It can be shown that the cone $\tilde{K}_{\max}(t)$ constructed in subsection 11.8 is the same for equation (13.1) as the variable cone defined in the $\{x, x'\}$ plane by the inequalities

$$x \geqslant 0, \qquad z_2(t)x + x' \geqslant 0. \qquad (13.23)$$

Thus, by Theorem 11.8, all solutions of equation (13.1) with initial values $\{x(0), x'(0)\}$ satisfying

$$x(0) > 0, \qquad z_2(0)x(0) + x'(0) > 0, \qquad (13.24)$$

converge to $x^*(t)$ as $t \to \infty$.

13.3. Lemma on positive solvability of the differential inequality

Lemma 13.1. Let the coefficients $p(t), q(t)$ of the ap-operator (13.4) satisfy inequality (13.16).

Then there exists an ap-function $u(t) \in B^2(R^1)$ such that

$$u(t) \gg 0, \qquad Lu(t) \gg 0. \qquad (13.25)$$

Proof. It follows from (13.16) that the equation

$$\lambda^2 - M(p)\lambda - M(q) + \frac{1}{4}M(p^2) = 0$$

has two distinct real roots. By (13.15), moreover, the larger of these two roots is positive. There exists a positive λ_0 such that

$$-\lambda_0^2 + M(p)\lambda_0 + M(q) - \frac{1}{4}M(p^2) > 0.$$

Choose $\varepsilon_0 > 0$ such that

$$a = -\lambda_0^2 + M(p)\lambda_0 + M(q) - \frac{1}{4}M(p^2) - \varepsilon_0 > 0.$$

Consider the ap-function

$$\nu(t) = \lambda_0^2 - \lambda_0 p(t) - q(t) + \frac{1}{4}p^2(t) + \varepsilon_0. \qquad (13.26)$$

Clearly $M(\nu) = -a$. Hence, $\nu(t)$ may be written as $\nu(t) = -a + \eta(t)$, where $\eta(t)$ is an ap-function with zero mean.

Every ap-function (see subsection 1.2) may be approximated arbitrarily closely by a trigonometric polynomial, uniformly on $(-\infty, \infty)$. In particular, if the mean of the ap-function is zero, it can be approximated by a trigonometric polynomial with zero mean. Hence one can construct a trigonometric polynomial $\eta_0(t)$ with zero mean such that

$$\| \eta_0(t) - \eta(t) \|_{B(R^1)} < a. \qquad (13.27)$$

We define

$$\xi_0(t) = \int_{-\infty}^{t} e^{-2\lambda_0(t-s)}\eta_0(s)\, ds. \qquad (13.28)$$

Clearly $\xi_0(t)$ is a trigonometric polynomial with zero mean. Thus there exists a constant c such that the trigonometric polynomial

$$\xi_1(t) = \int_{c}^{t} \xi_0(\tau)\, d\tau$$

also has zero mean (c is, of course, not unique).

We show that (13.25) holds for the ap-function

$$u(t) = e^{\xi_1(t)} \qquad (-\infty < t < \infty).$$

The first of these properties is obvious. The second follows from

$$Lu(t) \geqslant \frac{\varepsilon_0}{2}u(t) \qquad (-\infty < t < \infty), \qquad (13.29)$$

which we proceed to prove.

Let $p_n(t)$ be a sequence of trigonometric polynomials converging uniformly to $p(t)$. Then, for sufficiently large n,

$$\left\| [p(t) - p_n(t)] \frac{du(t)}{dt} \right\|_{B(R^1)} \leqslant \frac{\varepsilon_0}{2} \inf_{-\infty < t < \infty} u(t).$$

To prove (13.29) it thus suffices to show that for sufficiently large n we have

$$L_n u(t) \geqslant \varepsilon_0 u(t) \qquad (-\infty < t < \infty), \tag{13.30}$$

where

$$L_n x = \frac{d^2x}{dt^2} + p_n(t)\frac{dx}{dt} + q(t)x.$$

Set

$$\nu_n(t) = \lambda_0^2 - \lambda_0 p_n(t) - q(t) + \frac{1}{4}p_n^2(t) + \varepsilon_0.$$

We write these ap-functions in the form

$$\nu_n(t) = -a_n + \eta_n(t),$$

where $a_n = -M(\nu_n)$ and $\eta_n(t)$ are ap-functions with zero mean. Since the sequence $\nu_n(t)$ converges uniformly to the function (13.26), we have $a_n \to a$; by (13.28), we have, for sufficiently large n,

$$\| \eta_n(t) - \eta_0(t) \|_{B(R^1)} < a_n. \tag{13.31}$$

Define functions $f_n(t)$ by

$$f_n(t) = a_n + \eta_0(t) - \eta_n(t).$$

By (13.31), these functions are positive.

Since

$$\eta_0(t) = f_n(t) + \nu_n(t)$$

by definition, (13.28) may be written as

$$\xi_0(t) = \int_{-\infty}^{t} e^{-2\lambda_0(t-s)}[f_n(s) + \nu_n(s)]\,ds.$$

It is easy to check that

$$\int_{-\infty}^{t} e^{-2\lambda_0(t-s)}\nu_n(s)\,ds = \lambda_0 - \frac{1}{2}p_n(t) - \int_{-\infty}^{t} e^{-2\lambda_0(t-s)}q_n(s)\,ds,$$

where

$$q_n(t) = \lambda_0^2 + q(t) - \frac{1}{4}p_n^2(t) - \frac{1}{2}\frac{dp_n(t)}{dt} - \varepsilon_0.$$

Thus

$$\xi_0(t) = \lambda_0 - \frac{1}{2} p_n(t) - \int_{-\infty}^{t} e^{-2\lambda_0(t-s)} [q_n(s) - f_n(s)]\, ds$$

and, furthermore,

$$\xi_0(t) = \lambda_0 - \frac{1}{2} p_n(t) - z_n(t), \tag{13.32}$$

where $z_n(t)$ is an ap-solution of the equation

$$\frac{dz}{dt} = -2\lambda_0 z + q_n(t) - f_n(t).$$

A simple calculation using (13.32) yields

$$L_n u(t) \equiv [z_n^2(t) + f_n(t) + \varepsilon_0]\, u(t) \qquad (-\infty < t < \infty),$$

so that, by the positivity of $f_n(t)$ for large n, inequality (13.30) is valid for large n. Q. E. D.

13.4. Proof of Theorem 13.2

Consider the family of ap-operators

$$L(\beta)x = Lx - \beta x \qquad (0 < \beta < \infty). \tag{13.33}$$

Clearly,

$$L(\beta)1 = q(t) - \beta.$$

Thus, for

$$\beta > \sup_{-\infty < t < \infty} q(t) \tag{13.34}$$

Theorem 9.9 implies the regularity of the operators $L(\beta)$ and the nonpositivity of their Green's functions $G(t, s; \beta)$.

The problem of ap-solutions for equation (13.1) is equivalent to the fixed-point problem for the operators

$$\Pi(\beta)x(t) = -\int_{-\infty}^{\infty} G(t, s; \beta)\{\beta x(s) - f[s, x(s)]\}\, ds, \tag{13.35}$$

acting in the space $B(R^1)$. Naturally, only those β satisfying (13.34) are considered.

Let R_0 be a positive number such that for $R \geqslant R_0$

$$\frac{f(t, R)}{R} \geqslant q(t) \qquad (-\infty < t < \infty).$$

R_0 exists by (13.12). Then, for $R \geqslant R_0$,

$$L(\beta)R \leqslant -[\beta R - f(t, R)] \qquad (-\infty < t < \infty).$$

Multiplying this last inequality by the nonpositive function $G(t, s; \beta)$ and integrating from $-\infty$ to ∞, we obtain the inequality

$$R \geqslant \Pi(\beta)R \qquad (R \geqslant R_0, \ -\infty < t < \infty). \tag{13.36}$$

Let $u(t)$ be an ap-function satisfying (13.25); such a function exists by Lemma 13.1. By (13.11), there exists a positive r_0 such that for $0 < r \leqslant r_0$

$$Lu(t) \geqslant \frac{f[t, ru(t)]}{r} \qquad (-\infty < t < \infty).$$

From this inequality we obtain

$$L(\beta)[ru(t)] \geqslant -\{\beta ru(t) - f[t, ru(t)]\} \qquad (-\infty < t < \infty),$$

yielding the estimate

$$ru(t) \leqslant \Pi(\beta)[ru(t)] \qquad (0 < r \leqslant r_0, \ -\infty < t < \infty). \tag{13.36a}$$

We fix $r \in (0, r_0]$ and $R \in [R_0, \infty)$. Choose β so large that

$$f'_x(t, x) < \beta \qquad (0 < x \leqslant R, \ -\infty < t < \infty).$$

For this β, the operator $\Pi(\beta)$ is monotone and uniformly concave on the conic interval $\langle ru(t), R \rangle$. From (13.36) and (13.36a), it follows that the conic interval $\langle ru(t), R \rangle$ is invariant under $\Pi(\beta)$. By Theorem 10.3, $\Pi(\beta)$ has a unique fixed point $x^*(t)$ on the conic interval $\langle ru(t), R \rangle$. Clearly, $x^*(t) \gg 0$.

Now suppose $y(t)$ is a nonzero nonnegative ap-function, different from $x^*(t)$, which is also a solution of equation (13.1). Let β satisfy (13.34) and let the difference

$$g(t) = \beta y(t) - f[t, y(t)]$$

be a nonzero nonnegative ap-function. Then from the equality

$$y(t) = -\int_{-\infty}^{\infty} G(t, s; \beta)\, g(s)\, ds$$

we have $y(t) \gg 0$.

As r may be taken arbitrarily small and R arbitrarily large, we may assume that $y(t) \in \langle ru(t), R\rangle$. Hence $y(t) \equiv x^*(t)$.

We have shown that, under the hypotheses of Theorem 13.2, equation (13.1) has a unique nonzero nonnegative ap-solution $x^*(t)$. By Theorem 11.6, the solution $x^*(t)$ is bilaterally unstable, completing the proof of Theorem 13.2.

13.5. Proof of Theorem 13.3

We first consider the linear equation

$$\frac{d^2x}{dt^2} + p(t)\frac{dx}{dt} + q(t)x = 0. \qquad (13.37)$$

As usual, we reduce this equation to a system

$$\frac{du}{dt} + Q(t)u = 0 \qquad (13.38)$$

of two first-order equations. Here

$$u = \left\{x, \frac{dx}{dt}\right\}, \qquad Q(t) = \begin{Vmatrix} 0 & -1 \\ q(t) & p(t) \end{Vmatrix}. \qquad (13.39)$$

Let $U(t, s)$ denote the translation operator of system (13.38).

By the hypotheses of Theorem 13.3, the ap-operator L is regular and its Green's function is nonnegative. Moreover (see subsection 13.2)

$$U(t, s)\tilde{K}_{\max}(s) \subset \tilde{K}_{\max}(t) \qquad (t \geqslant s), \qquad (13.40)$$

where $\tilde{K}_{\max}(t)$ is the cone defined by inequalities (13.23):

$$\tilde{K}_{\max}(t) = \left\{u:\ x \geqslant 0,\ z_2(t)x + \frac{dx}{dt} \geqslant 0\right\}$$

$z_2(t)$ being the upper extremal solution of the Riccati equation (13.22).

As the vector $e_2 = \{0, 1\}$ belongs to the cone $\vec{K}_{\max}(t)$ for all $t \in (-\infty, \infty)$,

$$U(t, s) e_2 \in \vec{K}_{\max}(t) \qquad (t \geqslant s).$$

Let $K(t, s)$ denote the Cauchy function of equation (13.37); then

$$U(t, s) e_2 = \left\{ K(t, s), \frac{\partial K(t, s)}{\partial t} \right\}.$$

Thus

$$K(t, s) \geqslant 0, \quad z_2(t) K(t, s) + \frac{\partial K(t, s)}{\partial t} \geqslant 0 \quad (t \geqslant s). \tag{13.41}$$

By Theorem 9.3 and conditions (13.17)–(13.19), it follows that

$$\inf_{-\infty < t < \infty} [\alpha(t) - z_2(t)] > 0. \tag{13.42}$$

Hence it follows from (13.41) that the function

$$K_\alpha(t, s) = \alpha(t) K(t, s) + \frac{\partial K(t, s)}{\partial t}$$

is positive for $t > s$.

Set

$$\Pi_\alpha \sigma(t) = \int_{-\infty}^{t} K_\alpha(t, s) f[s, \sigma(s)]\, ds. \tag{13.43}$$

The operator (13.43) acts in $B(R^1)$ and is continuous there. The properties of the function $f(t, \sigma)$ entail that the operator Π_α has a unique nonzero fixed point $\sigma^*(t)$ in $\widehat{K}$, as may be seen by repeating the argument in the proof of Theorem 13.1.

Now consider the equation

$$\frac{dx}{dt} + \alpha(t) x = \sigma^*(t). \tag{13.44}$$

In the proof of Theorem 9.3 we saw that if the solutions of equation (13.37) are nonoscillatory and the ap-operator L is stable, then there exists an ap-function $\varphi(t)$ with nonnegative mean, such that $\varphi(t) < z_1(t)$ and, moreover,

$$\varphi(t) < z_2(t) \qquad (-\infty < t < \infty).$$

Thus (13.42) implies $M(\alpha) > 0$.

This last estimate implies that equation (13.44) has a unique ap-solution $x^*(t)$. This ap-solution is strictly positive.

We now show that $x^*(t)$ is an ap-solution of equation (13.2). Consider the equation

$$Lx = f[t, \sigma^*(t)]$$

and denote its ap-solution by $y^*(t)$. Clearly,

$$y^*(t) = \int_{-\infty}^{t} K(t, s) f[s, \sigma^*(s)]\, ds$$

and (since $K(t,t) \equiv 0$)

$$\frac{dy^*(t)}{dt} = \int_{-\infty}^{t} \frac{\partial K(t, s)}{\partial t} f[s, \sigma^*(s)]\, ds.$$

Therefore

$$\alpha(t) y^*(t) + \frac{dy^*(t)}{dt} = \int_{-\infty}^{t} K_\alpha(t, s) f[s, \sigma^*(s)]\, ds$$

and thus

$$\frac{dy^*(t)}{dt} + \alpha(t) y^*(t) = \sigma^*(t).$$

Hence

$$y^*(t) \equiv x^*(t).$$

Now we need only prove that the ap-solution $x^*(t)$ of equation (13.2) is stable. It is sufficient to show (see subsection 4.3) that the ap-operators

$$L(\mu) x = Lx - \mu \left[\frac{dx}{dt} + \alpha(t) x\right] f'_\sigma[t, \sigma^*(t)]$$

are regular for all $\mu \in [0, 1]$. In other words, it is sufficient to show that the equation

$$L(\mu) x = f(t) \qquad (13.45)$$

has a unique solution for $\mu \in [0, 1]$ and any ap-function $f(t)$. Equation (13.45) is equivalent to the integral equation

$$x(t) = \mu \int_{-\infty}^{t} K(t, s) f'_{\sigma}[s, \sigma^*(s)] \left[\frac{dx(s)}{ds} + \alpha(s) x(s)\right] ds + g(t),$$

where

$$g(t) = \int_{-\infty}^{t} K(t, s) f(s)\, ds.$$

Let

$$\sigma(t) = \frac{dx(t)}{dt} + \alpha(t) x(t).$$

Then we may clearly determine the ap-solutions $x(t)$ of equation (13.45) by finding the ap-solutions $\sigma(t)$ of the integral equations

$$\sigma(t) = \mu \int_{-\infty}^{t} K_{\alpha}(t, s) f'_{\sigma}[s, \sigma^*(s)] \sigma(s)\, ds + g_{\alpha}(t), \qquad (13.46)$$

where

$$g_{\alpha}(t) = \frac{dg(t)}{dt} + \alpha(t) g(t).$$

Equation (13.46) has a unique solution for every $\mu \in [0, 1]$ if the spectral radius of the integral operator

$$S\sigma(t) = \int_{-\infty}^{t} K_{\alpha}(t, s) f'_{\sigma}[s, \sigma^*(s)] \sigma(s)\, ds$$

is less than 1.

To estimate this radius, observe that S is positive with respect to the cone of nonnegative functions and, by (13.6) and (13.13),

$$S\sigma^*(t) \ll \sigma^*(t).$$

So indeed, by Theorem 5.1, the spectral radius of S is less than 1, completing the proof of Theorem 13.3.

Note that, under the hypotheses of Theorem 13.3, all solutions $x(t)$ of equation (13.2) whose initial values satisfy (13.24) approach the ap-solution $x^*(t)$ as $t \to \infty$.

13.6. Regularity of auxiliary ap-operators

Consider the family of ap-operators

$$L(\gamma)x = \frac{d^2x}{dt^2} + [p(t) - \gamma]\frac{dx}{dt} + [q(t) - \gamma\alpha(t)]x, \qquad (13.47)$$

where γ is a positive real parameter and $\alpha(t)$ an ap-function.

Lemma 13.2. Let $M(\alpha) > 0$.

Then there exists $\gamma_0 > 0$ *such that for* $\gamma \geqslant \gamma_0$ *the ap-operators (13.47) are regular and their Green's functions are nonpositive.*

We confine ourselves to the case that $p(t) \in B^1(R^1)$. As to the general case, one can use the constructions in the proof of Lemma 13.1.

By Theorem 9.9, the lemma will be established once we find ap-functions $v(t;\gamma) \in B^2(R^1)$ $(\gamma \geqslant \gamma_0)$ such that

$$v(t;\ \gamma) \gg 0 \qquad (\gamma \geqslant \gamma_0) \qquad (13.48)$$

and

$$L(\gamma)v(t;\ \gamma) < 0 \qquad (-\infty < t < \infty,\ \gamma \geqslant \gamma_0). \qquad (13.49)$$

Set

$$p(t;\ \gamma) = p(t) - \gamma, \qquad q(t;\ \gamma) = q(t) - \gamma\alpha(t)$$

and

$$r(t;\ \gamma) = q(t;\ \gamma) - \frac{1}{4}p^2(t;\ \gamma) - \frac{1}{2}p(t;\ \gamma)M[p(t;\ \gamma)] + \frac{3}{4}M^2[p(t;\ \gamma)].$$

It is easy to verify that

$$M[r(t;\ \gamma)] = M(q) - \frac{1}{4}M(p^2) + \frac{1}{4}M^2(p) - \gamma M(\alpha). \qquad (13.50)$$

Since $M(\alpha) > 0$ by assumption, (13.50) implies the existence of γ_1 such that

$$M[r(t;\ \gamma)] < 0 \qquad (\gamma \geqslant \gamma_1). \qquad (13.51)$$

We may take γ_1 to be large enough so that

$$M[p(t;\ \gamma)] = M(p) - \gamma < 0 \qquad (\gamma \geqslant \gamma_1). \qquad (13.52)$$

Let

$$\eta(t;\ \gamma) = r(t;\ \gamma) - M[r(t;\ \gamma)] \qquad (\gamma \geqslant \gamma_1). \tag{13.53}$$

Let $\eta_0(t;\gamma)$ denote a trigonometric polynomial with nonzero mean, such that

$$\| \eta_0(t;\ \gamma) - \eta(t;\ \gamma) \|_{B(R^1)} < -\frac{1}{2} M[r(t;\ \gamma)] \quad (\gamma \geqslant \gamma_1). \tag{13.54}$$

Set

$$\xi_0(t;\ \gamma) = \int_t^\infty e^{-M[p(t;\ \gamma)](t-s)} \eta_0(s;\ \gamma)\, ds \qquad (\gamma \geqslant \gamma_1)$$

and

$$\nu(t;\ \gamma) = e^{\int_{c(\gamma)}^{t} \xi_0(\tau;\ \gamma)\, d\tau} \qquad (\gamma \geqslant \gamma_1), \tag{13.55}$$

where $c(\gamma)$ are numbers such that

$$M\left[\int_{c(\gamma)}^{t} \xi_0(\tau;\ \gamma)\, d\tau\right] = 0.$$

The function (13.55) obviously satisfies (13.48). We have only to prove the validity of (13.49) for $\gamma \geqslant \gamma_0$, where $\gamma_0 > \gamma_1$.

Let

$$f(t;\ \gamma) = -r(t;\ \gamma) + \eta_0(t;\ \gamma) \qquad (\gamma \geqslant \gamma_1).$$

Then, obviously,

$$\xi_0(t;\ \gamma) = \int_t^\infty e^{-M[p(t,\ \gamma)](t-s)} [r(s;\ \gamma) + f(s;\ \gamma)]\, ds.$$

Now, it is easily checked that

$$\xi_0(t;\ \gamma) = \frac{1}{2} M[p(t;\ \gamma)] - \frac{1}{2} p(t;\ \gamma) - x(t;\ \gamma), \tag{13.56}$$

where $x(t;\gamma)$ denotes an ap-solution of the equation

$$\frac{dx}{dt} + M[p(t;\ \gamma)]\, x = \varphi(t;\ \gamma), \tag{13.57}$$

and

$$\varphi(t;\ \gamma) = q(t;\ \gamma) - \frac{1}{2}p^2(t;\ \gamma) - \frac{1}{2}\frac{dp(t)}{dt} + \\ + \frac{1}{4}M^2[p(t;\ \gamma)] + f(t;\ \gamma).$$

From (13.56),

$$L(\gamma)\nu(t;\ \gamma) \equiv [x^2(t;\ \gamma) - f(t;\ \gamma)]\,\nu(t;\ \gamma) \qquad (13.58)$$
$$(\gamma \geqslant \gamma_1,\ -\infty < t < \infty).$$

By (13.54), the identity (13.58) entails that

$$L(\gamma)\nu(t;\ \gamma) \leqslant \left\{x^2(t;\ \gamma) + \frac{1}{2}M[r(t;\ \gamma)]\right\}\nu(t;\ \gamma)$$
$$(\gamma \geqslant \gamma_1,\ -\infty < t < \infty).$$

To complete the proof, then, we must establish the existence of $\gamma_0 \geqslant \gamma_1$ such that

$$x^2(t;\ \gamma) + \frac{1}{2}M[r(t;\ \gamma)] < 0 \qquad (13.59)$$
$$(\gamma \geqslant \gamma_0,\ -\infty < t < \infty).$$

Since (13.50) implies that $M[r(t;\gamma)] \to -\infty$ as $\gamma \to \infty$, the inequalities (13.59) will follow if we show that the norms in $B(R^1)$ of the ap-solutions $x(t;\gamma)$ of equations (13.57) are bounded uniformly (in $\gamma \geqslant \gamma_1$).

Obviously,

$$x(t;\ \gamma) = -\int_t^\infty e^{[\gamma - M(p)](t-s)}\varphi(s;\ \gamma)\,ds \qquad (\gamma \geqslant \gamma_1)$$

and thus

$$\|x(t;\ \gamma)\|_{B(R^1)} \leqslant \frac{1}{\gamma - M(p)}\|\varphi(t;\ \gamma)\|_{B(R^1)}.$$

Hence, by the obvious estimate

$$\|\varphi(t;\ \gamma)\|_{B(R^1)} \leqslant a + b\gamma,$$

where a, b are positive constants, we have

$$\|x(t;\ \gamma)\|_{B(R^1)} \leqslant \frac{a + b\gamma}{\gamma - M(p)} \qquad (\gamma \geqslant \gamma_1).$$

Q. E. D.

13.7. Proof of Theorem 13.4

By Lemma 13.2, there exists $\gamma_0 > 0$ such that for $\gamma \geqslant \gamma_0$ the ap-operators (13.47) are regular and their Green's functions $G(t, s; \gamma)$ are nonpositive.

Consider the family of Riccati equations

$$\frac{dz}{dt} = z^2 - p(t;\ \gamma) z + q(t;\ \gamma) \qquad (\gamma \geqslant \gamma_0) \tag{13.60}$$

and let $z_1(t;\gamma)$ and $z_2(t;\gamma)$ denote the corresponding extremal solutions. Then by (9.65),

$$G(t,\ s;\ \gamma) = \begin{cases} \dfrac{1}{z_1(t;\ \gamma) - z_2(t;\ \gamma)} e^{-\int_s^t z_2(\tau;\ \gamma)\, d\tau}, & \text{if} \quad t \geqslant s, \\ \dfrac{1}{z_1(t;\ \gamma) - z_2(t;\ \gamma)} e^{-\int_s^t z_1(\tau;\ \gamma)\, d\tau}, & \text{if} \quad t < s. \end{cases}$$

From (13.18) and (13.19) it follows that

$$\frac{d\alpha(t)}{dt} \leqslant \alpha^2(t) - p(t;\ \gamma)\alpha(t) + q(t;\ \gamma) \qquad (-\infty < t < \infty),$$

where $\alpha(t)$ is not a solution of any of the equations (13.60). It follows by Theorem 9.3 that either

$$\inf_{-\infty < t < \infty} [\alpha(t) - z_2(t;\ \gamma)] > 0, \tag{13.61}$$

or

$$\inf_{-\infty < t < \infty} [z_1(t;\ \gamma) - \alpha(t)] > 0. \tag{13.61a}$$

The second inequality cannot hold, else it would follow that

$$M[z_1(t;\gamma)] > M(\alpha) > 0,$$

in contradiction to the inequality

$$M[z_1(t;\gamma)] < 0$$

shown in the proof of Theorem 9.9. Thus, inequality (13.61) holds.

Now set

$$G_\alpha(t,\ s;\ \gamma) = \alpha(t)\, G(t,\ s;\ \gamma) + \frac{\partial G(t,\ s;\ \gamma)}{\partial t}.$$

It follows from (13.61) that

$$G_\alpha(t,s;\gamma)<0 \qquad (-\infty<t,\ s<\infty). \tag{13.62}$$

Define integral operators

$$\Pi_\alpha(\gamma)\,\sigma(t)=-\int_{-\infty}^{\infty} G_\alpha(t,\ s;\ \gamma)\{\gamma\sigma(s)-f[s,\ \sigma(s)]\}\,ds \tag{13.63}$$

acting in $B(R^1)$ and continuous there. The existence of a unique ap-solution for equation (13.2) satisfying (13.20) is reducible (as in subsection 13.5, in the proof of Theorem 13.3) to the existence in $\hat{K}$ of a unique nonzero fixed point of any of the operators (13.69) for $\gamma \geqslant \gamma_0$.

By Lemma 13.1, there exists an ap-function $u(t) \in B^2(R^1)$ satisfying

$$u(t)\gg 0, \qquad Lu(t)\gg 0. \tag{13.64}$$

Let us introduce the ap-function

$$\sigma^0(t)=\frac{du(t)}{dt}+\alpha(t)\,u(t). \tag{13.65}$$

Since $M(\alpha)>0$ and $u(t)\gg 0$, there are points at which (13.65) is positive, and hence the nonnegative ap-function

$$\sigma_+^0(t)=\begin{cases}\sigma^0(t), & \text{if}\quad \sigma^0(t)\geqslant 0,\\ 0, & \text{if}\quad \sigma^0(t)\leqslant 0,\end{cases}$$

is not identically zero.

By (13.11) and (13.64), there exists $\delta_0>0$ such that

$$\begin{aligned} L(\gamma)[\delta u(t)] &= L[\delta u(t)]-\gamma\delta\sigma^0(t)\geqslant \\ &\geqslant -\{\gamma\delta\sigma^0(t)-f_0[t,\ \delta\sigma^0(t)]\} \qquad (-\infty<t<\infty) \end{aligned} \tag{13.66}$$

for all γ and $0<\delta\leqslant\delta_0$, where

$$f_0(t,\ \sigma)=\begin{cases} f(t,\ \sigma), & \text{if}\quad \sigma\geqslant 0,\\ 0 & \text{if}\quad \sigma<0.\end{cases}$$

By (13.66),

$$L(\gamma)[\delta u(t)]=-\{\gamma\delta\sigma^0(t)-f_0[t,\ \delta\sigma^0(t)]\}+\psi(t),$$

where $\psi(t)$ is a nonnegative ap-function. Thus

$$\delta u(t) = -\int_{-\infty}^{\infty} G(t, s; \gamma)\{\gamma\delta\sigma^0(s) - f_0[s, \delta\sigma^0(s)]\}\, ds + \\ + \int_{-\infty}^{\infty} G(t, s; \gamma)\psi(s)\, ds.$$

Therefore,

$$\delta\sigma^0(t) = -\int_{-\infty}^{\infty} G_a(t, s; \gamma)\{\gamma\delta\sigma^0(s) - f_0[s, \delta\sigma^0(s)]\}\, ds + \\ + \int_{-\infty}^{\infty} G_a(t, s; \gamma)\psi(s)\, ds,$$

whence, by (13.62),

$$\delta\sigma^0(t) \leqslant -\int_{-\infty}^{\infty} G_a(t, s; \gamma)\{\gamma\delta\sigma^0(s) - f_0[s, \delta\sigma^0(s)]\}\, ds. \qquad (13.67)$$

Moreover, (13.11) implies

$$\frac{\partial f_0(t, 0)}{\partial\sigma} = \frac{\partial f(t, 0)}{\partial\sigma} = 0.$$

We may thus suppose that for $\gamma \geqslant \gamma_0$ the function $\gamma\sigma - f_0(t, \sigma)$ is non-decreasing in σ for $\sigma \in (-\infty, \eta_0)$, where

$$\eta_0 = \delta_0 \sup_{-\infty < t < \infty} \sigma^0(t).$$

On the other hand, we conclude from (13.61) that for $\gamma \geqslant \gamma_0$, $0 < \delta \leqslant \delta_0$,

$$\delta\sigma^0(t) \leqslant \Pi_a(\gamma)[\delta\sigma^0_+(t)]$$

and, furthermore,

$$\delta\sigma^0_+(t) \leqslant \Pi_\alpha(\gamma)[\delta\sigma^0_+(t)] \qquad (\gamma \geqslant \gamma_0;\ 0 < \delta \leqslant \delta_0). \qquad (13.68)$$

Let $v_0(t)$ denote an ap-solution of the equation

$$\frac{dx}{dt} + \alpha(t)x = 1.$$

Clearly $v_0(t) \gg 0$.

Condition (13.12) guarantees the existence of $R_0 > 0$ such that for $R \geqslant R_0$

$$Lv_0(t) \leqslant \frac{f(t, R)}{R} \qquad (-\infty < t < \infty)$$

or, equivalently,

$$L(\gamma)[R\mathbf{v}_0(t)] = -[\gamma R - f(t, R)] + x(t) \qquad (-\infty < t < \infty),$$

where $x(t)$ is a nonpositive ap-function. We rewrite the last equation as

$$R\mathbf{v}_0(t) = -\int_{-\infty}^{\infty} G(t, s; \gamma)[\gamma R - f(s, R)]\,ds + \int_{-\infty}^{\infty} G(t, s; \gamma)\,x(s)\,ds$$

and thus

$$R = \Pi_\alpha(\gamma) R + \int_{-\infty}^{\infty} G_\alpha(t, s; \gamma)\,x(s)\,ds,$$

and finally, we have by (13.62) that

$$R \geqslant \Pi_\alpha(\gamma) R \qquad (R \geqslant R_0;\ \gamma \geqslant \gamma_0). \tag{13.69}$$

From (13.68) and (13.69) it follows that for sufficiently large γ the conic intervals $\langle \delta\sigma_+^0(t), R\rangle$ are invariant under the operators $\Pi_\alpha(\gamma)$. We may now complete the proof as in Theorems 13.2 and 13.3.

Chapter 4

EQUATIONS WITH A SMALL PARAMETER

§ 14. LINEAR EQUATIONS

14.1. Bogolyubov's lemma

Consider the regular operator

$$Lx = \frac{dx}{dt} + A(t)\,x \qquad (14.1)$$

in $B(R^n)$ and denote its Green's function by $G(t,s)$.

Let the vector-function $f(t;\varepsilon)$ be defined for $-\infty < t < \infty$ and $0 < \varepsilon < \varepsilon_0$. Let $f(t;\varepsilon)$ be almost periodic with respect to t for any fixed ε. Then the homogeneous linear equation

$$Lx = f(t;\varepsilon) \qquad (14.2)$$

has a unique ap-solution $x(t;\varepsilon)$ for any fixed $\varepsilon \in (0,\varepsilon_0)$. In this subsection we seek conditions such that

$$\lim_{\varepsilon\to 0} \| x(t;\ \varepsilon) \|_{B(R^n)} = 0. \qquad (14.3)$$

Clearly, (14.3) holds if

$$\lim_{\varepsilon\to 0} \| f(t;\ \varepsilon) \|_{B(R^n)} = 0. \qquad (14.4)$$

However, (14.4) is too restrictive a condition. For example, it is not valid in situations when the well-known a v e r a g i n g p r i n c i p l e is used.* We say that $f(t;\varepsilon)$ c o n v e r g e s i n t e g r a l l y t o z e r o as $\varepsilon \to 0$, if for every $T > 0$

$$\lim_{\varepsilon\to 0} \sup_{|t-s| \leqslant T} \left| \int_s^t f(\tau;\ \varepsilon)\, d\tau \right| = 0. \qquad (14.5)$$

* See, for example, Bogolyubov and Mitropol'skii /1/.

If in addition to (14.45)

$$\|f(t;\ \varepsilon)\|_{B(R^n)} < m \qquad (0 < \varepsilon < \varepsilon_0), \tag{14.6}$$

we say that $f(t;\varepsilon)$ converges regularly to zero as $\varepsilon \to 0$.

Lemma 14.1. Equation (14.3) holds whenever $f(t;\varepsilon)$ converges regularly to zero as $\varepsilon \to 0$. Conversely, given that (14.3) holds, $f(t;\varepsilon)$ converges integrally to zero as $\varepsilon \to 0$.

Proof. Necessity. It is clear that

$$x(t;\ \varepsilon) = \int_{-\infty}^{\infty} G(t,\ s) f(s;\ \varepsilon)\, ds. \tag{14.7}$$

By substituting $s = t + \tau$ in (14.7), we obtain

$$x(t;\ \varepsilon) = \int_{-\infty}^{\infty} G(t,\ t+\tau) f(t+\tau;\ \varepsilon)\, d\tau$$

or, equivalently,

$$x(t;\ \varepsilon) = \int_{-\infty}^{0} G(t,\ t+\tau)\, d_\tau \left[\int_{t}^{t+\tau} f(\sigma;\ \varepsilon)\, d\sigma\right] + \int_{0}^{\infty} G(t,\ t+\tau)\, d_\tau \left[\int_{t}^{t+\tau} f(\sigma;\ \varepsilon)\, d\sigma\right].$$

Integrating by parts in each term, we obtain

$$x(t;\ \varepsilon) = -\int_{-\infty}^{0} \frac{\partial G(t,\ t+\tau)}{\partial \tau} \left[\int_{t}^{t+\tau} f(\sigma;\ \varepsilon)\, d\sigma\right] d\tau - \int_{0}^{\infty} \frac{\partial G(t,\ t+\tau)}{\partial \tau} \left[\int_{t}^{t+\tau} f(\sigma;\ \varepsilon)\, d\sigma\right] d\tau. \tag{14.8}$$

In subsection 4.1 we proved that

$$G(t,\ s) = \begin{cases} U(t) P_+ U^{-1}(s), & \text{if} \quad t \geqslant s, \\ -U(t) P_- U^{-1}(s), & \text{if} \quad t < s. \end{cases} \tag{14.9}$$

Thus, for $\tau \neq 0$, we have the estimate

$$\left|\frac{\partial G(t,\ t+\tau)}{\partial \tau}\right| \leqslant m_1 e^{-\gamma_1 |\tau|} \qquad (-\infty < t < \infty), \tag{14.10}$$

where m_1, γ_1 are positive constants.

From (14.6) and (14.10), for arbitrary $T>0$,

$$|x(t;\ \varepsilon)| \leqslant mm_1 \int_{-\infty}^{-T} e^{-\gamma_1|\tau|}|\tau|\,d\tau +$$

$$+ m_1 \int_{-T}^{T} e^{-\gamma_1|\tau|} \left| \int_{t}^{t+\tau} f(\sigma;\ \varepsilon)\,d\sigma \right| d\tau + mm_1 \int_{T}^{\infty} e^{-\gamma_1|\tau|}\tau\,d\tau.$$

Given $\eta>0$, this last inequality implies the existence of $T>0$ such that

$$|x(t;\ \varepsilon)| < \frac{\eta}{2} + \frac{2m_1}{\gamma_1} \sup_{|\tau-s|\leqslant T} \left| \int_{s}^{\tau} f(\sigma;\ \varepsilon)\,d\sigma \right|. \tag{14.11}$$

By (14.5) and (14.11),

$$|x(t;\ \varepsilon)| < \eta \qquad (-\infty < t < \infty)$$

for small $\varepsilon \in (0,\ \varepsilon_0)$. Hence the sufficiency.

Necessity is readily proved. We need only integrate the identity

$$\frac{dx(t;\ \varepsilon)}{dt} + A(t)\,x(t;\ \varepsilon) \equiv f(t;\ \varepsilon)$$

from s to t to see that

$$\left| \int_{s}^{t} f(\sigma;\ \varepsilon)\,d\sigma \right| \leqslant (2 + T\,\|A(t)\|_{B(R^n)})\,\|x(t;\ \varepsilon)\|_{B(R^n)}.$$

Q. E. D.

An important example of a function converging regularly to zero as $\varepsilon \to 0$ is

$$f(t;\varepsilon) = f_0(t/\varepsilon), \tag{14.12}$$

where $f_0(t)$ is an ap-function with zero mean.

14.2. Dependence of the Green's function on a parameter

Consider the family of ap-operators

$$L(\varepsilon)\,x = \frac{dx}{dt} + A(t;\ \varepsilon)\,x \qquad (0 < \varepsilon < \varepsilon_0). \tag{14.13}$$

We say that the ap-matrices $A(t;\varepsilon)$ converge regularly to the ap-matrix $A(t)$ as $\varepsilon \to 0$ if

$$\| A(t;\ \varepsilon)\|_{B(R^n)} \leqslant m \qquad (0<\varepsilon<\varepsilon_0) \tag{14.14}$$

and for any $T>0$

$$\lim_{\varepsilon\to 0} \sup_{|t-s|\leqslant T} \left| \int_s^t [A(\tau;\ \varepsilon)-A(\tau)]\,d\tau \right| = 0. \tag{14.15}$$

Theorem 14.1. Let the operator (14.1) be regular. Let the ap-matrices $A(t;\varepsilon)$ converge regularly to the ap-matrix $A(t)$ as $\varepsilon \to 0$.

Then for sufficiently small $\varepsilon \in (0, \varepsilon_0)$ the ap-operators (14.13) are regular and

$$\lim_{\varepsilon\to 0} \sup_{-\infty<s,\,t<\infty} e^{\gamma_0|t-s|}|G(t,\ s;\ \varepsilon)-G(t,\ s)|=0, \tag{14.16}$$

where $G(t,s;\varepsilon)$ and $G(t,s)$ are, respectively, the Green's functions of the operators (14.13) and (14.1) and γ_0 is a positive number.

Proof. Set

$$H(t;\ \varepsilon)=\int_{-\infty}^{\infty} G(t,\ s)\,B(s;\ \varepsilon)\,ds, \tag{14.17}$$

where $B(t;\varepsilon)=A(t)-A(t;\varepsilon)$. By Lemma 14.1,

$$\lim_{\varepsilon\to 0}\| H(t;\ \varepsilon)\|_{B(R^n)}=0. \tag{14.18}$$

Let $f(t)$ be an ap-function. We seek an ap-solution $x(t)$ of the equation

$$\frac{dx}{dt}+A(t;\ \varepsilon)\,x=f(t)$$

of the form

$$x(t)=y(t)+H(t;\varepsilon)\,y(t). \tag{14.19}$$

A simple calculation shows that $y(t)$ must be an ap-solution of the equation

$$\frac{dy}{dt}+A(t)\,y+D(t;\ \varepsilon)\,y=[I+H(t;\ \varepsilon)]^{-1}f(t), \tag{14.20}$$

where

$$D(t;\ \varepsilon) = -[I + H(t;\ \varepsilon)]^{-1}\{H(t;\ \varepsilon)A(t) + B(t;\ \varepsilon)H(t;\ \varepsilon)\}.$$

By (14.18),

$$\lim_{\varepsilon\to 0}\| D(t;\ \varepsilon)\|_{B(R^n)} = 0.$$

Thus, for sufficiently small ε, every equation (14.20) has a unique ap-solution. Therefore, for sufficiently small ε, the ap-operators $L(\varepsilon)$ are regular.

For small ε, the left-hand sides of equations (14.20) define ap-operators; denote the corresponding Green's functions by $G_0(t, s; \varepsilon)$. By Theorem 4.3.

$$\lim_{\varepsilon\to 0}\sup_{-\infty < t,\, s < \infty} e^{\gamma_0|t-s|}|G(t,\ s) - G_0(t,\ s;\ \varepsilon)| = 0, \tag{14.21}$$

where γ_0 is a positive number.

As

$$y(t) = \int_{-\infty}^{\infty} G_0(t,\ s;\ \varepsilon)[I + H(s;\ \varepsilon)]^{-1} f(s)\,ds,$$

it follows from (14.19) that

$$x(t) = \int_{-\infty}^{\infty} [I + H(t;\ \varepsilon)]\,G_0(t,\ s;\ \varepsilon)[I + H(s;\ \varepsilon)]^{-1} f(s)\,ds,$$

and the uniqueness of the Green's function implies

$$G(t,\ s;\ \varepsilon) = [I + H(t;\ \varepsilon)]\,G_0(t,\ s;\ \varepsilon)[I + H(s;\ \varepsilon)]^{-1}$$
$$(-\infty < t,\ s < \infty).$$

The last equality and (14.21) yield (14.16). Q. E. D.

The only nontrivial point in this proof is the introduction of (14.19). It was proposed by Bogolyubov and Krylov and is widely applicable in different branches of oscillation theory.

Let us express the Green's function $G(t, s; \varepsilon)$ in a form analogous to (14.9):

$$G(t,\ s;\ \varepsilon) = \begin{cases} U(t;\ \varepsilon)P_+(\varepsilon)U^{-1}(s;\ \varepsilon), & \text{if } t \geqslant s, \\ -U(t;\ \varepsilon)P_-(\varepsilon)U^{-1}(s;\ \varepsilon), & \text{if } t < s. \end{cases} \tag{14.22}$$

Then

$$\lim_{\varepsilon\to 0} | P_+(\varepsilon) - P_+ | = \lim_{\varepsilon\to 0} |P_-(\varepsilon) - P_-| = 0,$$

yielding

Theorem 14.2. Under the hypotheses of Theorem 14.1, we have for small $\varepsilon \in (0, \varepsilon_0)$ that the ap-operators $L(\varepsilon)$ are stable if L is stable, and they are unstable if L is unstable.

Let $A_1(t)$ be an ap-matrix with mean value A_0:

$$A_0 = \lim_{T\to\infty} \frac{1}{2T} \int_{-T}^{T} A_1(\tau)\, d\tau. \tag{14.23}$$

Consider the equation

$$\frac{dx}{dt} = \varepsilon A_1(t)\, x + \varepsilon B(t;\ \varepsilon)\, x \qquad (0 < \varepsilon < \varepsilon_0), \tag{14.24}$$

where

$$\lim_{\varepsilon\to 0} \| B(t;\ \varepsilon) \|_{B(R^n)} = 0. \tag{14.25}$$

We are interested in the stability of the trivial solution of equation (14.24). The substitution $\tau = \varepsilon t$ converts equation (14.24) to the form

$$\frac{dx}{d\tau} = A_1\left(\frac{\tau}{\varepsilon}\right) x + B\left(\frac{\tau}{\varepsilon};\ \varepsilon\right) x. \tag{14.26}$$

As $\varepsilon \to 0$, the matrices $A_1\left(\frac{\tau}{\varepsilon}\right) + B\left(\frac{\tau}{\varepsilon};\ \varepsilon\right)$ clearly converge regularly to the constant matrix (14.23). Thus, Theorem 14.2 implies the simple but important

Theorem 14.3. If A_0 is a Hurwitz matrix, then, for sufficiently small $\varepsilon > 0$, the trivial solution of equation (14.24) is exponentially stable. But if A_0 has at least one eigenvalue in the right half-plane, then, for sufficiently small $\varepsilon > 0$, the trivial solution of equation (14.24) is unstable.

This theorem is due to Bogolyubov.

Suppose that A_0 has no eigenvalues in the right half-plane, but has eigenvalues on the imaginary axis. In this case, determining the stability of the trivial solution of equations (14.24) is a very complicated problem. This problem has been investigated by several authors (see, for example, Demidovich /2/, Hale /1/, Shtokalo /1/, Erugin /2/).

We shall need only one result, due to Shtokalo, which we present below.

14.3. The Bogolyubov-Shtokalo substitutions

Suppose equation (14.24) may be written as

$$\frac{dx}{dt} = \varepsilon A_1(t)\,x + \dots + \varepsilon^k A_k(t)\,x + \varepsilon^{k+1} C(t;\ \varepsilon)\,x, \qquad (14.27)$$

where $A_1(t), \dots, A_k(t)$ are matrices whose elements are trigonometric polynomials and $C(t; \varepsilon)$ are ap-matrices, continuous in $\varepsilon \in [0, \varepsilon_0]$ uniformly with respect to $t \in (-\infty, \infty)$.

In investigating the stability of the trivial solution of equation (14.27), it is natural to seek a new substitution such that the equation may be written

$$\frac{dy}{dt} = \varepsilon B_1 y + \dots + \varepsilon^k B_k y + \varepsilon^{k+1} D(t;\ \varepsilon)\,y \qquad (14.28)$$

where $B_1, \dots, B_k$ are constant matrices and the ap-matrix $D(t; \varepsilon)$ has the same properties as $C(t; \varepsilon)$.

Following Bogolyubov and Shtokalo, we seek a substitution of the form

$$x = [I + \varepsilon Y_1(t) + \dots + \varepsilon^k Y_k(t)]\,y, \qquad (14.29)$$

where $Y_1(t), \dots, Y_k(t)$ are ap-matrices.

The constant matrices $B_1, \dots, B_k$ and the ap-matrices $Y_1(t), \dots, Y_k(t)$ may be obtained simultaneously from the identity

$$\left[I + \sum_{i=1}^{k} \varepsilon^i Y_i(t)\right] \cdot \left[\sum_{i=1}^{k} \varepsilon^i B_i + \varepsilon^{k+1} D(t;\ \varepsilon)\right] + \sum_{i=1}^{k} \varepsilon^i \frac{dY_i(t)}{dt} =$$
$$= \left[\sum_{i=1}^{k} \varepsilon^i A_i(t) + \varepsilon^{k+1} C(t;\ \varepsilon)\right] \cdot \left[I + \sum_{i=1}^{k} \varepsilon^i Y_i(t)\right].$$

Expanding on both sides and comparing coefficients of $\varepsilon, \varepsilon^2, \dots \varepsilon^k$, we obtain a system of matrix equations

$$\left.\begin{aligned}
\frac{dY_1(t)}{dt} &= A_1(t) - B_1,\\
\frac{dY_2(t)}{dt} &= A_2(t) + A_1(t)\,Y_1(t) - B_2 - Y_1(t)\,B_1,\\
&\dots\dots\dots\dots\dots\dots\dots\dots\\
\frac{dY_k(t)}{dt} &= A_k(t) + A_{k-1}(t)\,Y_1(t) + \dots + A_1(t)\,Y_{k-1}(t) -\\
&\quad - B_k - Y_1(t)\,B_{k-1} - \dots - Y_{k-1}(t)\,B_1.
\end{aligned}\right\} \qquad (14.30)$$

The equations (14.30) have ap-solutions $Y_i(t)$ if and only if the mean value of the right-hand side is zero. Therefore, the following procedure may be used in solving (14.30).

The matrix B_1 is defined as the mean value of the ap-matrix $A_1(t)$:

$$B_1 = M[A_1(t)] = \lim_{T\to\infty} \frac{1}{2T} \int_{-T}^{T} A_1(\tau)\,d\tau.$$

Then $Y_1(t)$ is taken to be the ap-solution of the first equation of system (14.30) with zero mean. The elements of $Y_1(t)$ are clearly trigonometric polynomials with zero mean.

Having found B_1 and $Y_1(t)$, we proceed to find B_2 by stipulating that the right-hand side of the second equation in (14.30),

$$B_2 = M[A_2(t) + A_1(t)Y_1(t)],$$

have zero mean. Then $Y_2(t)$ is taken as the ap-solution with zero mean of the second equation of the system.

Continuing in this manner, we successively find $Y_3(t), \ldots, Y_k(t)$.

14.4. Formulas of perturbation theory

It is natural to expect that the solution of equation (14.28) in standard cases is defined by the distribution of the eigenvalues of

$$B(\varepsilon) = B_1 + \varepsilon B_2 + \ldots + \varepsilon^{k-1} B_k \qquad (0 < \varepsilon < \varepsilon_0). \qquad (14.31)$$

We confine ourselves here to the case that B_1 has a simple eigenvalue zero, all other eigenvalues lying in the left half-plane. It is known (see, for example, Gel'fand /1/) that the matrix (14.31) has a simple real eigenvalue $\lambda(\varepsilon)$ vanishing for $\varepsilon = 0$ and analytic in ε for sufficiently small ε. The other eigenvalues of (14.31) lie, as before, in the left half-plane for small ε.

The eigenvalue $\lambda(\varepsilon)$ may be represented by a series

$$\lambda(\varepsilon) = a_1\varepsilon + a_2\varepsilon^2 + \ldots + a_j\varepsilon^j + \ldots \qquad (14.32)$$

The sign of $\lambda(\varepsilon)$ for small ε depends only on the sign of the first nonvanishing coefficient a_j; we denote the corresponding index by j_0. The value of j_0 will be essential in the next subsections.

Let e_0 and g_0 denote eigenvectors of the matrix B_1 and of its adjoint B_1^*, respectively, corresponding to the zero eigenvalue:

$$B_1 e_0 = 0, \qquad B_1^* g_0 = 0.$$

We assume e_0 and g_0 normalized so that

$$(e_0, g_0) = 1.$$

Similarly, we denote by $e_0(\varepsilon)$ and $g_0(\varepsilon)$ eigenvectors of $B(\varepsilon)$ and $B^*(\varepsilon)$ corresponding to the eigenvalue $\lambda(\varepsilon)$:

$$B(\varepsilon)\, e_0(\varepsilon) = \lambda(\varepsilon)\, e_0(\varepsilon), \quad B^*(\varepsilon)\, g_0(\varepsilon) = \lambda(\varepsilon)\, g_0(\varepsilon), \tag{14.33}$$

normalized so that

$$(e_0(\varepsilon),\, g_0(\varepsilon)) = 1. \tag{14.34}$$

It follows from perturbation theory (see Gel'fand /1/) that for small ε

$$e_0(\varepsilon) = e_0 + \varepsilon e_1 + \varepsilon^2 e_2 + \ldots + \varepsilon^l e_l + \ldots \tag{14.35}$$

and

$$g_0(\varepsilon) = g_0 + \varepsilon g_1 + \varepsilon^2 g_2 + \ldots + \varepsilon^l g_l + \ldots, \tag{14.36}$$

where e_i, g_i are vectors satisfying

$$(e_i, g_0) = (e_0, g_i) = 0 \qquad (i = 1, 2, \ldots). \tag{14.37}$$

The coefficients $a_1, a_2, \ldots$ and vectors $e_1, e_2, \ldots; g_1, g_2, \ldots$ may be derived by the method of undetermined coefficients. We are interested only in j_0 and a_{j_0}.

Substituting (14.31), (14.32) and (14.35) into the first equation of (14.33), we obtain

$$\left(\sum_{i=1}^{k} \varepsilon^{i-1} B_i\right) \sum_{l=0}^{\infty} \varepsilon^l e_l = \left(\sum_{i=1}^{\infty} a_i \varepsilon^i\right) \sum_{l=0}^{\infty} \varepsilon^l e_l. \tag{14.38}$$

Clearing parentheses and comparing coefficients of like powers of ε on both sides of (14.38), we obtain an infinite system of equations for the numbers $a_1, a_2, \ldots$ and vectors $e_1, e_2, \ldots$.

We are interested only in the case that the index j_0 of the first nonvanishing coefficient of (14.32) does not exceed $k-1$. Therefore we shall write down only those equations needed to determine the coefficients $a_1, \ldots, a_{k-1}$. These equations are clearly

$$\left.\begin{aligned} B_1 e_1 &= a_1 e_0 - B_2 e_0, \\ B_1 e_2 &= a_1 e_1 + a_2 e_0 - B_3 e_0 - B_2 e_1, \\ &\cdots\cdots\cdots\cdots \\ B_1 e_{k-1} &= a_1 e_{k-2} + a_2 e_{k-3} + \ldots + a_{k-1} e_0 - \\ &\quad - B_k e_0 - B_{k-1} e_1 - \ldots - B_2 e_{k-2}. \end{aligned}\right\} \tag{14.39}$$

The solvability condition for the first of equations (14.39) implies that

$$a_1 = (B_2 e_0, \ g_0). \tag{14.40}$$

If a_1 vanishes, we must evaluate a_2, i. e., we must find e_1 satisfying (14.37) from the first of equations (14.39) and then determine a_2 from the solvability condition for the second equation:

$$a_2 = (B_3 e_0 + B_2 e_1, \ g_0). \tag{14.41}$$

If a_2 also vanishes, we proceed to the third equation, and so on.

14.5. Shtokalo's theorem

We return to our examination of system (14.27).

Suppose that the substitution (14.29) gives system (14.28). Let zero be a simple eigenvalue of the matrix B_1 and suppose all other eigenvalues lie in the left half-plane. We evaluate the eigenvalue (14.32) of the matrix (14.31) by the procedure described in subsection 14.4. Let a_{j_0} be the first nonvanishing coefficient of (14.32), $j_0 \leqslant k-1$.

Theorem 14.4. If

$$a_{j_0} < 0, \tag{14.42}$$

then the ap-operators

$$L(\varepsilon)\, x = \frac{dx}{dt} - [\varepsilon A_1(t) + \varepsilon^2 A_2(t) + \ldots + \varepsilon^k A_k(t) + \varepsilon^{k+1} C(t; \varepsilon)]\, x \tag{14.43}$$

are regular and stable for small positive ε.

But if

$$a_{j_0} > 0, \tag{14.44}$$

then the ap-operators (14.43) are regular and unstable for small positive ε.

Proof. The regularity and stability of the ap-operators (14.43) are clearly equivalent to the regularity and stability of the ap-operators

$$L_1(\varepsilon)\,x = \frac{dx}{dt} - \left[B_1 + \varepsilon B_2 + \ldots + \varepsilon^{k-1}B_k + \varepsilon^k D\left(\frac{t}{\varepsilon};\varepsilon\right)\right]x. \quad (14.45)$$

Consider the auxiliary family of ap-operators

$$L_1(\varepsilon;\mu)\,x = \frac{dx}{dt} - \left[B_1 + \varepsilon B_2 + \ldots + \varepsilon^{k-1}B_k + \mu\varepsilon^k D\left(\frac{t}{\varepsilon};\varepsilon\right)\right]x, \quad (14.46)$$

where μ takes values in $[0, 1]$. When $\mu = 1$, the operators (14.46) coincide with the ap-operators (14.45), while when $\mu = 0$ they are regular ap-operators

$$L_1(\varepsilon;0)\,x = \frac{dx}{dt} - [B_1 + \varepsilon B_2 + \ldots + \varepsilon^{k-1}B_k]\,x, \quad (14.47)$$

whose stability for small positive ε depends on the sign of a_{j_0}.

For some fixed ε, let all the ap-operators (14.46) be regular. Then, by the results of subsection 4.3, the operators (14.45) and (14.47) are either both stable or both unstable. Hence, to complete the proof, we must establish the existence of some $\varepsilon_0 > 0$ such that for $0 < \varepsilon < \varepsilon_0$, $0 \leqslant \mu \leqslant 1$, the operators (14.46) are regular. In other words, it is sufficient to prove that for $0 < \varepsilon < \varepsilon_0$, $0 \leqslant \mu \leqslant 1$ and for an arbitrary ap-function $f(t)$ the equation

$$L_1(\varepsilon;\mu)\,x = f(t) \quad (14.48)$$

has an ap-solution.

Equation (14.48) is equivalent to the integral equation

$$x(t) = \mu\varepsilon^k \int_{-\infty}^{\infty} G(t, s; \varepsilon)\, D\left(\frac{s}{\varepsilon};\varepsilon\right) x(s)\,ds + \int_{-\infty}^{\infty} G(t, s; \varepsilon)\, f(s)\,ds,$$

where $G(t, s; \varepsilon)$ is the Green's function of (14.47). It remains only to show, then, that for $0 < \varepsilon < \varepsilon_0$, where ε_0 is some (positive) number, the norm of the integral operator

$$S(\varepsilon)\,x(t) = \varepsilon^k \int_{-\infty}^{\infty} G(t, s; \varepsilon)\, D\left(\frac{s}{\varepsilon};\varepsilon\right) x(s)\,ds \quad (14.49)$$

is less than 1.

Let $e_0(\varepsilon)$ and $g_0(\varepsilon)$ be the eigenvectors (14.38) and (14.36) of $B(\varepsilon)$ and $B^*(\varepsilon)$. Each element $x \in R^n$ may be represented in the form

$$x = P(\varepsilon)x + Q(\varepsilon)x, \tag{14.50}$$

where

$$P(\varepsilon)x = \frac{(x, g_0(\varepsilon))}{(e_0(\varepsilon), g_0(\varepsilon))} e_0(\varepsilon) = (x, g_0(\varepsilon))\, e_0(\varepsilon),$$

and $Q(\varepsilon)$ is the projection operator onto the $B(\varepsilon)$-invariant subspace $R^{n-1}(\varepsilon) \subset R^n$ orthogonal to the vector $g_0(\varepsilon)$.

We introduce a family of norms in R^n.

$$|x|_\varepsilon = |P(\varepsilon)x| + |Q(\varepsilon)x|. \tag{14.51}$$

Since the operators $P(\varepsilon)$ and $Q(\varepsilon)$ are continuous in ε (for small ε), there exist $\varepsilon_0 > 0$ and $m_0 > 0$ such that for $0 \leqslant \varepsilon \leqslant \varepsilon_0$

$$|x| \leqslant |x|_\varepsilon \leqslant m_0 |x| \qquad (x \in R^n). \tag{14.52}$$

Further estimates depend on the obvious inequality

$$|e^{B(\varepsilon)t} Q(\varepsilon)| \leqslant m e^{-\gamma_0 t} \qquad (t \geqslant 0), \tag{14.53}$$

where $m, \gamma_0 > 0$; this estimate is valid for sufficiently small ε.

We first consider the case that (14.42) holds. Then

$$G(t, s; \varepsilon) = \begin{cases} e^{B(\varepsilon)(t-s)}, & \text{if} \quad t \geqslant s, \\ 0, & \text{if} \quad t < s. \end{cases}$$

Therefore, for any $x \in R^n$ and $t \geqslant s$,

$$\begin{aligned} |G(t, s; \varepsilon)x| &\leqslant |e^{B(\varepsilon)(t-s)} P(\varepsilon)x| + |e^{B(\varepsilon)(t-s)} Q(\varepsilon)x| \leqslant \\ &\leqslant e^{\lambda(\varepsilon)(t-s)} |P(\varepsilon)x| + m e^{-\gamma_0 (t-s)} |Q(\varepsilon)x|, \end{aligned}$$

so that for sufficiently small $\varepsilon > 0$

$$|G(t, s; \varepsilon)x| \leqslant m_1 e^{\lambda(\varepsilon)(t-s)} |x|_\varepsilon \qquad (x \in R^n;\ t \geqslant s).$$

This last inequality yields a simple estimate for the norm of the operator (14.49):

$$\| S(\varepsilon) \|_{B(R^n)} \leqslant \frac{m_0 m_1}{|\lambda(\varepsilon)|} \varepsilon^k \cdot \| D(t; \varepsilon) \|_{B(R^n)},$$

from which it follows that

$$\lim_{\varepsilon\to 0}\| S(\varepsilon)\|_{B(R^n)} = 0. \tag{14.54}$$

Thus the first assertion of Theorem 14.4 is proved.
Now suppose that (14.44) holds. Then

$$G(t, s; \varepsilon) = \begin{cases} e^{B(\varepsilon)(t-s)}Q(\varepsilon), & \text{if} \quad t \geqslant s, \\ -e^{\lambda(\varepsilon)(t-s)}P(\varepsilon), & \text{if} \quad t < s. \end{cases}$$

By (14.53), we have that for small $\varepsilon>0$

$$|G(t, s; \varepsilon)x| \leqslant m_2 e^{-\lambda(\varepsilon)|t-s|}|x|,$$

which implies

$$\| S(\varepsilon)\|_{B(R^n)} \leqslant \frac{2m_0m_2}{\lambda(\varepsilon)}\varepsilon^k \| D(t; \varepsilon)\|_{B(R^n)}.$$

Thus, in this case too (14.54) holds, and the proof of Theorem 14.4 is complete.

§ 15. BIFURCATION OF ALMOST-PERIODIC SOLUTIONS

15.1. Statement of the problem

Our subject in this section is ap-solutions originating from an equilibrium state when the parameters of the differential equation are varied. Suppose that for every value of μ the system

$$\frac{dx}{dt} = f(t, x; \mu), \tag{15.1}$$

whose right-hand side is almost periodic in t, is satisfied by the trivial solution ($f(t, 0; \mu) \equiv 0$). A value μ_0 of the parameter μ is called a *bifurcation point* if for any $\eta>0$ there exists $\mu \in (\mu_0 - \eta, \mu_0 + \eta)$ such that system (15.1) has a nontrivial ap-solution satisfying

$$\| x(t; \mu)\|_{B(R^n)} < \eta. \tag{15.2}$$

Suppose the right-hand side of equation (15.1) is sufficiently smooth with respect to the space variables. Consider the ap-operators

$$L(\mu)x = \frac{dx}{dt} - f'_x(t, 0; \mu)x. \tag{15.3}$$

Let the operator $L(\mu_0)$ be regular for some μ_0 and let the ap-matrices $f'_x(t, 0; \mu_0)$ converge regularly to $f'_x(t, 0; \mu)$ as $\mu \to \mu_0$. Then equation (15.1) has no small nontrivial ap-solutions for μ close to μ_0, as we may see by applying the contracting-mapping principle to the integral operator

$$\Pi(\mu)x(t) = \int_{-\infty}^{\infty} G_0(t, s; \mu)\{f[s, x(s); \mu] - f'_x(s, 0; \mu)x(s)\}\,ds,$$

where $G_0(t, s; \mu)$ is the Green's function of the operator (15.3).

It follows from the foregoing discussion that bifurcation points can occur only at values of μ for which the ap-operator (15.3) is not regular. Let μ_0 be such a value. Set $\mu = \mu_0 + \varepsilon$. Then equation (15.1) becomes

$$\frac{dx}{dt} = g(t, x; \varepsilon). \tag{15.4}$$

This reduces the problem to examination of small nontrivial ap-solutions of equation (15.4) for small ε.

Suppose that equation (15.4) has an ap-solution $x^*(t; \varepsilon)$ continuous in ε and converging to the trivial solution as $\varepsilon \to 0$. Consider the family of ap-operators

$$L(\varepsilon)x = \frac{dx}{dt} - g'_x[t, x^*(t; \varepsilon); \varepsilon]x. \tag{15.5}$$

These operators are, in general, regular for small ε. Thus, the ap-operators

$$L_1(\varepsilon)x = \frac{dx}{dt} - g'_x[t, x_0(t; \varepsilon); \varepsilon]x \tag{15.6}$$

are also regular, provided the ap-functions $x_0(t; \varepsilon)$ are sufficiently "close" to $x^*(t; \varepsilon)$.

When the ap-functions $x_0(t; \varepsilon)$ are known and the ap-solutions $x^*(t; \varepsilon)$ unknown, one may pass from the differential equation (15.4) to the integral equation

$$x(t) = \int_{-\infty}^{\infty} G(t, s; \varepsilon)\{g[s, x(s); \varepsilon] - g'_x[s, x_0(s; \varepsilon); \varepsilon]x(s)\}\,ds, \tag{15.7}$$

and then try to use successive approximations.

This technique is well known. It may be used in proving the existence of ap-solutions $x^*(t; \varepsilon)$. In this case the ap-functions $x_0(t; \varepsilon)$ must be chosen in such a way that equation (15.7) has solutions for small ε. In the cases considered below, equation (15.4) as a rule has several families of ap-solutions; each family is to be determined using a specific ap-function $x_0(t;\varepsilon)$.

Let us assume equation (15.4) is of the form

$$\frac{dx}{dt} = \varepsilon A_1(t)\,x + \ldots + \varepsilon^k A_k(t)\,x + \varepsilon^{k+1} A(t;\varepsilon)\,x + \\ + \varepsilon F(t, x; \varepsilon) + \varepsilon\omega(t, x; \varepsilon). \qquad (15.8)$$

Here $A_1(t), \ldots, A_k(t)$ are square $n \times n$ matrices whose elements are trigonometric polynomials; $A(t;\varepsilon)$ is a matrix almost periodic in t, continuous in ε uniformly with respect to $t \in (-\infty, \infty)$; $F(t, x; \varepsilon)$ is a form of degree $m\,(m \geqslant 2)$ in the space variables; its coefficients are trigonometric polynomials, continuous in ε uniformly with respect to t; the remainder term $\omega(t, x; \varepsilon)$ contains terms that vanish with x to order greater than m: $\omega(t, 0; \varepsilon) \equiv 0$ and

$$|\omega(t, x; \varepsilon) - \omega(t, y; \varepsilon)| \leqslant q(r)\,|x - y| \\ (|x|, |y| \leqslant r; \ -\infty < t < \infty), \qquad (15.9)$$

where

$$\lim_{r \to 0} r^{-m+1} q(r) = 0,$$

$\omega(t, x; \varepsilon)$ is almost periodic in t and continuous in ε uniformly with respect to $t \in (-\infty, \infty)$.

To simplify matters, we shall restrict ourselves to the case that ε is positive.

The results presented below were first announced in a paper by Burd, Zabreiko, Kolesov and Krasnosel'skii, read at the Fifth International Conference on Nonlinear Oscillations.

15.2. Passage to an equation of special form

By substituting (14.29) into equation (15.8), we obtain an equation of the form

$$\frac{dy}{dt} = \varepsilon B_1 y + \ldots + \varepsilon^k B_k y + \varepsilon^{k+1} B(t;\varepsilon)\,y + \\ + \varepsilon F(t, y; \varepsilon) + \varepsilon^2 F_1(t, y; \varepsilon) + \varepsilon\omega_1(t, y; \varepsilon), \qquad (15.10)$$

where $B_1, \ldots, B_k$ are constant matrices, the matrix $B(t;\varepsilon)$ has the same properties as $A(t;\varepsilon)$; $F_1(t, y; \varepsilon)$ is a form of degree m with the same properties as $F(t, y; \varepsilon)$; and the remainder term $\omega_1(t, y; \varepsilon)$ has the same properties as $\omega(t, y; \varepsilon)$. It is evident from (14.29) that finding the ap-solutions of equation (15.8) for small ε is equivalent to finding the ap-solutions of equation (15.10).

We define an operator J on the set of all trigonometric polynomials

$$f(t) = a_0 + \sum_{i=1}^{l} a_i \sin \lambda_i t + b_i \cos \lambda_i t,$$

where $\lambda_i \neq 0$, by

$$J[f(t)] = \sum_{i=1}^{l} \frac{-a_i \cos \lambda_i t + b_i \sin \lambda_i t}{\lambda_i}. \qquad (15.11)$$

Apply the Bogolyubov-Krylov substitution

$$y = z + \varepsilon J[F(t, z; \varepsilon)] \qquad (15.12)$$

to (15.10), where $J[F(t, z; \varepsilon)]$ is the form obtained by applying the operator (15.11) to the coefficients of $F(t, y; \varepsilon)$. Obviously

$$\frac{dz}{dt} + \varepsilon F(t, z; \varepsilon) - \varepsilon \bar{F}(z; \varepsilon) + \varepsilon J\left[F'_z(t, z; \varepsilon)\right] \frac{dz}{dt} =$$
$$= \varepsilon B_1 z + \ldots + \varepsilon^k B_k z + \varepsilon^{k+1} B(t; \varepsilon) z + \varepsilon F(t, z; \varepsilon) +$$
$$+ \varepsilon^2 F_2(t, z; \varepsilon) + \varepsilon \omega_2(t, z; \varepsilon),$$

where $F_2(t, z; \varepsilon)$ and $\omega_2(t, z; \varepsilon)$ have the same properties as $F_1(t, z; \varepsilon)$, $\omega_1(t, z; \varepsilon)$ and

$$\bar{F}(z; \varepsilon) = \lim_{T \to \infty} \frac{1}{T} \int_0^T F(s, z; \varepsilon)\, ds \qquad (z \in R^n). \qquad (15.13)$$

From the last equation, we have

$$\frac{dz}{dt} = \varepsilon B_1 z + \ldots + \varepsilon^k B_k z + \varepsilon^{k+1} B(t; \varepsilon) z + \varepsilon \bar{F}(z; \varepsilon) +$$
$$+ \varepsilon^2 F_3(t, z; \varepsilon) + \varepsilon \omega_3(t, z; \varepsilon). \qquad (15.14)$$

Let us now introduce a new time variable $\tau = t\varepsilon$. Then equation (15.14) assumes the form

$$\frac{dz}{d\tau} = B_1 z + \ldots + \varepsilon^{k-1} B_k z + \bar{F}(z; \varepsilon) +$$
$$+ \varepsilon^k B\left(\frac{\tau}{\varepsilon}; \varepsilon\right) z + \varepsilon F_3\left(\frac{\tau}{\varepsilon}, z; \varepsilon\right) + \omega_3\left(\frac{\tau}{\varepsilon}, z; \varepsilon\right). \qquad (15.15)$$

As a result of several substitutions, we have reduced the problem of finding small ap-solutions of equation (15.8) to the same problem for equation (15.5). The advantage of the latter is that the "principal' terms on its right do not depend on time.

Of course, equation (15.15) is meaningful only for $\varepsilon \neq 0$. For $\varepsilon = 0$, a natural choice for the equation is (see, e. g., Bogolybov and Mitropol'skii /1/)

$$\frac{dz}{dt} = B_1 z + \bar{F}(z; 0) + \bar{\omega}_3(z; 0), \tag{15.16}$$

where

$$\bar{\omega}_3(z; \varepsilon) = \lim_{T \to \infty} \frac{1}{T} \int_0^T \omega_3(s, z; \varepsilon)\, ds \qquad (z \in R^n).$$

Equation (15.16) is known as the averaged equation of the first approximation.

15.3. Fundamental theorem

As shown in subsection 15.1, equation (15.15) has small nontrivial ap-solutions (for small ε) only if the ap-operator

$$Lz = \frac{dz}{dt} - B_1 z$$

is not regular or, equivalently, if the matrix B_1 has eigenvalues on the imaginary axis. For simplicity's sake, we confine ourselves to the case that zero is a simple eigenvalue of B_1, all other eigenvalues lying in the left half-plane.

Let

$$\lambda(\varepsilon) = a_1\varepsilon + a_2\varepsilon^2 + \ldots + a_j\varepsilon^j + \ldots \tag{15.17}$$

be the eigenvalue (14.32) of the matrix

$$B(\varepsilon) = B_1 + \varepsilon B_2 + \ldots + \varepsilon^{k-1} B_k. \tag{15.18}$$

Let a_{j_0} be the first nonvanishing coefficient in (15.17). As in subsection 14.5 we assume that

$$j_0 \leqslant k - 1. \tag{15.19}$$

As in subsection 14.4, we denote by e_0 and g_0, respectively, eigenvectors of the matrices B_1 and B_1^* corresponding to the zero eigenvalue, normalized so that

$$(e_0, g_0) = 1.$$

We set

$$\alpha_0 = (\bar{F}(e_0, 0), g_0) \tag{15.20}$$

and assume henceforth that

$$\alpha_0 \neq 0. \tag{15.21}$$

Let $V(r)$ denote the ball $\|x(t)\| \leqslant r$ in $B(R^n)$.

Theorem 15.1. Let conditions (15.19) and (15.21) be satisfied. Then there exist $\varepsilon_0, r_0 > 0$ such that:

1. Equation (15.16) has no nontrivial ap-solutions in the ball $V(r_0)$.

2. If m is even, then for $0 < \varepsilon < \varepsilon_0$ equation (15.15) has a unique ap-solution in $V(r_0)$; this solution is exponentially stable if $a_{j_0} > 0$ and unstable if $a_{j_0} < 0$.

3. If m is odd and $\alpha_0 a_{j_0} < 0$ then for $0 < \varepsilon < \varepsilon_0$ equation (15.15) has exactly two nontrivial ap-solutions in $V(r_0)$, which are exponentially stable if $a_{j_0} > 0$ and unstable if $a_{j_0} < 0$.

4. If m is odd and $\alpha_0 a_{j_0} > 0$, then for $0 < \varepsilon < \varepsilon_0$ equation (15.15) has no nontrivial ap-solutions in $V(r_0)$.

15.4. Proof of the existence of ap-solutions

Consider the algebraic equation

$$B_1 z + \varepsilon B_2 z + \ldots + \varepsilon^{k-1} B_k z + \bar{F}(z; 0) = 0. \tag{15.22}$$

We seek an approximate solution in the form

$$z_0(\varepsilon) = \beta(\varepsilon) e_0(\varepsilon), \tag{15.23}$$

where $e_0(\varepsilon)$ is an eigenvector of the matrix (14.31). Substitution of (15.23) into (15.22) and scalar multiplication on the left by g_0 yield an equation for $\beta(\varepsilon)$:

$$\lambda(\varepsilon) + [\beta(\varepsilon)]^{m-1} (\bar{F}[e_0(\varepsilon); 0], g_0) = 0,$$

whence

$$\beta(\varepsilon) = \left\{ \frac{-\lambda(\varepsilon)}{(\bar{F}[e_0(\varepsilon); 0], g_0)} \right\}^{\frac{1}{m-1}}, \tag{15.24}$$

and so

$$z_0(\varepsilon) = \left\{ \frac{-\lambda(\varepsilon)}{(\bar{F}[e_0(\varepsilon); 0], g_0)} \right\}^{\frac{1}{m-1}} e_0(\varepsilon). \tag{15.25}$$

The denominator in (15.25) does not vanish for small ε, since for $\varepsilon = 0$ it is the number (15.20). Clearly, (15.24) defines a unique real function if m is even. If m is odd, then, under the hypotheses of part 3 of Theorem 15.1, it defines two continuous real functions $\beta(\varepsilon)$ and $\beta(\varepsilon)$.

We seek an ap-solution $z(\tau; \varepsilon)$ of equation (15.15), in the form

$$z = u + z_0(\varepsilon). \tag{15.26}$$

It is convenient to write the differential equation for u in the form

$$\frac{du}{d\tau} - B(\varepsilon)u - \bar{F}'_z[z_0(\varepsilon); 0]u = D(\tau, u; \varepsilon), \tag{15.27}$$

where

$$\begin{aligned} D(\tau, u; \varepsilon) = \lambda(\varepsilon) z_0(\varepsilon) + \varepsilon^k B\left(\frac{\tau}{\varepsilon}; \varepsilon\right)[u + z_0(\varepsilon)] + \\ + \bar{F}[u + z_0(\varepsilon); \varepsilon] - \bar{F}'_z[z_0(\varepsilon); 0]u + \\ + \varepsilon F_3\left[\frac{\tau}{\varepsilon}, u + z_0(\varepsilon); \varepsilon\right] + \omega_3\left[\frac{\tau}{\varepsilon}, u + z_0(\varepsilon); \varepsilon\right]. \end{aligned}$$

We first consider the ap-operators

$$L(\varepsilon)u = \frac{du}{dt} - B(\varepsilon)u - \bar{F}'_z[z_0(\varepsilon); 0]u \tag{15.28}$$

and show that they are regular for small nonzero ε. To this end, we first observe that the matrices $B(\varepsilon) + \bar{F}'_z[z_0(\varepsilon); 0]$ are analytic in ε and when $\varepsilon=0$ they coincide with B_1, which, by assumption, has a simple eigenvalue zero and all its other eigenvalues are situated in the left half-plane. Thus, for small ε, the matrices $B(\varepsilon) + \bar{F}'_z[z_0(\varepsilon); 0]$ have a simple eigenvalue $\eta(\varepsilon)$ which vanishes for $\varepsilon = 0$, the remaining eigenvalues lying in the left half-plane. To prove the regularity of the ap-operators (15.28) we must show that $\eta(\varepsilon) \neq 0$ for small nonzero ε.

To evaluate $\eta(\varepsilon)$, we use the technique of perturbation theory as described in subsection 14.4. Since

$$\begin{aligned} \bar{F}'_z[z_0(\varepsilon); 0] = \frac{-\lambda(\varepsilon)}{(\bar{F}[e_0(\varepsilon); 0], g_0)} \bar{F}'_z[e_0(\varepsilon); 0] = \\ = -\alpha_0^{-1} a_{j_0} \varepsilon^{j_0} \bar{F}'_z(e_0; 0) + o(\varepsilon^{j_0}), \end{aligned}$$

it follows from (14.39) that

$$\eta(\varepsilon) = \{a_{j_0} - \alpha_0^{-1} a_{j_0} (\overline{F}'_z(e_0; 0) e_0, g_0)\} \varepsilon^{j_0} + o(\varepsilon^{j_0}). \tag{15.29}$$

By Euler's formula for homogeneous functions,

$$\overline{F}'_z(e_0; 0) e_0 = m\overline{F}(e_0; 0)$$

and thus (15.29) may be written

$$\eta(\varepsilon) = -(m-1) a_{j_0} \varepsilon^{j_0} + o(\varepsilon^{j_0}). \tag{15.30}$$

By (15.30), $\eta(\varepsilon) \neq 0$ for small nonzero ε, so that the operators (15.28) are indeed regular for these ε. Their Green's function $G(\tau, \sigma; \varepsilon)$, as shown in the proof of Theorem 14.4, satisfies the inequality

$$|G(\tau, \sigma; \varepsilon)| \leqslant M_1 e^{-|\eta(\varepsilon)| \cdot |\tau - \sigma|} \qquad (-\infty < \tau, \sigma < \infty).$$

This inequality together with (15.30) show that for small positive ε

$$|G(\tau, \sigma; \varepsilon)| \leqslant M_1 e^{-\gamma \varepsilon^{j_0} |\tau - \sigma|} \qquad (-\infty < \tau, \sigma < \infty), \tag{15.31}$$

where γ is positive.

Set

$$S(\varepsilon) u(\tau) = \int_{-\infty}^{\infty} G(\tau, \sigma; \varepsilon) u(\sigma) \, d\sigma. \tag{15.32}$$

This operator acts in $B(R^n)$. By (15.31), it follows that

$$\| S(\varepsilon) \|_{B(R^n)} \leqslant \frac{2M_1}{\gamma} \varepsilon^{-j_0}. \tag{15.33}$$

Now consider the nonlinear operator

$$D(\varepsilon) u(\tau) = D[\tau, u(\tau); \varepsilon] \tag{15.34}$$

acting in $B(R^n)$. Clearly,

$$D(\varepsilon) 0 = \lambda(\varepsilon) z_0(\varepsilon) + \overline{F}[z_0(\varepsilon); 0] + \varepsilon^k B\left(\frac{\tau}{\varepsilon}, \varepsilon\right) z_0(\varepsilon) +$$
$$+ \{\overline{F}[z_0(\varepsilon); \varepsilon] - \overline{F}[z_0(\varepsilon); 0]\} + \varepsilon F_3\left[\frac{\tau}{\varepsilon}, z_0(\varepsilon); \varepsilon\right] +$$
$$+ \omega_3\left[\frac{\tau}{\varepsilon}, z_0(\varepsilon); \varepsilon\right],$$

so that

$$D(\varepsilon)0 = \lambda(\varepsilon) z_0(\varepsilon) + \bar{F}[z_0(\varepsilon); 0] + o(\varepsilon^{j_0 + j_0/(m-1)}). \qquad (15.35)$$

By (15.33) and (15.35),

$$S(\varepsilon) D(\varepsilon) 0 = S(\varepsilon)\{\lambda(\varepsilon) z_0(\varepsilon) + \bar{F}[z_0(\varepsilon); 0]\} + o(\varepsilon^{j_0/(m-1)})$$

or, equivalently,

$$S(\varepsilon) D(\varepsilon) 0 = -\{B(\varepsilon) + \bar{F}'_z[z_0(\varepsilon); 0]\}^{-1} \times \\ \times \{\lambda(\varepsilon) z_0(\varepsilon) + \bar{F}[z_0(\varepsilon); 0]\} + o(\varepsilon^{j_0/(m-1)}). \qquad (15.36)$$

Let $e^*(\varepsilon)$ and $g^*(\varepsilon)$, respectively, denote the eigenvectors of $B(\varepsilon) + \bar{F}'_z[z_0(\varepsilon); 0]$ and its adjoint, corresponding to the eigenvalue $\eta(\varepsilon)$ which reduce to e_0 and g_0 for $\varepsilon = 0$ and are normalized so that

$$(e^*(\varepsilon), g^*(\varepsilon)) = 1.$$

Then every vector $x \in R^n$ may be represented in the form

$$x = (x, g^*(\varepsilon)) e^*(\varepsilon) + Q_*(\varepsilon) x,$$

where $Q_*(\varepsilon)$ is the projection operator onto the subspace orthogonal to $g^*(\varepsilon)$, which is invariant under $B(\varepsilon) + \bar{F}'_z[z_0(\varepsilon); 0]$. For small ε we have the obvious inequality

$$|\{B(\varepsilon) + \bar{F}'_z[z_0(\varepsilon); 0]\}^{-1} Q_*(\varepsilon)| \leqslant M_2,$$

where M_2 is a constant, and again for small ε,

$$|\{B(\varepsilon) + \bar{F}'_z[z_0(\varepsilon); 0]\}^{-1} Q_*(\varepsilon)\{\lambda(\varepsilon) z_0(\varepsilon) + \bar{F}[z_0(\varepsilon); 0]\}| \leqslant \\ \leqslant M_2 |\lambda(\varepsilon) z_0(\varepsilon) + \bar{F}[z_0(\varepsilon); 0]| \leqslant M_3 \varepsilon^{j_0 + j_0/(m-1)}. \qquad (15.37)$$

Now, by the definition of $z_0(\varepsilon)$,

$$|\{B(\varepsilon) + \bar{F}'_z[z_0(\varepsilon); 0]\}^{-1} (\lambda(\varepsilon) z_0(\varepsilon) + \bar{F}[z_0(\varepsilon); 0], g^*(\varepsilon)) e^*(\varepsilon)| \leqslant \\ \leqslant \frac{1}{|\eta(\varepsilon)|} |(\lambda(\varepsilon) z_0(\varepsilon) + \bar{F}[z_0(\varepsilon); 0], g^*(\varepsilon) - g_0)| |e^*(\varepsilon)| = \\ = o(\varepsilon^{j_0/(m-1)}). \qquad (15.38)$$

As a result, (15.36)—(15.38) yield the estimate

$$\| S(\varepsilon) D(\varepsilon) 0 \|_{B(R^n)} \leqslant \alpha(\varepsilon) \cdot \varepsilon^{j_0/(m-1)}, \qquad (15.39)$$

where $\alpha(\varepsilon)$ is positive for $\varepsilon > 0$ and

$$\lim_{\varepsilon \to 0} \alpha(\varepsilon) = 0.$$

The operator $S(\varepsilon)D(\varepsilon)$ satisfies a Lipschitz condition on every ball $V(r) = \{u(\tau): \|u\|_{B(R^n)} \leqslant r\}$. We claim that the corresponding Lipschitz constant $q(\varepsilon)$ converges to 0 as $\varepsilon \to 0$, if the radius of the ball is given by

$$r = \varepsilon^{j_0/(m-1)} \sqrt{\alpha(\varepsilon)}. \tag{15.40}$$

Let $u(\tau), v(\tau) \in V(r)$. Then

$$D(\varepsilon) u(\tau) - D(\varepsilon) v(\tau) = \Delta_1 + \Delta_2 + \Delta_3 + \Delta_4 + \Delta_5, \tag{15.41}$$

where

$$\Delta_1 = \varepsilon^k B\left(\frac{\tau}{\varepsilon}; \varepsilon\right)[u(\tau) - v(\tau)],$$

$$\Delta_2 = \bar{F}[z_0(\varepsilon) + u(\tau); \varepsilon] - \bar{F}[z_0(\varepsilon) + v(\tau); \varepsilon] - \\ - \bar{F}'_z[z_0(\varepsilon); \varepsilon][u(\tau) - v(\tau)],$$

$$\Delta_3 = \{\bar{F}'_z[z_0(\varepsilon); \varepsilon] - \bar{F}'_z[z_0(\varepsilon); 0]\}[u(\tau) - v(\tau)],$$

$$\Delta_4 = \varepsilon F_3\left[\frac{\tau}{\varepsilon}, z_0(\varepsilon) + u(\tau); \varepsilon\right] - \varepsilon F_3\left[\frac{\tau}{\varepsilon}, z_0(\varepsilon) + v(\tau); \varepsilon\right],$$

$$\Delta_5 = \omega_3\left[\frac{\tau}{\varepsilon}, z_0(\varepsilon) + u(\tau); \varepsilon\right] - \omega_3\left[\frac{\tau}{\varepsilon}, z_0(\varepsilon) + v(\tau); \varepsilon\right].$$

As $k > j_0$,

$$\|\Delta_1\|_{B(R^n)} \leqslant M\varepsilon^k \|u - v\|_{B(R^n)} = o(\varepsilon^{j_0})\|u - v\|_{B(R^n)}.$$

The equality

$$\Delta_2 = \int_0^1 \{\bar{F}'_z[z_0(\varepsilon) + \theta u(\tau) + (1 - \theta) v(\tau); \varepsilon] - \bar{F}'_z[z_0(\varepsilon); \varepsilon]\} \times \\ \times [u(\tau) - v(\tau)]\, d\theta$$

and (15.40) yield the estimate

$$\|\Delta_2\|_{B(R^n)} \leqslant M(r^{m-1} + r\varepsilon^{j_0(m-2)/(m-1)})\|u - v\|_{B(R^n)} = \\ = o(\varepsilon^{j_0})\|u - v\|_{B(R^n)}.$$

By the continuity of the form $\bar{F}(z; \varepsilon)$ with respect to ε,

$$\|\Delta_3\|_{B(R^n)} = o(\varepsilon^{j_0})\|u - v\|_{B(R^n)}.$$

Finally, it is clear that

$$\| \Delta_4 \|_{B(R^n)} = o(\varepsilon^{j_0}) \| u - v \|_{B(R^n)},$$
$$\| \Delta_5 \|_{B(R^n)} = o(\varepsilon^{j_0}) \| u - v \|_{B(R^n)}.$$

Thus

$$\| D(\varepsilon) u(\tau) - D(\varepsilon) v(\tau) \|_{B(R^n)} = o(\varepsilon^{j_0}) \| u - v \|_{B(R^n)},$$

and (15.33) gives the estimate

$$\| S(\varepsilon) D(\varepsilon) u(\tau) - S(\varepsilon) D(\varepsilon) v(\tau) \|_{B(R^n)} \leqslant q(\varepsilon) \| u - v \|_{B(R^n)}$$
$$(u(\tau),\ v(\tau) \in V(r)),$$

where

$$\lim_{\varepsilon \to 0} q(\varepsilon) = 0.$$

The problem of finding the ap-solutions of equation (15.27) is equivalent to the integral equation

$$u(\tau) = S(\varepsilon) D(\varepsilon) u(\tau).$$

As we have shown, for small positive ε, $S(\varepsilon)D(\varepsilon)$ is a contraction operator ($q(\varepsilon) < 1$) on the ball $V(r)$ of radius (15.40). From (15.39) and (15.40) it follows that for small ε the operator $S(\varepsilon)D(\varepsilon)$ maps the ball $V(r)$ into itself: for $u(\tau) \in V(r)$

$$\| S(\varepsilon) D(\varepsilon) u(\tau) \|_{B(R^n)} \leqslant$$
$$\leqslant \| S(\varepsilon) D(\varepsilon) u(\tau) - S(\varepsilon) D(\varepsilon) 0 \|_{B(R^n)} + \| S(\varepsilon) D(\varepsilon) 0 \|_{B(R^n)} \leqslant$$
$$\leqslant q(\varepsilon) r + \varepsilon^{j_0/(m-1)} \alpha(\varepsilon) = r \left[q(\varepsilon) + \sqrt{\alpha(\varepsilon)} \right] < r.$$

By the contracting-mapping principle, the operator $S(\varepsilon)D(\varepsilon)$ has a unique fixed point in $V(r)$.

We have shown that, under the hypotheses of part 2 or 3, respectively, equation (15.5) has at least one or two nontrivial ap-solutions. We leave the examination of their stability to the reader (it is convenient to use Theorems 11.1 and 11.2).

The remaining parts of Theorem 15.1 will be proved in subsequent subsections.

15.5. Invariant cones

Recall that $e_0(\varepsilon)$ and $g_0(\varepsilon)$ denote eigenvectors of $B(\varepsilon)$ and $B^*(\varepsilon)$ respectively, which reduce to e_0 and g_0 when $\varepsilon = 0$ and are so normalized that $(e_0(\varepsilon), g_0(\varepsilon)) = 1$. Let $Q(\varepsilon)$ denote the projection operator onto the subspace $R(\varepsilon)$ of vectors orthogonal to $g_0(\varepsilon)$:

$$Q(\varepsilon)x = x - (x, g_0(\varepsilon))e_0(\varepsilon) \qquad (x \in R^n).$$

Let $U(\varepsilon)$ be the matrix whose columns are the vectors

$$e_0(\varepsilon),\ Q(\varepsilon)f_1,\ \ldots,\ Q(\varepsilon)f_{n-1},$$

where $f_1, \ldots, f_{n-1}$ is an arbitrary but fixed basis of $R(0)$. $U(\varepsilon)$ is clearly continuous in ε and nonsingular for sufficiently small ε.

We set

$$z = U(\varepsilon)w. \tag{15.42}$$

Then equation (15.15) may be rewritten

$$\frac{dw}{d\tau} = C_0(\varepsilon)w + \varepsilon^k C\left(\frac{\tau}{\varepsilon}; \varepsilon\right)w + \Phi_0(w; \varepsilon) + \\ + \varepsilon\Phi\left(\frac{\tau}{\varepsilon}, w; \varepsilon\right) + \Omega\left(\frac{\tau}{\varepsilon}, w; \varepsilon\right), \tag{15.43}$$

where

$$C_0(\varepsilon) = U^{-1}(\varepsilon)B(\varepsilon)U(\varepsilon), \qquad \Phi_0(w; \varepsilon) = U^{-1}(\varepsilon)\bar{F}[U(\varepsilon)w; \varepsilon],$$

while the other terms on the right have the same properties as the corresponding terms in equation (15.15). A simple check shows that $C_0(\varepsilon)$ is of the form

$$C_0(\varepsilon) = \begin{Vmatrix} \lambda(\varepsilon) & 0 \\ 0 & D_0(\varepsilon) \end{Vmatrix}, \tag{15.44}$$

where $D_0(\varepsilon)$ is a square matrix of order $n-1$.

Since the eigenvalues of $C_0(\varepsilon)$ coincide with those of $B(\varepsilon)$, all the eigenvalues of $D_0(\varepsilon)$ are situated in the left half-plane.

It is known (see, for example, Malkin /1/), that there exists a positive-definite symmetric matrix H of order $n-1$ such that

$$HD_0(0) + D_0^*(0)H = -I.$$

It is clear that, for sufficiently small ε,

$$HD_0(\varepsilon)+D_0^*(\varepsilon)H<-\frac{1}{2}I. \tag{15.45}$$

Recall that (15.45) is equivalent to

$$(HD_0(\varepsilon)x,\,x)\leqslant-\frac{1}{4}(x,\,x)\qquad(x\in R^{n-1}).$$

We shall find it convenient to write a vector $w=\{\zeta_1,\,\ldots,\,\zeta_n\}\in R^n$ as

$$w=\{\zeta,\,x\},$$

where $\zeta=\zeta_1$ and $x=\{\zeta_2,\,\ldots,\,\zeta_n\}$ is a point of R^{n-1}.

Let $K(\rho)$ denote the cone in R^n selected by the functional

$$q(w)=-\rho\zeta+\sqrt{(Hx,\,x)}, \tag{15.46}$$

where ρ is positive, and let $K(\rho;r)$ be the intersection of $K(\rho)$ with the ball $V(r)$. $\Gamma(\rho)$ and $\Gamma(\rho,r)$, respectively, will be the boundary of $K(\rho)$ and of its intersection with $V(r)$. We now show that, for any $\rho>0$, there exist positive $\varepsilon_0=\varepsilon_0(\rho)$ and $r_0=r_0(\rho)$ such that for $0<\varepsilon<\varepsilon_0$ every solution $w(t)=\{\zeta(t),x(t)\}$ of equation (15.43) having a nonzero initial value $w(t_0)\in\Gamma(\rho,r_0)$ lies within $K(\rho)$ for values of t close to and greater than t_0. To see this, it suffices to prove the inequality

$$\frac{dq(t_0)}{dt}<0, \tag{15.47}$$

where

$$q(t)=q[w(t)]=-\rho\zeta(t)+\sqrt{(Hx(t),\,x(t))}.$$

Clearly,

$$\frac{dq(t)}{dt}=(\operatorname{grad}q[w(t)],\,C_0(\varepsilon)w(t))+\Delta(t),$$

where

$$\Delta(t)=\Big(\operatorname{grad}q[w(t)],\,\varepsilon^kC\left(\frac{\tau}{\varepsilon};\,\varepsilon\right)w(t)+\Phi_0[w(t);\,\varepsilon]+\\+\varepsilon\Phi\left[\frac{\tau}{\varepsilon},\,w(t);\,\varepsilon\right]+\Omega\left[\frac{\tau}{\varepsilon},\,w(t);\,\varepsilon\right]\Big).$$

Since

$$(\operatorname{grad}q[w(t)],\,C_0(\varepsilon)w(t))=\\=\frac{\big((HD_0(\varepsilon)+D_0^*(\varepsilon)H)x(t),\,x(t)\big)}{2\sqrt{(Hx(t),\,x(t))}}-\rho\lambda(\varepsilon)\zeta(t),$$

it follows from the equality

$$\rho\xi(t_0) = \sqrt{(Hx(t_0), x(t_0))} \tag{15.48}$$

that

$$(\operatorname{grad} q[w(t_0)], C_0(\varepsilon) w(t_0)) = \\ = \frac{(\{H[D_0(\varepsilon) - \lambda(\varepsilon) I] + [D_0^*(\varepsilon) - \lambda(\varepsilon) I] H\} x(t_0), x(t_0))}{2\sqrt{(Hx(t_0), x(t_0))}}.$$

Hence, by the positive-definiteness of the matrix H and by the inequality (15.45), we have that for small ε

$$(\operatorname{grad} q[w(t_0)], C_0(\varepsilon) w(t_0)) \leqslant -a_0 \sqrt{(x(t_0), x(t_0))} \quad (a_0 > 0)$$

and (15.48) gives

$$(\operatorname{grad} q[w(t_0)], C_0(\varepsilon) w(t_0)) \leqslant -b_0(\rho) | w(t_0) |, \tag{15.49}$$

where $b_0(\rho)$ is a positive constant. Since, except for $C_0(\varepsilon)w$, all terms on the right of equation (15.43) either vanish with w to order greater than unity or are linear in w and involve ε as a factor, it follows that there exist ε_0, $r_0 > 0$ such that for $0 < \varepsilon < \varepsilon_0$ and $|w(t_0)| \leqslant r_0$

$$|\Delta(t_0)| \leqslant \frac{1}{2} b_0(\rho).$$

Thus, for $0 < \varepsilon < \varepsilon_0$ and $|w(t_0)| \leqslant r_0$,

$$\frac{dq(t_0)}{dt} \leqslant -\frac{1}{2} b_0(\rho) | w(t_0) |,$$

implying (15.47).

Suppose that $w(t)$ is a nontrivial ap-solution of equation (15.43) lying entirely in the ball $V(r_0)$ for $\varepsilon \in (0, \varepsilon_0)$. If there exists $t_0 \in (-\infty, \infty)$ such that $w(t_0) \in K(\rho, r_0)$, it follows from the above argument that $w(t) \in K(\rho, r_0)$ for $t \geqslant t_0$ and therefore for all $t \in (-\infty, \infty)$. Similarly, if for some t_0 the point $w(t_0)$ belongs to $-K(\rho, r_0)$, then $w(t) \in -K(\rho, r_0)$ for all $t \in (-\infty, \infty)$.

Let us show that for sufficiently small ε equations (15.43) have no solutions in $V(r_0)$ not contained in the cones $K(\rho)$ and $K(\rho)$. Suppose, on the contrary, that there exists a sequence of positive numbers ε_j converging to zero such that the equations

$$\frac{dw}{d\tau} = C_0(\varepsilon_j) w + \varepsilon_j^k C\left(\frac{\tau}{\varepsilon_j}; \varepsilon_j\right) w + \Phi_0(w; \varepsilon_j) + \\ + \varepsilon_j \Phi\left(\frac{\tau}{\varepsilon_j}, w; \varepsilon_j\right) + \Omega\left(\frac{\tau}{\varepsilon_j}, w; \varepsilon_j\right) \tag{15.50}$$

have nontrivial ap-solutions $w_j(\tau)$ satisfying

$$\lim_{j\to\infty} \| w_j(\tau) \|_{B(R^n)} = 0$$

and lying entirely outside the cones $K(\rho)$ and $-K(\rho)$. Select a sequence τ_j such that

$$| w_j(\tau_j) | \geqslant \frac{1}{2} \| w_j(\tau) \|_{B(R^n)}, \qquad (15.51)$$

and set

$$h_j(\tau) = \frac{w_j(\tau + \tau_j)}{\| w_j(\tau) \|_{B(R^n)}} \qquad (j = 1, 2, \ldots).$$

Clearly,

$$\frac{dh_j(\tau)}{d\tau} = C_0(\varepsilon_j) h_j(\tau) + \delta_j(\tau),$$

where

$$\lim_{j\to\infty} \| \delta_j(\tau) \|_{B(R^n)} = 0.$$

Hence the functions $h_j(\tau)$ are equicontinuous, and we may assume without loss of generality that they converge uniformly on every finite interval to a function $h_0(\tau) \in C(R^n)$. It is clear that $h_0(\tau)$ is a solution of the differential equation

$$\frac{dh}{d\tau} = C_0(0)\, h. \qquad (15.52)$$

The matrix $C_0(0)$ has an eigenvalue zero with eigenvector $e_1 = \{1, 0, \ldots, 0\}$ and all other eigenvalues lie in the left half-plane. It follows that all solutions of equation (15.52) bounded on $(-\infty, \infty)$ have the form $h(t) = \varkappa e_1$, where $\varkappa$ is some number. Hence,

$$h_0(\tau) = \varkappa_1 e_1,$$

and, by (15.51), $\varkappa_1 \neq 0$. Since e_1 is an interior point of $K(\rho)$, we have that for all sufficiently large j, the vectors $w_j(\tau_j)$ belong either to $K(\rho)$ or to $-K(\rho)$; this is a contradiction.

For sufficiently small ε, then, the small ap-solutions of equation (15.43) lie either in $K(\rho)$ or in $-K(\rho)$.

Below we shall need the fact that, for small positive ε, any two small ap-solutions $w_1(t; \varepsilon)$ and $w_2(t; \varepsilon)$ of equation (15.43) are comparable, in the sense that either $w_1(t; \varepsilon) - w_2(t; \varepsilon) \in K(\rho)$ or $w_2(t; \varepsilon) - w_1(t; \varepsilon) \in K(\rho)$. To see this, it suffices to write down the equation satisfied by the difference $w_1(t;\varepsilon)-w_2(t;\varepsilon)$ and to repeat the reasoning of the preceding paragraphs.

To fix ideas, suppose that $w_1(t;\varepsilon)-w_2(t;\varepsilon)\in K(\rho)$. If the ap-solutions $w_1(t;\varepsilon)$ and $w_2(t;\varepsilon)$ are distinct, a more detailed analysis shows that for any $t\in(-\infty,\infty)$ the difference $w_1(t;\varepsilon)-w_2(t;\varepsilon)$ has a spherical neighborhood in $K(\rho)$ with radius independent of t (but, of course, dependent on ε).

15.6. Uniqueness of ap-solution in cone

Let $K(\rho)$ be the cone in the phase space R^n, defined in subsection 15.5. As usual, we denote by $\hat{K}(\rho)$ the cone in $B(R^n)$ of functions $w(t)$ with values in $K(\rho)$.

Consider the ap-operators

$$L(\varepsilon)w=\frac{dw}{d\tau}-C_0(\varepsilon)w-\varepsilon^kC\left(\frac{\tau}{\varepsilon};\varepsilon\right)w+(w,e_1)\varkappa e_1, \qquad (15.53)$$

where $\varkappa$ is positive. It is evident that, for every fixed $\varkappa$, the operators (15.53) are regular and stable, provided ε is sufficiently small. Let

$$G(\tau,\sigma;\varepsilon)=\begin{cases}U(\tau,\sigma;\varepsilon), & \text{if } \tau\geqslant\sigma,\\ 0, & \text{if } \tau<\sigma,\end{cases} \qquad (15.54)$$

where $U(\tau,\sigma;\varepsilon)$ is the fundamental matrix of the equation $L(\varepsilon)w=0$, denote the Green's function of the ap-operators (15.53). We shall later need the fact that, for fixed small $\varkappa$ and all sufficiently small ε, the Green's function (15.54) is strongly positive with respect to all cones $K(\rho)$ for sufficiently large ρ. Indeed, the positive constant $b_0(\rho)$ in (15.49) may be selected simultaneously for all sufficiently large ρ. Henceforth we assume that $\varkappa$ has already been fixed.

Condition (15.21) may be written

$$\alpha_0=(\Phi_0(e_1;0),e_1)\neq 0. \qquad (15.55)$$

This enables us to select ρ_0 in such a way that either $\Phi_0(e_1;0)$ or $-\Phi_0(e_1;0)$ is an interior point of the cone $K(\rho_0)$ (depending on the sign of α_0). In all subsequent constructions, we assume that the semi-orders are defined in the spaces R^n and $B(R^n)$ by the cones $K(\rho_0)$ and $\hat{K}(\rho_0)$, respectively.

The problem of ap-solutions of equation (15.43) reduces to the integral equation

$$w(\tau)=S(\varepsilon)D(\varepsilon)w(\tau),$$

where

$$S(\varepsilon)\,w(\tau) = \int_{-\infty}^{\infty} G(\tau, \sigma; \varepsilon)\,w(\sigma)\,d\sigma,$$

and

$$D(\varepsilon)\,w(\tau) = (w(\tau), e_1)\,\varkappa e_1 + \Phi_0[w(\tau); \varepsilon] + \\ + \varepsilon\Phi\left[\frac{\tau}{\varepsilon}, w(\tau); \varepsilon\right] + \Omega\left[\frac{\tau}{\varepsilon}, w(\tau); \varepsilon\right].$$

We claim that the operator $S(\varepsilon)D(\varepsilon)$ is monotone on the ball $V(r)$ in $B(R^n)$, provided r is sufficiently small.

Let $w_1(\tau) = \{\zeta_1(\tau), x_1(\tau)\}$, $w_2(\tau) = \{\zeta_2(\tau), x_2(\tau)\}$; $w_1(\tau)$, $w_2(\tau) \in V(r)$ and

$$w_1(\tau) \leqslant w_2(\tau).$$

This means that

$$\zeta_1(\tau) \leqslant \zeta_2(\tau) \qquad (-\infty < \tau < \infty)$$

and

$$\sqrt{(H[x_2(\tau) - x_1(\tau)], x_2(\tau) - x_1(\tau))} \leqslant \rho_0[\zeta_2(\tau) - \zeta_1(\tau)] \\ (-\infty < \tau < \infty).$$

Let Δ be the difference

$$\Delta = D(\varepsilon)\,w_1(\tau) - D(\varepsilon)\,w_2(\tau).$$

To afford a more concise notation, we define a projection operator in R^n:

$$Qw = w - (w, e_1)\,e_1. \qquad (15.56)$$

Then for each $\tau \in (-\infty, \infty)$, clearly,

$$|Q\Delta| \leqslant q(r)\,|w_1(\tau) - w(\tau)|,$$

where $q(r) \to 0$ as $r \to 0$; thus

$$\sqrt{(HQ\Delta, Q\Delta)} \leqslant M_1 q(r)\,|w_1(\tau) - w_2(\tau)|. \qquad (15.57)$$

On the other hand,

$$|(\Delta, e_1)| \geqslant \varkappa[\zeta_2(\tau) - \zeta_1(\tau)] - q(r)\,|w_1(\tau) - w_2(\tau)|.$$

Since the operators $S(\varepsilon)$ are positive, the operator $S(\varepsilon)D(\varepsilon)$ will be proven monotone if we show that the inequality

$$\rho_0\varkappa(\zeta_2-\zeta_1)\geqslant(M_1+\rho_0)\,q(r)\,|w_2-w_1| \tag{15.58}$$

holds for sufficiently small r, uniformly with respect to the difference $w_2-w_1\in K(\rho_0)$. But for all $w\in K(\rho_0)$, we have the inequality $(w, e_1)\geqslant m(\rho_0)\,|w|$, where $m(\rho_0)>0$. So (15.58), in turn, is valid provided

$$\rho_0\varkappa m(\rho_0)\geqslant(M_1+\rho_0)\,q(r).$$

Finally, we have that $S(\varepsilon)D(\varepsilon)$ is monotone on $V(r)$ provided

$$q(r)\leqslant\frac{\rho_0\varkappa m(\rho_0)}{M_1+\rho_0}. \tag{15.59}$$

The cases $\alpha_0>0$ and $\alpha_0<0$ must now be handled separately.

Let $\alpha_0>0$. Then, for sufficiently small ρ_1 and r, the operator $S(\varepsilon)D(\varepsilon)$ has the following property in $\widehat{K}(\rho_1, r)$: if $w_0\in\widehat{K}_{\text{in}}(\rho_1, r)$ and $\beta\in(0,1)$, there exists $\eta=\eta(w_0,\beta)>0$ such that

$$S(\varepsilon)\,D(\varepsilon)\,(\beta w_0)\leqslant\beta(1-\eta)\,S(\varepsilon)\,D(\varepsilon)\,w_0. \tag{15.60}$$

To prove this, we need only check that the elements

$$w=\beta S(\varepsilon)D(\varepsilon)\,w_0-S(\varepsilon)D(\varepsilon)\,(\beta w_0) \tag{15.61}$$

are interior points of the cone $\widehat{K}(\rho_0)$. The simple calculation called for is tedious, however, and is omitted here.

Suppose that for sufficiently small ε equation (15.45) has two distinct nontrivial ap-solutions in $\widehat{K}(\rho_1)$. As shown in subsection 15.5, these solutions w_1 and w_2 are then comparable, say $w_1\leqslant w_2$. Let β_0 be the smallest β such that $w_1\leqslant\beta w_2$. Since w_2-w_1 is an interior point of the cone $\widehat{K}(\rho_1)$, we have $\beta_0<1$. Thus, by (15.60),

$$S(\varepsilon)\,D(\varepsilon)\,(\beta_0 w_2)\leqslant\beta_0(1-\eta)\,S(\varepsilon)\,D(\varepsilon)\,w_2=\beta_0(1-\eta)\,w_2$$

and by the monotonicity of $S(\varepsilon)D(\varepsilon)$,

$$w_1=S(\varepsilon)\,D(\varepsilon)\,w_1\leqslant S(\varepsilon)\,D(\varepsilon)\,(\beta_0 w_2)\leqslant\beta_0(1-\eta)\,w_2,$$

contradicting the minimality of β_0.

Consequently, for small ε, equation (15.43) can have at most one nontrivial ap-solution in the cone $\widehat{K}(\rho_1)$.

A similar argument proves uniqueness for $\alpha_0 < 0$. Here (15.60) is replaced by the inequality

$$S(\varepsilon) D(\varepsilon) (\beta w_0) \geqslant \beta (1+\eta) S(\varepsilon) D(\varepsilon) w_0.$$

Then, if $w_1 = S(\varepsilon) D(\varepsilon) w_1$ and $w_2 = S(\varepsilon) D(\varepsilon) w_2$, $w_1 \neq w_2$ and $w_1 \leqslant w_2$, we let β_0 be the largest β such that $\beta w_2 \leqslant w_1$. Then the inequality

$$w_1 = S(\varepsilon) D(\varepsilon) w_1 \geqslant S(\varepsilon) D(\varepsilon) (\beta_0 w_2) \geqslant \geqslant \beta_0 (1+\eta) S(\varepsilon) D(\varepsilon) w_2 = \beta_0 (1+\eta) w_2$$

contradicts the maximality of β_0.

In concluding this subsection we note that equation (15.43) can have at most one nontrivial ap-solution for small ε in the cones $-\hat{K}(\rho_1)$ as well. To see this, merely substitute $v = -w$ for w in (15.43).

15.7. End of the proof of Theorem 15.1

Let us begin with part 1. Equation (15.16), upon the substitution (15.42)(with $\varepsilon = 0$), takes the form

$$\frac{dw}{d\tau} = C_0(0) w + \Phi_0(w; 0) + \overline{\Omega}(w; 0), \tag{15.62}$$

where

$$\overline{\Omega}(w; 0) = \lim_{T \to \infty} \frac{1}{T} \int_0^T \Omega(t, w; 0)\, dt.$$

By the arguments of subsection 15.5, for any $\rho > 0$ one can find $r = r(\rho) > 0$ such that the small nontrivial ap-solutions of equation (15.62) lie entirely either in $K(\rho)$ or in $-K(\rho)$. Therefore, part 1 will be proved if we show that, for small ρ and r, no nontrivial ap-solution of equation (15.62) lies entirely in $K(\rho, r)$ or $-K(\rho, r)$.

Suppose the contrary. To fix ideas, we assume that for small ρ and r there is at least one nontrivial ap-solution $w(\tau)$ lying entirely in $K(\rho, r)$. Consider the scalar almost periodic function

$$v(\tau) = (w(\tau), e_1) \qquad (-\infty < \tau < \infty). \tag{15.63}$$

Clearly,

$$\frac{dv(\tau)}{d\tau} = (\Phi_0[w(\tau); 0], e_1) + (\overline{\Omega}[w(\tau); 0], e_1).$$

We write $w(\tau)$ in the form $w(\tau)=\zeta(\tau)e_1+x(\tau)$, where $(x(\tau),e_1)=0$. Then

$$\frac{dv(\tau)}{d\tau}=\alpha_0\zeta^m(\tau)+(\Phi_0[\zeta(\tau)e_1+x(\tau);0]-\Phi_0[\zeta(\tau)e_1;0],e_1)+\\ +(\bar{\Omega}[w(\tau);0],e_1).$$

Since $\alpha_0\neq 0$, it follows that for sufficiently small ρ and r

$$\operatorname{sign}\frac{dv(\tau)}{d\tau}=\operatorname{sign}\alpha_0\qquad(-\infty<\tau<\infty).$$

Therefore $v(\tau)$ is strictly monotone, contradicting the fact that it is almost periodic.

We now prove part 2, examining the case that

$$\alpha_0\cdot a_{j_0}>0. \tag{15.64}$$

Part 2 will be proved if we show that equation (15.41) has no nontrivial ap-solutions in $K(\rho,r)$ for small ρ and r and for sufficiently small ε. Indeed, the results of subsection 15.5 imply that all small nontrivial ap-solutions lie in $-K(\rho)$; moreover, by the results of subsection 15.4, at least one such solution exists in $-K(\rho)$, and its uniqueness was established in subsection 15.6.

If we assume that (15.43) has small nontrivial ap-solutions $w(\tau;\varepsilon)$ lying in $K(\rho,r)$, then, corresponding to each of these, we have a scalar function

$$v(\tau;\varepsilon)=(w(\tau;\varepsilon),e_1)\qquad(-\infty<\tau<\infty).$$

This function is almost periodic. On the other hand, it is easy to check that it is monotone (by the method used in proving that (15.63) is monotone), contradicting the almost periodicity.

If the inequality

$$\alpha_0\cdot a_{j_0}<0$$

holds rather than (15.64), then equations (15.43) have no small nontrivial ap-solutions in $-K(\rho,r)$.

Let us now prove part 3. Under the hypotheses of part 3, two families of ap-solutions, $w_1(\tau;\varepsilon)$ and $w_2(\tau;\varepsilon)$, were constructed in subsection 15.4. The solutions in the first family lie entirely within balls of radius $r(\varepsilon)$ as given by (15.40), centered at $\beta(\varepsilon)e(\varepsilon)$, where $\beta(\varepsilon)$ is given by (15.24). Those in the second family all lie within balls of the same radius centered at $-\beta(\varepsilon)e(\varepsilon)$. It is easy to see that these balls do not intersect for small ε. Thus the solutions constructed in subsection 15.4 are distinct.

To prove part 4, we repeat the argument used in completing the proof of part 2. This completes the proof of Theorem 15.1.

15.8. Pendulum with vibrating point of suspension

As an example, consider the equation

$$\ddot{\theta} + 2\alpha\varepsilon\dot{\theta} + [\varepsilon^2 k^2 - \varepsilon p''(t)]\sin\theta = 0 \tag{15.65}$$

describing the oscillations of a pendulum with vibrating point of suspension. Here $\alpha > 0$, $p(t)$ is a trigonometric polynomial with zero mean and ε is a positive parameter. For small ε, the lower equilibrium position ($\theta = 0$) is always stable. The upper equilibrium position ($\theta = \pi$) may be stable or unstable, depending on $p(t)$, as shown by Bogolyubov /1/ and Kapitsa /1/ (see also Landau and Lifshits /1/).

We are interested in the case that ap-solutions of equation (15.65) originate from the upper equilibrium position.

In order to employ Theorem 15.1, we first need to transform (15.65) to the form of (15.8).

Toward this end, we replace the unknown function θ by two unknown functions x_1 and x_2 via Bogolyubov's substitution

$$\left.\begin{aligned} \theta &= x_1 + \pi - \varepsilon p(t)\sin x_1, \\ \dot{\theta} &= \varepsilon x_2 - \varepsilon p'(t)\sin x_1. \end{aligned}\right\} \tag{15.66}$$

Equation (15.65) thus becomes a system of two equations

$$x_1' = \frac{\varepsilon x_2}{1 - \varepsilon p(t)\cos x_1}, \tag{15.67}$$

$$\begin{aligned} x_2' = p''(t)\{\sin x_1 - \sin[x_1 - \varepsilon p(t)\sin x_1]\} &+ \\ + \frac{\varepsilon p'(t)\, x_2 \cos x_1}{1 - \varepsilon p(t)\cos x_1} - 2\alpha\varepsilon x_2 + 2\alpha\varepsilon p'(t)\sin x_1 &+ \\ + \varepsilon k^2 \sin[x_1 - \varepsilon p(t)\sin x_1]&. \end{aligned} \tag{15.68}$$

Expanding the right-hand sides in powers of ε and introducing vector notation, we obtain the equation

$$\frac{dx}{dt} = \varepsilon A_1(t)\,x + \varepsilon^2 A_2(t)\,x + \ldots + \varepsilon F(t, x) + \ldots, \tag{15.69}$$

where

$$A_1(t) = \begin{Vmatrix} 0 & 1 \\ p(t)\,p''(t) + 2\alpha p'(t) + k^2 & p'(t) - 2\alpha \end{Vmatrix},$$

$$A_2(t) = \begin{Vmatrix} 0 & p(t) \\ -\,p(t)\,k^2 & p'(t)\,p(t) \end{Vmatrix},$$

$$F(t, x) = \\ = \left\{0, -\frac{1}{6}[k^2 + 4p(t)p''(t) + 2\alpha p'(t)]x_1^3 - \frac{1}{2}p'(t)x_1^2x_2\right\}.$$

Using the substitutions (14.29) and (15.12), we pass from equation (15.69) to the equation

$$\frac{dz}{dt} = \varepsilon B_1 z + \varepsilon^2 B_2 z + \ldots + \varepsilon \bar{F}(z) + \ldots, \tag{15.70}$$

where

$$B_1 = \begin{Vmatrix} 0 & 1 \\ k^2 - M[(p')^2] & -2\alpha \end{Vmatrix}, \quad B_2 = \begin{Vmatrix} 0 & 0 \\ 2M[p(p')^2] & 0 \end{Vmatrix},$$

$$\bar{F}(z) = \left\{0, \frac{2}{3}M[(p')^2]z_1^3 - \frac{1}{6}k^2z_1^3\right\}.$$

Clearly, when $M[(p')^2] > k^2$, the upper equilibrium position is stable. If $M[(p')^2] < k^2$, then the upper equilibrium position is unstable. In the latter situation, for sufficiently small ε, there are no almost periodic oscillations in the neighborhood of the upper equilibrium position.

We are interested in what happens when

$$M[(p')^2] = k^2. \tag{15.71}$$

In this case, in order to examine the stability of the upper equilibrium position, we must construct (see subsection 14.5) the sequence of matrices

$$B_1 + \varepsilon B_2, \quad B_1 + \varepsilon B_2 + \varepsilon^2 B_3, \quad B_1 + \varepsilon B_2 + \varepsilon^2 B_3 + \varepsilon^3 B_4, \ldots \tag{15.72}$$

continuing until one of them is nonsingular for small nonzero ε. Thus, for example, if $M[p(p')^2] \neq 0$, this will be the matrix $B_1 + \varepsilon B_2$. The stability of the trivial solution depends on the first nonsingular matrix in (15.72). For example, if

$$M[p(p')^2] < 0, \tag{15.73}$$

then the upper equilibrium position is stable, but if

$$M[p(p')^2] > 0, \tag{15.74}$$

then it is unstable. Henceforth we shall assume that one of the matrices (15.72) is nonsingular for small nonzero ε.

Theorem 15.1 implies

Theorem 15.2. Let condition (15.71) hold.

Then:

1. If the upper equilibrium position of the pendulum is unstable, there are no ap-oscillations in its neighborhood for small positive ε.

2. If the upper equilibrium position is stable, then in each of its neighborhoods there are, for small ε, exactly two ap-oscillations, both unstable.

A more detailed analysis shows that under the hypotheses of part 2, there exists a circular neighborhood $V(r_0)$ of the equilibrium position (in the phase plane) with the property: for any other neighborhood $V(r)$ there exists $\varepsilon_0 = \varepsilon_0(r) > 0$ such that for $0 < \varepsilon < \varepsilon_0$ each of equations (15.65) has solutions with initial values in $V(r)$, and as t increases these solutions leave $V(r_0)$. This means that if (15.71) holds the upper equilibrium position is practically always unstable.

15.9. Further remarks

In the theory of equations with a small parameter, significant progress has been made in regard to extending ap-solutions with respect to the parameter in the case that linearization leads to an equation with a regular ap-operator. This case has been described in detail in several books (see, for example, Demidovich /1/, Zubov /1/, Hale /1/). The case in which the limiting equation with regular ap-operator is obtained by averaging has also been examined thoroughly (see Bogolyubov and Mitropol'skii /1/).

Branching of ap-solutions has been examined only for very special classes of equations. We mention here the sophisticated investigations of Moser /1, 2, 4/, Burd /1/, and Burd and Sabirov /1/.

§ 16. BIFURCATION OF AP-SOLUTIONS OF SINGULARLY PERTURBED SECOND-ORDER EQUATIONS*

16.1. Statement of the problem. Necessary condition

In this section we examine the problem of small nontrivial ap-solutions of the scalar equation

$$\varepsilon \frac{d^2x}{dt^2} + \frac{dx}{dt} + q(t)\, x = a(t)\, x^n + \omega(t, x), \tag{16.1}$$

* Burd, Kolesov, Krasnosel'skii /4/.

where ε is a small positive parameter, $q(t) \in B^2(R^1)$ $a(t) \in B^1(R^1)$, n is a positive integer, $n \geqslant 2$, $\omega(t, x)$ is almost periodic in t, $\omega(t, 0) \equiv 0$ and

$$|\omega(t, x_1) - \omega(t, x_2)| \leqslant \varkappa(r) |x_1 - x_2| \quad (|x_1|, |x_2| \leqslant r), \qquad (16.2)$$

where

$$\lim_{r \to 0} r^{-n+1} \varkappa(r) = 0. \qquad (16.3)$$

Equation (16.1) describes the oscillations of a small mass suspended on an elastic element with nonlinear restoring force. Let us first find necessary conditions for the existence of small nontrivial ap-solutions of equation (16.1) for small ε. Suppose such solutions $x(t; \varepsilon)$ exist; the function $x(t; \varepsilon)$ converges uniformly to zero as $\varepsilon \to 0$. Consider the function

$$y(t; \varepsilon) = \frac{x(t + t_\varepsilon; \varepsilon)}{\| x(t; \varepsilon) \|_{B(R^1)}},$$

where t_ε is chosen so that

$$|y(0; \varepsilon)| > 1/2. \qquad (16.4)$$

The function $y(t; \varepsilon)$ is a solution of the equation

$$\varepsilon \frac{d^2 y}{dt^2} + \frac{dy}{dt} = -q(t + t_\varepsilon) y + f(t; \varepsilon), \qquad (16.5)$$

where $f(t; \varepsilon)$ is almost periodic with respect to t and converges to zero uniformly with respect to t as $\varepsilon \to 0$. By (16.5),

$$\frac{dy(t; \varepsilon)}{dt} = \frac{1}{\varepsilon} \int_{-\infty}^{t} e^{-\frac{1}{\varepsilon}(t-s)} [f(s; \varepsilon) - q(s + t_\varepsilon) y(s; \varepsilon)] \, ds,$$

and thus

$$\left\| \frac{dy(t; \varepsilon)}{dt} \right\|_{B(R^1)} \leqslant c, \qquad (16.6)$$

where the constant c does not depend on ε. By (16.6), there exists a sequence $\varepsilon_j \to 0$ such that the ap-functions $y_j(t) = y(t; \varepsilon_j)$ converge uniformly on any finite interval to a function $u(t)$ which is bounded on the whole real line and not identically zero (by (16.4)), and the ap-functions $q(t + t_{\varepsilon_j})$ converge uniformly on $(-\infty, \infty)$ to an ap-function $q^*(t)$. The right-hand side of the equality

$$\frac{dy_j(t)}{dt} = \frac{1}{\varepsilon_j} \int_{-\infty}^{t} e^{-\frac{1}{\varepsilon_j}(t-s)} \left[f(s;\, \varepsilon_j) - q(s + t_{\varepsilon_j})\, y_j(s)\right] ds,$$

clearly converges to $-q^*(t)u(t)$ uniformly on every finite interval. Thus, the derivatives $y_j'(t)$ converge to $u'(t)$ and therefore $u(t)$ is a solution, bounded on the real line, of the differential equation

$$\frac{dx}{dt} + q^*(t)\, x = 0. \tag{16.7}$$

Thus, the ap-operator on the left of (16.7) is not regular, implying (see subsection 4.4) that $M(q^*) = 0$. Hence

$$M(q) = \lim_{T \to \infty} \frac{1}{T} \int_0^T q(s)\, ds = 0. \tag{16.8}$$

We have thus proved:

Equality (16.8) is a necessary condition for the existence of small nontrivial ap-solutions of equation (16.1) for small ε.

16.2. Fundamental theorem

In what follows we assume that (16.8) holds and also that $q(t) \not\equiv 0$. Suppose also that

$$\int_0^t q(s)\, ds \in B(R^1) \tag{16.9}$$

and

$$\int_0^t [q^2(s) - M(q^2)]\, ds \in B(R^1). \tag{16.10}$$

Set

$$b(t) = a(t) \exp\left[-(n-1) \int_0^t q(s)\, ds\right]; \tag{16.11}$$

we assume that $M(b) \neq 0$ and

$$\int_0^t [b(s) - M(b)]\, ds \in B(R^1). \tag{16.12}$$

If two ap-functions $x_1(t)$ and $x_2(t)$ satisfy either $x_1(t) \geqslant x_2(t)$ or $x_1(t) \leqslant x_2(t)$, we say that they are comparable.

Theorem 16.1. There exist $\varepsilon_0 > 0$ *and* $r_0 > 0$ *such that*

1. For $0 < \varepsilon < \varepsilon_0$, *the trivial solution of equation (16.1) is exponentially stable.*

2. If n *is even, then for* $0 < \varepsilon < \varepsilon_0$ *equation (16.1) has at least one nontrivial unstable ap-solution in the ball* $V(r_0)$. *All ap-solutions* $x(t; \varepsilon)$ *belonging to* $V(r_0)$ *are comparable and satisfy the inequality*

$$M(b)x(t; \varepsilon) > 0 \qquad (-\infty < t < \infty). \tag{16.13}$$

3. If n *is odd and* $M(b) > 0$, *then for* $0 < \varepsilon < \varepsilon_0$ *equation (16.1) has at least one positive and at least one negative unstable solution in the ball* $V(r_0)$. *All solutions in* $V(r_0)$ *are of constant sign and comparable.*

4. If n *is odd and* $M(b) < 0$, *then for* $0 < \varepsilon < \varepsilon_0$ *equation (16.1) has no nontrivial ap-solutions in the ball* $V(r_0)$.

The proof occupies the next three subsections.

16.3. Proof of part 1

By Theorem 11.1, it suffices to show that, for small ε, the ap-operator

$$L(\varepsilon)x = \varepsilon \frac{d^2x}{dt^2} + \frac{dx}{dt} + q(t)x \tag{16.14}$$

is regular and stable. In order to apply Theorem 9.8, we must show that the solutions of the equation

$$\frac{d^2x}{dt^2} + \frac{1}{\varepsilon}\frac{dx}{dt} + \frac{q(t)}{\varepsilon}x = 0 \tag{16.15}$$

are nonoscillatory and that there exists an ap-function $v(t; \varepsilon) \gg 0$ satisfying the inequality

$$L(\varepsilon)v(t; \varepsilon) \geqslant \gamma_0 \varepsilon v(t; \varepsilon), \tag{16.16}$$

where $\gamma_0 > 0$.

To investigate equation (16.15), let us substitute

$$x = e^{-t/2\varepsilon}y.$$

The equation then takes the form

$$\frac{d^2y}{dt^2} + \left[\frac{q(t)}{\varepsilon} - \frac{1}{4\varepsilon^2}\right]y = 0.$$

The solutions of this equation are nonoscillatory for ε such that

$$\frac{q(t)}{\varepsilon} - \frac{1}{4\varepsilon^2} < 0 \qquad (-\infty < t < \infty).$$

Define a function $v(t; \varepsilon)$ by

$$v(t; \varepsilon) = u_1(t) + \varepsilon u_2(t), \tag{16.17}$$

where

$$u_1(t) = \exp\left[-\int_0^t q(s)\,ds\right] \tag{16.18}$$

and $u_2(t)$ is some fixed ap-solution of the equation

$$\frac{dx}{dt} + q(t)x = M(q^2)u_1(t) - \frac{d^2u_1(t)}{dt^2}. \tag{16.19}$$

It follows from (16.9) that $u_1(t)$ is almost periodic, and from (16.10) that all solutions of equation (16.19) are almost periodic. A simple check shows that

$$L(\varepsilon)v(t; \varepsilon) = \varepsilon M(q^2)u_1(t) + \varepsilon^2 \frac{d^2u_2(t)}{dt^2}.$$

Since $u_1(t)$ is strictly positive and $\frac{d^2u_2(t)}{dt^2}$ is almost periodic, this inequality implies the validity of (16.16) for small ε.

16.4. Existence of nontrivial ap-solutions in parts 2 and 3

We confine ourselves to proving the existence of a small nontrivial ap-solution under the hypotheses of part 2. The proof is basically the same as in subsection 15.1.

Set

$$\alpha = \left[\frac{M(q^2)}{M(b)}\right]^{1/(n-1)}$$

and let $v_1(t)$ denote an ap-solution of the equation

$$\frac{dx}{dt} + q(t)x = \alpha^n a(t)u_1^n(t) - \alpha\frac{d^2u_1(t)}{dt^2},$$

where $u_1(t)$ is given by (16.18). The existence of $v_1(t)$ is guaranteed by (16.10) and (16.12).

The substitution

$$x = \alpha\varepsilon^{1/(n-1)}u_1(t) + \varepsilon^{n/(n-1)}v_1(t) + y \tag{16.20}$$

puts equation (16.1) in the form

$$L_1(\varepsilon)y = D(t, y; \varepsilon), \tag{16.21}$$

where

$$\begin{aligned} D(t, y; \varepsilon) = \varepsilon^{(2n-1)/(n-1)}\varphi_1(t; \varepsilon) + \varepsilon^{(n-2)/(n-1)}\varphi_2(t; \varepsilon)y^2 + \dots \\ \dots + \varepsilon^{1/(n-1)}\varphi_{n-1}(t; \varepsilon)y^{n-1} + a(t)y^n + \\ + \omega[t, \alpha\varepsilon^{1/(n-1)}u_1(t) + \varepsilon^{n/(n-1)}v_1(t) + y] \end{aligned} \tag{16.22}$$

and

$$\begin{aligned} L_1(\varepsilon)y = \varepsilon\frac{d^2y}{dt^2} + \frac{dy}{dt} + \\ + \{q(t) - na(t)[\alpha\varepsilon^{1/(n-1)}u_1(t) + \varepsilon^{n/(n-1)}v_1(t)]^{n-1}\}y, \end{aligned} \tag{16.23}$$

the functions $\varphi_i(t; \varepsilon)$ $(i = 1, \dots, n-1)$ are almost periodic in t and continuous in ε $(0 \leqslant \varepsilon \leqslant \varepsilon_0)$ uniformly with respect to t.

Let us show that the ap-operators (16.23) are regular for small nonzero ε and that their inverses satisfy the estimates

$$\| L_1^{-1}(\varepsilon)\|_{B(R^1)} \leqslant \frac{c}{\varepsilon}. \tag{16.24}$$

Let $w(t; \varepsilon) = u_1(t) + \varepsilon w_1(t)$, where $u_1(t)$ is given by (16.18) and $w_1(t)$ is some ap-solution of the equation

$$\frac{dx}{dt} + q(t)x = -(n-1)M(q^2)u_1(t) - \frac{d^2u_1(t)}{dt^2} + n\alpha^{n-1}a(t)u_1^{n-1}(t).$$

The existence of $w_1(t)$ follows from (16.10) and (16.12). It is easily verified that

$$L_1(\varepsilon)w(t; \varepsilon) = -\varepsilon(n-1)M(q^2)u_1(t) + o(\varepsilon).$$

Therefore, for sufficiently small ε,

$$L_1(\varepsilon)w(t; \varepsilon) \leqslant -\varepsilon\gamma w(t; \varepsilon) \qquad (-\infty < t < \infty), \tag{16.25}$$

where $\gamma > 0$. Since $w(t; \varepsilon) \gg 0$ for small ε, Theorem 9.9 implies that the operators $L_1(\varepsilon)$ are regular and that their Green's functions are negative. Thus, (16.25) implies the inequality

$$-L_1^{-1}(\varepsilon)\,w(t;\varepsilon) \leqslant \frac{1}{\varepsilon\gamma} w(t;\varepsilon) \qquad (-\infty < t < \infty),$$

from which the estimate (16.24) follows.

Since the operators (16.23) are regular, we may pass from the differential equation (16.21) to the integral equation

$$y = \Pi(\varepsilon)\,y, \tag{16.26}$$

where

$$\Pi(\varepsilon)\,y(t) = L_1^{-1}(\varepsilon)\,D[t, y(t); \varepsilon]. \tag{16.27}$$

A simple check shows that (16.24) implies the estimate

$$\|\Pi(\varepsilon)\,0\|_{B(R^1)} = o(\varepsilon^{1/(n-1)}).$$

Therefore, there exists a function $\beta(\varepsilon)$, continuous for $\varepsilon \geqslant 0$, positive for $\varepsilon > 0$, $\beta(0) = 0$, such that

$$\|\Pi(\varepsilon)\,0\|_{B(R^1)} \leqslant \varepsilon^{1/(n-1)}\beta(\varepsilon).$$

Then, by (16.24), the operator (16.27) satisfies the hypotheses of the contracting-mapping principle in a ball $V(r)$ of radius

$$r = \varepsilon^{1/(n-1)}\sqrt{\beta(\varepsilon)}.$$

Thus, equation (16.21) has an ap-solution $y(t;\varepsilon)$ in $V(r)$. By (16.20), there is a corresponding ap-solution of constant sign for equation (16.1), whose sign coincides with that of α or, equivalently, of $M(b)$.

It was shown above that theGreen's functions of the operators $L_1(\varepsilon)$ are negative. It follows from Theorem 9.9 that the operators $L_1(\varepsilon)$ are unstable. Then, Theorem 11.2, yields that the solutions established above are unstable.

16.5. End of the proof of Theorem 16.1

We first show that for small $\varepsilon > 0$ every small nontrivial ap-solution $x(t;\varepsilon)$ is of constant sign. To this end, we define a cone $K(N)$ in the $\{x, x'\}$ plane by the inequalities $x \geqslant 0$ and $Nx + x' \geqslant 0$, where N is so chosen that $\varepsilon_0 N^2 - N - q(t) \leqslant -c < 0$ (we assume that $\varepsilon_0 < 1$). We may assume that $x(0;\varepsilon) = \sup_t x(t;\varepsilon)$. If $x(0;\varepsilon) > 0$, then $\{x(0;\varepsilon),\ x'(0;\varepsilon)\} \in K(N)$ and so $\{x(t;\varepsilon),\ x'(t;\varepsilon)\} \in K(N)$ for all t, i. e., $x(t;\varepsilon) > 0$.

To prove the comparability of any two small ap-solutions $x_1(t;\varepsilon)$ and $x_2(t;\varepsilon)$, it suffices to substitute $x = x_1(t;\varepsilon) + y$ in equation (16.1) and to repeat the argument of the previous paragraph.

It remains to prove that under the hypotheses of part 2 equation (16.1) has no ap-solutions not satisfying (16.13), and that under the hypotheses of part 4 there are no small nontrivial ap-solutions. As the proofs of these two propositions are similar, we prove only part 4.

By (16.12), all solutions of the equation

$$\frac{dx}{dt} + q(t)x = [b(t) - M(b)]\, e^{-\int\limits_0^t q(s)\,ds}$$

are almost periodic. Let $g(t)$ be such a solution. Set

$$z(t;\tau;\varepsilon) = \tau v(t\cdot\varepsilon) + \tau^n g(t), \tag{16.28}$$

where $v(t;\varepsilon)$ is the strictly positive ap-function (16.17), and τ is a scalar variable. Clearly,

$$L(\varepsilon) z(t;\tau;\varepsilon) = \tau L(\varepsilon) v(t;\varepsilon) + \tau^n L(\varepsilon) g(t),$$

and, therefore, we have by (16.16) that for positive τ,

$$L(\varepsilon) z(t;\tau;\varepsilon) \geqslant \tau\gamma_0\varepsilon v(t;\varepsilon) + \\ + \tau^n \frac{d^2 g(t)}{dt^2} + \tau^n [b(t) - M(b)] \exp\left[-\int\limits_0^t q(s)\,ds\right].$$

On the other hand,

$$a(t) z^n(t;\tau;\varepsilon) + \omega[t, z(t;\tau;\varepsilon)] = \\ = \tau^n b(t) \exp\left[-\int\limits_0^t q(s)\,ds\right] + o(\tau)[\varepsilon + o(\varepsilon)] + o(\tau^n).$$

Hence,

$$L(\varepsilon) z(t;\tau;\varepsilon) \geqslant \\ \geqslant a(t) z^n(t;\tau;\varepsilon) + \omega[t, z(t;\tau,\varepsilon)] + \tau\gamma_0\varepsilon v(t;\varepsilon) = \\ = \tau^n M(b) \exp\left[-\int\limits_0^t q(s)\,ds\right] + o(\tau)[\varepsilon + o(\varepsilon)] + o(\tau^n).$$

Since $M(b) < 0$, by assumption this last inequality implies the existence of positive ε_0 and τ_0 such that for $0 < \varepsilon < \varepsilon_0$, $0 < \tau \leqslant \tau_0$

$$L(\varepsilon) z(t;\tau;\varepsilon) \gg a(t) z^n(t;\tau;\varepsilon) + \omega[t, z(t;\tau;\varepsilon)]. \tag{16.29}$$

Let r_0 be so small that the ball $V(r_0)$ is contained in all conic intervals $\langle -z(t;\tau_0;\varepsilon),\ z(t;\tau_0;\varepsilon)\rangle$ for $0<\varepsilon<\varepsilon_0$.

Suppose that for some $\varepsilon^* \in (0,\varepsilon_0)$ equation (16.1) has an ap-solution $x(t;\varepsilon^*) \in V(r_0)$ which is positive for some t. Let τ^* be the smallest τ such that

$$x(t;\ \varepsilon^*) \leqslant z(t;\tau;\varepsilon^*) \qquad (-\infty<t<\infty). \tag{16.30}$$

Clearly, $0<\tau^*\leqslant\tau_0$.

Consider the ap-operator

$$L_2(\varepsilon)x = L(\varepsilon)x + x. \tag{16.31}$$

An argument as in subsection 16.3 readily shows that for small ε the operators (16.31) satisfy the hypotheses of Theorem 9.8. Hence they are regular and their Green's functions are positive. We may assume that this is the case for all $\varepsilon \in (0,\varepsilon_0]$.

Consider the integral operator

$$\Pi_* x(t) = L_2^{-1}(\varepsilon^*)\{x(t) + a(t)x^n(t) + \omega[t, x(t)]\}$$

on the ball $V(r_*)$ of radius

$$r_* = \sup_{0<\varepsilon<\varepsilon_0;\ 0<\tau\leqslant\tau_0} \left[\sup_{-\infty<t<\infty} z(t;\tau;\varepsilon)\right].$$

If ε_0 and τ_0 are sufficiently small, Π_* is monotone on $V(r_*)$.

Clearly,

$$x(t;\varepsilon^*) = \Pi_* x(t;\varepsilon^*)$$

and, by the monotonicity of Π_*,

$$x(t;\varepsilon^*) \leqslant \Pi_* z(t;\tau^*;\varepsilon^*).$$

By (16.29) we have

$$z(t;\tau^*;\varepsilon^*) \gg \Pi_* z(t;\tau^*;\varepsilon^*).$$

Therefore

$$x(t;\varepsilon^*) \ll z(t;\tau^*;\varepsilon^*),$$

contradicting our choice of τ^*.

We have shown that, under the hypotheses of part 4, equation (16.1) can have no small ap-solutions which are positive at certain points.

If equation (16.1) had nontrivial nonpositive small ap-solutions, then the equation

$$L(\varepsilon)x = a(t)x^n - \omega(t, -x)$$

would have nontrivial and nonnegative small ap-solutions. But this is impossible by the proposition just proved, completing the proof of Theorem 16.1.

16.6. Functions $a(t)$ of constant sign

If the coefficients $q(t)$ and $a(t)$ and the function $\omega(t, x)$ are periodic with period T_0, then, rather than seeking ap-solutions of equation 16.1, it is natural to seek -periodic solutions. The construction of these solutions is similar to that of ap-solutions. Note that, under the hypotheses of part 2 of Theorem 16.1, equation (16.1) has exactly one nontrivial small T_0-periodic solution, and, under the hypotheses of part 3, exactly two such solutions. We have been able to prove this proposition in the general case only under the further assumption that $a(t) \gg 0$ or $a(t) \ll 0$ (the constructions of § 13, for example, maybe used in proving this).

16.7. General equation

The technique developed in this section allows us to examine small ap-solutions of equations more general than (16.1):

$$\varepsilon[m_0(t) + m(t; \varepsilon)]\frac{d^2x}{dt^2} + \frac{dx}{dt} + [q_0(t) + \varepsilon q_1(t) + \varepsilon q(t; \varepsilon)]x = \\ = [a_0(t) + a(t; \varepsilon)]x^n + \omega(t, x; \varepsilon). \tag{16.32}$$

Let us assume that all functions appearing in the equation are almost periodic in t. Let

$$m_0(t),\ q_1(t),\ a_0(t) \in B^1(R^1), \quad q_0(t) \in B^2(R^1). \tag{16.33}$$

We assume that $m(t; \varepsilon)$, $q(t; \varepsilon)$ and $a(t; \varepsilon)$ converge to zero uniformly in t as $\varepsilon \to 0$. The remainder term $\omega(t, x; \varepsilon)$ has the same properties as $\omega(t, x)$ in (16.1). We also assume that

$$m_0(t) \gg 0. \tag{16.34}$$

A necessary condition for the existence of small ap-solutions for (16.32), analogous to condition (16.8) for equation (16.1), is

$$M(q_0) = 0. \tag{16.35}$$

We assume henceforth that this condition holds.

Consider the functions

$$b_0(t) = a_0(t) \exp\left[-(n-1)\int_0^t q_0(s)\,ds\right] \tag{16.36}$$

and

$$\lambda_0(t) = q_1(t) - m_0(t)\frac{dq_0(t)}{dt} + m_0(t)\,q_0^2(t). \tag{16.37}$$

Let us assume that the following conditions are valid:

$$\int_0^t q_0(s)\,ds \in B(R^1), \tag{16.38}$$

$$\int_0^t [b_0(s) - M(b_0)]\,ds \in B(R^1) \tag{16.39}$$

and

$$\int_0^t [\lambda_0(s) - M(\lambda_0)]\,ds \in B(R^1) \tag{16.40}$$

(compare (16.9), (16.10) and (16.12)). In studying equation (16.1), we assumed that $M(b) \neq 0$ and $q(t) \not\equiv 0$; now the appropriate conditions are

$$M(b_0) \neq 0 \tag{16.41}$$

and

$$M(\lambda_0) \neq 0. \tag{16.42}$$

By the method used in proving Theorem 16.1, we may obtain the following more general proposition.

Theorem 16.2. There exist $\varepsilon_0 > 0$ *and* $r_0 > 0$ *with the following properties:*

1. For $0<\varepsilon<\varepsilon_0$, the trivial solution of equation (16.32) is exponentially stable if

$$M(\lambda_0)>0 \tag{16.43}$$

and unstable if

$$M(\lambda_0)<0. \tag{16.44}$$

2. If n is even, then for $0<\varepsilon<\varepsilon_0$ equation (16.32) has at least one nontrivial ap-solution (exponentially stable if $M(\lambda_0)<0$, unstable if $M(\lambda_0)>0$) in the ball $V(r_0)$. All other nontrivial solutions in $V(r_0)$ are comparable and satisfy the inequality

$$M(\lambda_0)\,M(b_0)\,x(t;\varepsilon)>0 \qquad (-\infty<t<\infty). \tag{16.45}$$

3. If n is odd and

$$M(\lambda_0)M(b_0)>0, \tag{16.46}$$

then for $0<\varepsilon<\varepsilon_0$ equation (16.32) has at least one positive and at least one negative ap-solution in the ball $V(r_0)$. These ap-solutions are exponentially stable if $M(\lambda_0)<0$ and unstable if $M(\lambda_0)>0$. All solutions in $V(r_0)$ are of constant sign and pairwise comparable.

4. If n is odd and

$$M(\lambda_0)M(b_0)<0, \tag{16.47}$$

then for $0<\varepsilon<\varepsilon_0$ equation (16.61) has no nontrivial ap-solutions in $V(r_0)$.

16.8. Example

Suppose a pendulum of slowly varying length $l(\varepsilon t)$ is oscillating in the horizontal plane. Its position is defined by the angle θ between the pendulum and some fixed direction. If the suspended mass is equal to 1, the equation describing its oscillation is

$$\frac{d}{dt}\left[l^2(\varepsilon t)\frac{d\theta}{dt}\right]+2\chi\frac{d}{dt}\left[l(\varepsilon t)\,\theta\right]=0,$$

where the friction coefficient χ is positive.

Suppose the point of suspension of the pendulum oscillates along the straight line defined by the above-mentioned direction according to the law $\frac{1}{\varepsilon} p(\varepsilon t)$. Then (Mitropol'skii /1/) θ satisfies the equation

$$\frac{d}{dt}\left[l^2(\varepsilon t)\frac{d\theta}{dt}\right]+2\chi\frac{d}{dt}[l(\varepsilon t)\theta]+\varepsilon p''(\varepsilon t)\,l(\varepsilon t)\sin\theta=0.$$

Setting $\tau=\varepsilon t$, we rewrite this equation

$$\varepsilon\frac{d}{d\tau}\left[l^2(\tau)\frac{d\theta}{d\tau}\right]+2\chi\frac{d}{d\tau}[l(\tau)\theta]+p''(\tau)\,l(\tau)\sin\theta=0$$

or, equivalently,

$$\varepsilon\frac{l(\tau)}{2[\chi+\varepsilon l'(\tau)]}\frac{d^2\theta}{d\tau^2}+\frac{d\theta}{d\tau}+\frac{2\chi l'(\tau)+l(\tau)p''(\tau)}{2[\chi+\varepsilon l'(\tau)]\,l(\tau)}\theta=$$
$$=\frac{p''(\tau)}{12[\chi+\varepsilon l'(\tau)]}\theta^3+\omega(\tau,\theta;\varepsilon), \tag{16.48}$$

where $\omega(\tau,\theta;\varepsilon)$ vanishes to order 5 with θ.

Suppose that the functions $l(\tau)$ and $p(\tau)$ are almost periodic and sufficiently smooth. Then Theorem 16.2 may be used in studying the small nontrivial ap-oscillations of equation (16.48). In this case

$$m_0(\tau)=\frac{1}{2\chi}l(\tau),\qquad q_0(\tau)=\frac{l'(\tau)}{l(\tau)}+\frac{p''(\tau)}{2\chi},$$
$$q_1(\tau)=-\frac{[l'(\tau)]^2}{\chi l(\tau)}-\frac{l'(\tau)\,p''(\tau)}{2\chi^2},\qquad a_0(\tau)=\frac{p''(\tau)}{12\chi}.$$

Conditions (16.34), (16.35) and (16.38) are clearly satisfied. The function (16.36) differs by a positive factor from the function

$$b_{00}(\tau)=\frac{p''(\tau)}{l^2(\tau)}e^{-\frac{1}{\chi}p'(\tau)}, \tag{16.49}$$

and the function (16.37) is

$$\lambda_0(\tau)=-\frac{l(\tau)}{4\chi^2}p'''(\tau)+\frac{l(\tau)}{8\chi^3}[p''(\tau)]^2-\frac{l''(\tau)}{2\chi}. \tag{16.50}$$

We assume that these functions satisfy the condition (16.39)—(16.42).

From (16.49) and (16.50), it is clear that the numbers $M(b_0)$ and $M(\lambda_0)$ may have different signs (in all combinations) depending on $l(\tau)$ and $p(\tau)$. Thus, depending on these functions, any of the alternatives in parts 1, 3, 4 of Theorem 16.2 may occur.

*SUMMARY OF CHAPTERS 1 – 4**

CHAPTER 1. LINEAR DIFFERENTIAL OPERATORS WITH ALMOST PERIODIC COEFFICIENTS

§ 1. Almost periodic functions

1.1. We first state the definitions.

R^n denotes n-dimensional euclidean space; $|x|$ denotes the euclidean norm of a vector $x \in R^n$.

The space of continuous functions uniformly bounded on $(-\infty, \infty)$ with values in R^n is denoted by $C(R^n)$. The norm in $C(R^n)$ is defined by

$$\|x(t)\|_{C(R^n)} = \sup_{-\infty < t < \infty} |x(t)| \qquad (x(t) \in C(R^n)).$$

The functions $x(t)$ of finite norm

$$\| x(t) \|_{C^m(R^n)} = \| x(t) \|_{C(R^n)} + \| x'(t) \|_{C(R^n)} + \ldots + \| x^{(m)}(t) \|_{C(R^n)}$$

form the space $C^m(R^n)$.

A function $x(t) \in C(R^n)$ is said to be almost periodic (or an ap-function) if the set of functions

$$x_h(t) = x(t+h) \qquad (-\infty < h < \infty)$$

is compact in $C(R^n)$. The ap-functions form a subspace $B(R^n)$ of $C(R^n)$. $B^m(R^n)$ denotes the subspace of $C^m(R^n)$, consisting of all ap-functions having almost periodic derivatives up to order m.

Every ap-function $x(t)$ has a mean value defined by

$$M[x] = \lim_{T \to \infty} \frac{1}{2T} \int_{-T}^{T} x(s)\, ds.$$

* Throughout this summary the formulas are numbered independently.

1.2. The trigonometric polynomials

$$x(t) = \sum_{i=1}^{k} a_i \cos \lambda_i t + b_i \sin \lambda_i t \qquad (-\infty < t < \infty)$$

are important examples of ap-functions (k is a positive integer, λ_i are arbitrary real numbers and the coefficients, a_i, b_i are vectors in R^n).

The spaces $B(R^n)$ and $B^m(R^n)$ are nonseparable. The trigonometric polynomials are dense in both spaces.

1.3. A function $f(t, x)$ of the two variables $-\infty < t < \infty$, $x \in R^n$ is said to be *uniformly almost periodic* if it is periodic in t for every fixed x, and if it is continuous in x on every ball $|x| \leqslant r$ uniformly in t.

Every function $f(t, x)$ induces a superposition operator

$$\mathfrak{f}x(t) = f[t, x(t)].$$

If $f(t, x)$ is uniformly almost periodic, then the superposition operator acts in $B(R^n)$ and is continuous and bounded there.

1.4. The notion of ap-function may be generalized in a natural way for functions taking values in an arbitrary metric space $\mathfrak{R}$. The space $C(\mathfrak{R})$ of continuous functions uniformly bounded on $(-\infty, \infty)$ and the space of ap-functions with values in $\mathfrak{R}$ are metric spaces under the metric

$$\rho_{C(\mathfrak{R})}[x(t), y(t)] = \sup_{-\infty < t < \infty} \rho_{\mathfrak{R}}[x(t), y(t)].$$

§ 2. Regular ap-operators

2.1. The main constructions of this chapter concern the linear differential operator

$$Lx = \frac{d^m x}{dt^m} + A_1(t) \frac{d^{m-1} x}{dt^{m-1}} + \ldots + A_m(t) x, \tag{2.1}$$

where $x \in R^n$ and the $A_i(t)$ are square matrices of order m whose elements are ap-functions. The operators (2.1) are called *ap-operators*.

An ap-operator is said to be *regular* if the equation

$$Lx = f(t) \tag{2.2}$$

has a unique solution $x(t) \in C^m(R^n)$ for any $f(t) \in B(R^n)$ and this solution belongs to $B^m(R^n)$ if $f(t) \in C(R^n)$.

Regularity of ap-operators is invariant under small perturbations of the coefficients.

An ap-operator L is said to be weakly regular if equation (2.2) has at least one solution when the function on the right is in $B(R^n)$. It turns out that *weak regularity of an ap-operator implies its regularity.*

2.2. Equations (2.2), may be more conveniently written as first-order equations in the phase space R^{mn}.

Let $x = \{x_1, \ldots, x_n\}$. Set $u = \{u_1, \ldots, u_m\}$, where

$$u_1 = \{x_1, \ldots, x_n\}, \quad u_2 = \{x_1', \ldots, x_n'\}, \ldots, u_m = \{x_1^{(m-1)}, \ldots, x_n^{(m-1)}\}.$$

Then equation (2.2) is written as

$$\tilde{L}u = \tilde{f}(t),$$

where

$$\tilde{f}(t) = \{0, \ldots, 0, f(t)\}$$

and

$$\tilde{L}u = \frac{du}{dt} + Q(t)u, \tag{2.3}$$

with

$$Q(t) = \begin{Vmatrix} 0 & -I & 0 & \ldots & 0 \\ 0 & 0 & -I & \ldots & 0 \\ \cdot & \cdot & \cdot & \cdot & \cdot \\ 0 & 0 & 0 & & -I \\ A_m(t) & A_{m-1}(t) & A_{m-2}(t) & \ldots & A_1(t) \end{Vmatrix}. \tag{2.4}$$

The ap-operator (2.3) is regular if and only if the ap-operator (2.1) is regular.

2.3. In examining ap-operators, a well-known and importnat theorem of Esclangon plays an essential role.

Esclangon's theorem. If the right-hand side of equation (2.2) is bounded on the whole real line and $x(t)$ is a solution bounded on the whole real line, then the derivatives $x'(t), \ldots, x^{(m)}(t)$ are bounded on $(-\infty, \infty)$.

2.4. Consider the ap-operator (2.1). Let $H(L)$ denote the set of ap-operators

$$L_* x = \frac{d^m x}{dt^m} + A_1^*(t)\frac{d^{m-1}x}{dt^{m-1}} + \ldots + A_m^*(t)x,$$

to each of which there corresponds a sequence h_k such that

$$\lim_{k\to\infty} \sup_{-\infty<t<\infty} \left| A_i(t+h_k) - A_i^*(t) \right| = 0 \qquad (i=1, \ldots, m).$$

The regularity of the ap-operator L implies the regularity of all ap-operators L_* in $H(L)$. Moreover, the regularity of one operator in $H(L)$ implies the regularity of all operators in $H(L)$.

§ 3. Behavior of solutions of the homogeneous equation

3.1. Consider the homogeneous equation

$$Lx = 0 \tag{3.1}$$

in R^n, where L is a regular first-order ap-operator given by

$$Lx = \frac{dx}{dt} + A(t)\,x. \tag{3.2}$$

Let $U(t)$ be the translation operator of equation (3.1). ($U(t)$ is the fundamental matrix of solutions of system (3.1) satisfying the initial condition $U(0) = I$).

Let $E_+(s)$ denote the set of initial values (at $t = s$) in R^n for which the corresponding solutions of equation (3.1) are bounded on $[s, \infty)$. Let $E_-(s)$ denote the set of initial values for which the corresponding solutions are bounded on $(-\infty, s]$. Clearly, for any t and s,

$$E_+(t) = U(t)\,E_+(0)\,U^{-1}(s), \qquad E_-(t) = U(t)\,E_-(0)\,U^{-1}(s).$$

It turns out that for each t, $E_+(t) \dotplus E_-(t)$ is a direct sum decomposition of R^n.

This decomposition of R^n is equivalent to decomposition of the solution set X of equation (3.1) into a direct sum $X = X_+ \dotplus X_-$, where X_+ is the subspace of solutions bounded as $t\to\infty$, and X_- the subspace of solutions bounded as $t\to-\infty$. This decomposition is known as a *dichotomy* of solutions.

3.2. A dichotomy of solutions of equation (3.1) is said to be *exponential* if there exist positive constants M_+, M_-, γ_+, γ_-, such that for $x(t)\in X_+$

$$|x(t)| \leqslant M_+ e^{-\gamma_+(t-s)}\,|x(s)| \qquad (-\infty < s \leqslant t < \infty),$$

and for $x(t) \in X_-$

$$|x(t)| \geqslant M_- e^{\gamma_-(t-s)}\,|x(s)| \qquad (-\infty < s \leqslant t < \infty).$$

If the ap-operator (3.2) is regular, then the dichotomy $X = X_+ \dot{+} X_-$ *of the solutions of equation (3.1) is exponential.*

3.3. The angle between two subspaces E_1 and E_2 is defined by

$$\alpha(E_1, E_2) = \min_{x \in E_1,\ y \in E_2,\ |x| = |y| = 1} |x - y|.$$

If the ap-operator (3.2) is regular, then

$$\alpha[E_+(t), E_-(t)] \geqslant \alpha_0 > 0 \qquad (-\infty < t < \infty). \tag{3.3}$$

A dichotomy satisfying (3.3) is said to be **uniform**.

3.4. The decomposition of R^n as a direct sum of subspaces $E_+(t)$ and $E_-(t)$ entails that every element $x \in R^n$ may be uniquely represented as

$$x = u + v \qquad (u \in E_+(t),\ v \in E_-(t)).$$

This representation defines projection operators

$$P_+(t)x = u, \quad P_-(t)x = v \qquad (x \in R^n,\ -\infty < t < \infty)$$

onto the subspaces $E_+(t)$ and $E_-(t)$, respectively.

Clearly, $P_+(t) + P_-(t) = I$ and

$$P_+(t) = U(t)P_+(0)U^{-1}(t), \qquad P_-(t) = U(t)P_-(0)U^{-1}(t). \tag{3.4}$$

The operators $P_+(t)$ *and* $P_-(t)$ *are almost periodic with respect to* t. In particular, their norms are uniformly bounded.

§ 4. The Green's function

4.1. Let the ap-operator

$$Lx = \frac{d^m x}{dt^m} + A_1(t)\frac{d^{m-1}x}{dt^{m-1}} + \ldots + A_m(t)x \tag{4.1}$$

be regular. Then, as we know, the inverse L^{-1} is defined; it acts from $C(R^n)$ into $C^m(R^n)$ and from $B(R^n)$ into $B^m(R^n)$. The operator L^{-1} has an integral representation

$$L^{-1}f(t) = \int_{-\infty}^{\infty} G(t, s) f(s)\, ds. \tag{4.2}$$

The kernel matrix $G(t, s)$ is known as the Green's function of the ap-operator L.

The Green's function is fully determined by the following properties;

1. For $t \neq s$,

$$\frac{\partial^m G(t, s)}{\partial t^m} + A_1(t) \frac{\partial^{m-1} G(t, s)}{\partial t^{m-1}} + \ldots + A_m(t) G(t, s) \equiv 0.$$

2. The matrix-functions

$$G(t, s), \; \frac{\partial G(t, s)}{\partial t}, \; \ldots, \; \frac{\partial^{m-2} G(t, s)}{\partial t^{m-2}}$$

are jointly continuous in $t, s \in (-\infty, \infty)$, the derivative $\frac{\partial^{m-1} G(t, s)}{\partial t^{m-1}}$ is jointly continuous in t, s for $t \neq s$ and

$$\frac{\partial^{m-1} G(t+0, t)}{\partial t^{m-1}} - \frac{\partial^{m-1} G(t-0, t)}{\partial t^{m-1}} = I.$$

3.

$$|G(t, s)|, \; \left| \frac{\partial G(t, s)}{\partial t} \right|, \; \ldots, \; \left| \frac{\partial^{m-1} G(t, s)}{\partial t^{m-1}} \right| \leqslant M e^{-\gamma |t-s|}$$
$$(-\infty < t, s < \infty; \; t \neq s),$$

where $M, \gamma > 0$.

4.2. The Green's function $G(t, s)$ of a regular operator (4.2) can be constructed quite simply in a general setting.

Let $\tilde{L}$ be the operator (2.3). The regularity of the ap-operator L implies the regularity of $\tilde{L}$. Thus the set of solutions of the homogeneous equation

$$\tilde{L} u = 0 \qquad (4.3)$$

admits a dichotomy. Let $U(t)$ be the translation operator of equation (4.3), $P_+(t)$ and $P_-(t)$ the operators (3.4). Then the Green's function of the operator (2.3) is given by

$$\tilde{G}(t, s) = \begin{cases} U(t) P_+(0) U^{-1}(s), & \text{if} \quad t \geqslant s, \\ -U(t) P_-(0) U^{-1}(s), & \text{if} \quad t < s. \end{cases} \qquad (4.4)$$

This matrix-function can be written

$$\widetilde{G}(t, s) = \begin{Vmatrix} g_{11}(t, s) & g_{12}(t, s) & \dots & g_{1m}(t, s) \\ g_{21}(t, s) & g_{22}(t, s) & \dots & g_{2m}(t, s) \\ \cdot & \cdot & \cdot & \cdot \\ g_{m1}(t, s) & g_{m2}(t, s) & \dots & g_{mm}(t, s) \end{Vmatrix}, \tag{4.5}$$

where $g_{ij}(t, s)$ are square matrices of order n.

The matrix-function $g_{1m}(t, s)$ *is the Green's function of the regular ap-operator (4.1).*

4.3. Consider the family of ap-operators

$$L_\mu x = \frac{dx}{dt} + A(t; \mu)x, \tag{4.6}$$

where μ ranges over some compact set in a metric space.

Suppose that the ap-operators (4.6) are regular for all μ and $A(t; \mu)$ are continuous in μ uniformly in t. Then the Green's function $G(t, s; \mu)$ of the operator (4.6) is continuous in μ and, moreover, there exist positive M_0 and γ_0 such that for any μ_1, μ_2

$$|G(t, s; \mu_1) - G(t, s; \mu_2)| \leqslant \\ \leqslant M_0 e^{-\gamma_0 |t-s|} \sup_{-\infty < \tau < \infty} |A(\tau; \mu_1) - A(\tau; \mu_2)|.$$

4.4. An ap-operator with constant coefficients is regular if and only if its characteristic polynomial has no roots on the imaginary axis. The ap-operator

$$Lx = x' + a(t)x, \tag{4.7}$$

where $a(t)$ is a scalar ap-function, is regular if and only if $M[a] \neq 0$. (If, say, $M[a] < 0$, then $G(t, s) = 0$ for $t \geqslant s$ and

$$G(t, s) = -\exp\left[-\int_s^t a(\tau)\, d\tau\right] \quad \text{for} \quad t < s.)$$

§ 5. Positivity of the Green's function

5.1. A closed convex set K in a Banach space E is called a w e d g e if $\alpha x \in K$ whenever $x \in K$ and $\alpha \geqslant 0$. A wedge is called a c o n e, if $x, -x \in K$ implies $x = 0$.

Every cone $K \subset E$ induces a semi-order in E: we write $x \leqslant y$ if $y - x \in K$. The relation $\leqslant$ has all the properties of the usual order $\leqslant$.

A cone is said to be solid if it contains interior points. If $y-x$ is an interior element of a solid cone K, we write $x \ll y$. A cone K is said to be normal if $0 \leqslant x \leqslant y$ implies $\|x\| \leqslant N\|y\|$, where N does not depend on x and y. Only solid and normal cones will be considered henceforth.

The simplest example of a normal solid cone K in $B(R^n)$ is the set of nonnegative functions. All cones in finite-dimensional spaces are normal.

A cone K is said to be minihedral if, for every pair $x, y \in E$, there exists an element u such that $x \leqslant u$ and $y \leqslant u$ and, furthermore, $x \leqslant v$ and $y \leqslant v$ imply $u \leqslant v$. We write $u = \sup\{x, y\}$. In the finite-dimensional space R^n, minihedral cones are cones of vectors with nonnegative components relative to some basis.

Let K be a wedge in E. A linear functional l is said to be positive if $l(x) \geqslant 0$ for $x \in K$. The set K^* of positive linear functionals is also a wedge.

5.2. A continuous functional $q(x)$ $(x \in E)$ is said to be sublinear if

$$q(\alpha x) = \alpha q(x) \qquad (\alpha \geqslant 0,\ x \in E)$$

and

$$q(x+y) \leqslant q(x) + q(y) \qquad (x, y \in E).$$

Consider linear functionals $q(x)$ for which the set $K = \{x: q(x) \leqslant 0\}$ is a solid cone and $K_{\text{in}} = \{x: q(x) < 0\}$ is its interior. $q(x)$ is said to select the cone K.

Every solid cone may be selected by various sublinear functionals. It is convenient to use the functionals

$$q_0(x) = \min_{x+ru_0 \in K} r \qquad (x \in E), \tag{5.1}$$

where u_0 is a fixed interior element of the cone K, and

$$q_1(x) = -\min_{l \in K^*,\ \|l\|=1} l(x) \qquad (x \in E). \tag{5.2}$$

For various specific cones, an appropriate sublinear functional can be written down immediately. As an example, the cone K of nonnegative ap-functions in $B(R^1)$ is selected by the functional

$$q(x) = -\inf_{-\infty < t < \infty} x(t) \qquad (x(t) \in B(R^1)).$$

5.3. The set $\mathfrak{K}$ of all cones and wedges in R^n becomes a complete metric space if we set

$$\rho(K_1, K_2) = \max\left\{\sup_{x \in \Omega_1} \rho(x, \Omega_2), \quad \sup_{y \in \Omega_2} \rho(y, \Omega_1)\right\},$$

where

$$\Omega_1 = \{x: x \in K_1, |x| \leqslant 1\}, \quad \Omega_2 = \{y: y \in K_2, |y| \leqslant 1\}.$$

Thus we may speak of ap-functions $K(t)$ with values in $\mathfrak{K}$ and, in particular, of cone-valued functions.

Let $K(t)$ be a cone-valued ap-function. Let $\widehat{K}(t)$ denote the set of ap-functions $x(t)$ with values in R^n such that $x(t) \in K(t)$ $(-\infty < t < \infty)$. A cone-valued ap-function $K(t)$ is said to be solid if the closure of its range in $\mathfrak{K}$ consists of solid cones. If a cone-valued ap-function $K(t)$ is solid, it turns out that the cone $\widehat{K}(t)$ is both solid and normal in $B(R^n)$. If all the cones $K(t)$ are minihedral, then $\widehat{K}(t)$ is also minihedral.

Let $q(t, x)$ be a functional which is almost periodic in t and sublinear in x $(-\infty < t < \infty,\ x \in R^n)$, satisfying a Lipschitz condition

$$|q(t, x) - q(t, y)| \leqslant a|x - y| \qquad (x, y \in R^n),$$

where the constant a does not depend on t. Suppose there exists a function $x_0(t) \in B(R^n)$ such that

$$q[t, x_0(t)] \leqslant -\varepsilon_0 < 0 \qquad (-\infty < t < \infty).$$

Finally, assume that

$$\max\{q(t, x), q(t, -x)\} \geqslant \varepsilon_1 > 0 \qquad (-\infty < t < \infty, |x| = 1).$$

Under these conditions, *the cones* $K(t)$ *selected by the functional* $q(t, x)$ *form a solid cone-valued ap-function.*

5.4. An operator A acting in a Banach space E is said to be positive with respect to a cone K if $AK \subset K$. If K is solid and $x \in K$ and $x \neq 0$ imply $Ax \gg 0$, then A is said to be strongly positive. A is said to be monotone if $x \leqslant y$ implies $Ax \leqslant Ay$.

For linear operators, the notions of positivity and monotonicity are equivalent.

Let $K(t)$ be a solid cone-valued ap-function. The Green's function $G(t, s)$ of the regular ap-operator

$$Lx = \frac{d^m x}{dt^m} + A_1(t)\frac{d^{m-1}x}{dt^{m-1}} + \ldots + A_m(t)x \qquad (5.3)$$

is said to be nonnegative with respect to $K(t)$ if

$$G(t, s) K(s) \subset K(t) \qquad (-\infty < t,\ s < \infty). \tag{5.4}$$

If the Green's function is nonnegative, then the operator

$$L^{-1}f(t) = \int_{-\infty}^{\infty} G(t, s) f(s)\, ds \tag{5.5}$$

is positive in $B(R^n)$ *with respect to the cone* $\hat{K}(t)$.

Let L be a regular ap-operator. Its Green's function is said to be strongly positive with respect to a solid cone-valued ap-function $K(t)$ if it is nonnegative and, for every Green's function $G_*(t, s)$ corresponding to an operator in $H(L)$ (see subsection 2.4) and every t, there exists a set $\mathfrak{M}$, dense in some infinite interval, such that, for nonzero $x \in K_*(s)$ and $s \in \mathfrak{M}$, the elements $G_*(t, s)x$ are interior points of the cone $K_*(t)$.

If $G(t, s)$ *is strongly positive, then the operator* L^{-1} *is strongly positive with respect to* $\hat{K}(t)$.

Nonpositivity and strong negativity of the Green's function are similarly defined; they imply, respectively, the positivity and strong positivity of the operator L^{-1}.

Note, in conclusion, that in examining ap-operators L acting on scalar functions $x(t)$, we consider only one cone in $B(R^1)$, namely, the cone of nonnegative ap-functions. In this important case we are concerned with the sign (in the ordinary sense) of the scalar Green's function $G(t, s)$. Thus the general notions are significant only in the case of ap-operators acting on vector-functions.

CHAPTER 2. ANALYSIS OF SPECIAL TYPES OF AP-OPERATORS

§ 6. First-order systems of equations

6.1. We examine the ap-operator

$$Lx = \frac{dx}{dt} + A(t)\, x, \tag{6.1}$$

where $A(t)$ is a square matrix of order n with almost periodic elements. Conditions are found for the operator (16.1) to be regular and for its Green's function to be of constant sign with respect to a cone-valued function.

It turns out that the operator (6.1) may be regular and the Green's function of constant sign only if the trivial solution of the homogeneous equation

$$Lx = 0 \tag{6.2}$$

is exponentially stable, either as t increases or as t decreases. In addition, the Green's function may be nonnegative only if the trivial solution is stable as t increases, and it may be nonpositive only if the trivial solution is stable as t decreases (tending to $-\infty$).

We now consider conditions for the existence of a nonnegative Green's function. Let $U(t)$ denote the fundamental matrix of solutions of equation (6.2) having initial value $U(0) = I$. The expression $U(t, s) = U(t)U^{-1}(s)$ defines the translation operator along the trajectories of equation (6.2) (or, simply: the translation operator of system (6.2)) over the interval from s to t. The translation operator is positive with respect to a cone-valued function $K(t)$ if $U(t, s)K(s) \subset K(t)$ for $t \geqslant s$, strongly positive if $U(t, s)x \gg 0$ for $t > s, x \in K(s)$ and $x \neq 0$.

Theorem 6.1. The ap-operator (6.1) is regular and its Green's function is nonnegative with respect to a solid cone-valued function $K(t)$ if and only if the following two conditions hold:

1. *The translation operator $U(t, s)$ is positive with respect to $K(t)$.*
2. *The translation operator $U(t, s)$ satisfies the inequalities*

$$|U(t, s)| \leqslant Me^{-\gamma(t-s)} \qquad (-\infty < s \leqslant t < \infty),$$

where $M, \gamma > 0$.

If the Green's function exists, then it is of the form

$$G(t, s) = \begin{cases} U(t, s) & \text{for } t \geqslant s, \\ 0 & \text{for } t < s. \end{cases}$$

This theorem reduces the problem of the regularity of (6.1) and of the nonnegativity of its Green's function to an analysis of the translation operator $U(t, s)$. We confine ourselves to the case $K(t) \equiv K$.

6.2. Let $q(x)$ be a sublinear functional defined on E. Then it has right and left Gâteaux derivatives, defined by

$$q'_+(x, h) = \lim_{t \to +0} \frac{q(x + th) - q(x)}{t} \qquad (x, h \in E)$$

and

$$q'_-(x, h) = \lim_{t \to -0} \frac{q(x + th) - q(x)}{t} \qquad (x, h \in E).$$

Clearly,

$$q'_-(x, h) = -q'_+(x, -h) \qquad (x, h \in E).$$

Thus, it suffices to evaluate right derivatives of sublinear functionals. We cite some simple examples.

a) Let

$$q(x) = |\xi_1| + \ldots + |\xi_n| \qquad (x = \{\xi_1, \ldots, \xi_n\} \in R^n).$$

Then

$$q'_+(x, h) = s(\xi_1, h_1) + \ldots + s(\xi_n, h_n) \qquad (h = \{h_1, \ldots, h_n\} \in R^n),$$

where

$$s(\xi_i, h_i) = \begin{cases} h_i \operatorname{sign} \xi_i & \text{for } \xi_i \neq 0, \\ |h_i| & \text{for } \xi_i = 0. \end{cases}$$

b) Let

$$q(x) = \max_{i=1,\ldots,n} |\xi_i| \qquad (x = \{\xi_1, \ldots, \xi_n\} \in R^n).$$

Then

$$q'_+(x, h) = \max_{i \in i(x)} s(\xi_i, h_i) \qquad (h = \{h_1, \ldots, h_n\} \in R^n),$$

where $i(x)$ is the set of indices i such that $q(x) = |\xi_i|$.

c) Let

$$q(x) = -\min_{i=1,\ldots,n} \xi_i \qquad (x = \{\xi_1, \ldots, \xi_n\} \in R^n).$$

Then

$$q'_+(x, h) = -\min_{i \in i(x)} h_i \qquad (h = \{h_1, \ldots, h_n\} \in R^n),$$

where $i(x)$ is the set of indices i such that $q(x) = -\xi_i$.

d) Let C be the space of continuous scalar functions on [0.1] and

$$q(x) = \|x\| = \max_{0 \leqslant t \leqslant 1} |x(t)| \qquad (x \in C).$$

Then

$$q'_+(x, h) = \max\left\{\max_{\tau \in \Omega_+} h(\tau), \ -\min_{\tau \in \Omega_-} h(\tau)\right\},$$

where

$$\Omega_+ = \{\tau:\ \tau \in [0, 1],\ x(\tau) = \|x\|\}$$

and

$$\Omega_- = \{\tau:\ \tau \in [0, 1],\ x(\tau) = -\|x\|\}.$$

6.3. Let $q(x)$ be a sublinear functional selecting a solid cone K with boundary Γ. In checking for positivity and strong positivity of the translation operator $U(t, s)$ of equation (6.2), conditions of the following kind are used:

$$q'_-[x, -A(t)x] \leqslant 0 \qquad (x \in \Gamma,\ -\infty < t < \infty), \qquad (6.3)$$

$$q'_+[x, -A(t)x] \leqslant 0 \qquad (x \in \Gamma,\ -\infty < t < \infty) \qquad (6.4)$$

and

$$q'_-[x, -A(t)x] < 0 \qquad (x \in \Gamma,\ x \neq 0,\ t \in \mathfrak{M}), \qquad (6.5)$$

where $\mathfrak{M}$ is a dense subset of $(-\infty, \infty)$.

Theorem 6.2. For the operator $U(t, s)$ to be positive with respect to the cone K, it is necessary that (6.3) hold and sufficient that (6.4) hold. A sufficient condition for $U(t, s)$ to be strongly positive is that (6.5) hold.

When the operator (6.1) has constant coefficients and K is a faceted cone in R^n (K is said to be faceted if it is the intersection of a finite number of half-spaces), condition (6.5) is not only sufficient but also necessary for the translation operator to be strongly positive.

6.4. Let K be the cone of vectors $x = \{\xi_1, \ldots, \xi_n\} \in R^n$ with nonnegative components. Let $A = (a_{ij})$ be a constant matrix. Then: *The translation operator of equation (6.2) is positive if and only if all off-diagonal elements of the matrix A are nonpositive:*

$$a_{ij} \leqslant 0 \qquad (i \neq j;\ i, j = 1, \ldots, n). \qquad (6.6)$$

The sign of the diagonal elements a_{ii} is not important.

The condition for strong positivity can be conveniently formulated with the help of an additional concept.

A sequence

$$a_{i_1 i_2},\ a_{i_2 i_3},\ \ldots,\ a_{i_{m-1} i_m},\ a_{i_m i_1}$$

is called a nondegeneracy path of the matrix A if none of its terms vanish, if

$$i_1 \neq i_2,\ i_2 \neq i_3,\ \ldots,\ i_{m-1} \neq i_m,\ i_m \neq i_1$$

and the sequence $i_1, i_2, \ldots, i_m$ includes all the numbers $1, 2\ldots, n$ (possibly with repetitions).

The translation operator of an equation (6.2) with constant matrix A is strongly positive with respect to the cone K of vectors with nonnegative components if and only if condition (6.6) is valid and the matrix A has at least one nondegeneracy path.

If the matrix $A(t) = (a_{ij}(t))$ is variable, the above conditions are only sufficient for the translation operator to be positive. Condition (6.6) must hold for all t and the nondegeneracy paths of $A(t)$ may be distinct for distinct t.

6.5. *Theorem 6.3. Let the translation operator $U(t, s)$ be positive with respect to a solid cone-valued ap-function $K(t)$. Suppose there exists a function $x_0(t) \in C^1(R^n)$ such that*

$$x_0(t) \gg 0, \quad Lx_0(t) \gg 0. \tag{6.7}$$

Then condition 2 of Theorem 6.1 is satisfied.

Thus our examination of the operator (6.1) may be pursued as follows. First we must find a cone K satisfying (6.3). Then we seek a function $x_0(t)$ satisfying (6.7).

Having found such K and $x_0(t)$, we have that the operator (6.1) is regular and that its Green's function is nonnegative.

If inequality (6.7), which is stronger than (6.3), holds, then the Green's function is strongly positive. Note, moreover, that if the cone K is faceted, then the operator L^{-1} is strongly positive with respect to the cone $\hat{K}$, provided (6.3) holds for all t and (6.5) for some t_0.

Theorem 6.3 is a simple example of the method of test functions, widely applied in investigations of various differential equations. By this method, the general properties of an operator are deduced from the existence of a function satisfying certain inequalities.

We note one obvious corollary of Theorem 6.3:

If there exists a function satisfying (6.7) and the translation operator is positive, then the trivial solution of equation (6.2) is exponentially stable.

§7. Second-order systems

7.1. We first examine the operator

$$Lx = \frac{d^2x}{dt^2} + Ax, \tag{7.1}$$

where A is a constant matrix. The regularity of the operator (7.1) is equivalent to the non-existence of nonnegative real eigenvalues of the matrix A. If this condition holds, then A can be uniquely represented as

$$A = -B^2,$$

where B is a real Hurwitz matrix (i.e., a matrix with eigenvalues in the left half-plane). The Green's function is given by

$$G(t, s) = \begin{cases} \frac{1}{2} B^{-1} e^{B(t-s)} & \text{for } t \geqslant s, \\ \frac{1}{2} B^{-1} e^{-B(t-s)} & \text{for } t < s. \end{cases} \quad (7.2)$$

The construction of the matrix B and analysis of the matrix-function $B^{-1}e^{B\tau}$ are rather tedious. It is therefore desirable to formulate conditions for the Green's functions to be nonpositive or nonnegative directly, in terms of the matrix A. Formula (7.2) implies that the Green's function of the operator (7.1) cannot be nonnegative with respect to any solid cone $K \subset R^n$. Simple tests for nonpositivity can be derived for operators (7.1) with Hurwitz matrices A.

Let K be a solid cone with boundary Γ. Suppose A is a Hurwitz matrix. Then:

a) If $e^{At}K \subset K$ $(t > 0)$, the ap-operator (7.1) is regular and its Green's function is nonpositive.

b) If

$$q'_{-}(x, Ax) < 0 \qquad (x \in \Gamma, \ x \neq 0),$$

where $q(x)$ is a sublinear functional selecting K, then the ap-operator (7.1) is regular and its Green's function is strongly negative.

c) If K is a faceted cone and $e^{At}x \gg 0$ $(x \in K, \ x \neq 0)$, the ap-operator (7.1) is regular and its Green's function is strongly negative.

7.2. We now ask when the ap-operator

$$Lx = \frac{d^2x}{dt^2} + A(t)\,x \qquad (7.3)$$

is regular and when its Green's function $G(t, s)$ is nonpositive. The most important technique for examining this case is the method of comparison. We compare the operator (7.3) with some ap-operator

$$L_0 x = \frac{d^2x}{dt^2} + A_0(t)\,x, \tag{7.4}$$

known to be regular, whose Green's function $G_0(t, s)$ is nonpositive. For example, L_0 could be an ap-operator (7.1) with constant matrix.

Theorem 7.1. Let the Green's function $G_0(t,s)$ *of the ap-operator (7.4) be nonpositive with respect to a solid cone-valued ap-function* $K(t)$. *Let*

$$A(t)\,x \geqslant A_0(t)\,x \qquad (x \in K(t),\ -\infty < t < \infty).$$

Finally, suppose there exists a function $x_0(t) \in B^2(R^n)$ *such that*

$$x_0(t) \gg 0, \qquad Lx_0(t) \ll 0. \tag{7.5}$$

Then the ap-operator (7.3) is regular, its Green's function $G(t, s)$ *is nonpositive with respect to* $K(t)$ *and, moreover,*

$$G_0(t, s)\,x - G(t, s)\,x \in K(t) \qquad (x \in K(s)). \tag{7.6}$$

From (7.6) we have

Theorem 7.2. Suppose the conditions of Theorem (7.1) are valid and the Green's function of the ap-operator (7.4) is strongly negative.

Then the Green's function $G_0(t, s)$ *of the operator (7.3) is also strongly negative.*

7.3. It turns out that the regularity of the operator (7.3) and the properties of its Green's function are closely related to the translation operator $U(t, s)$ of the linear first-order system

$$\frac{dx}{dt} - A(t)\,x = 0. \tag{7.7}$$

Theorem 7.3. Let K *be a solid cone such that* $U(t, s)K \subset K$ *for* $t \geqslant s$. *Suppose there exists a function* $x_0(t) \in B^2(R^n)$ *satisfying inequalities (7.5).*

Then the ap-operator (7.3) is regular and its Green's function is nonpositive with respect to K.

Recall that conditions for the translation operator of a first-order equation to be positive were established in § 6 (see Theorem 6.2).

7.4. If the Green's function $G(t, s)$ of the operator (7.3) is nonpositive with respect to the cone K, then the integral operator

$$Sf(t) = -L^{-1}f(t) = -\int_{-\infty}^{\infty} G(t, s) f(s)\, ds \tag{7.8}$$

is positive with respect to the cone $\hat{K} \subset B(R^n)$.

We turn now to investigation of the strong positivity of the operator (7.8).

Theorem 7.4. Suppose there exists an ap-function $x_0(t) \in B^2(R^n)$, satisfying inequalities (7.5). Let

$$\sup_{-\infty < t < \infty} q'_-[x, A(t)x] < 0 \qquad (x \in \Gamma,\ x \neq 0),$$

where $q(x)$ is a sublinear functional selecting K and Γ is the boundary of K.

Then the integral operator (7.8) is strongly positive with respect to $\hat{K}$.

For a faceted cone K, an essentially stronger assertion can be proved: Under the hypotheses of Theorem 7.3, if the cone K is faceted, then the operator (7.8) is strongly positive if there exists t_0 such that

$$q'_-[x, A(t_0)x] < 0 \qquad (x \in \Gamma,\ x \neq 0).$$

§ 8. Higher-order scalar equations

8.1. We investigate the ap-operator

$$Lx = \frac{d^m x}{dt^m} + a_1(t)\frac{d^{m-1}x}{dt^{m-1}} + \ldots + a_m(t)x \tag{8.1}$$

acting on scalar functions $x(t)$. In this section and the next, we are interested in the sign (in the usual sense) of the values of the Green's function $G(t, s)$.

Theorem 8.1. If the Green's function $G(t, s)$ of a regular ap-operator (8.1) is nonnegative, then it is strongly positive; if it is nonpositive, then it is strongly negative.

Therefore, we need only discuss conditions for the regularity of the ap-operator (8.1) and give criteria for the nonnegativity or nonpositivity of its Green's function.

8.2. We begin with an analysis of the ap-operator with constant coefficients:

$$Lx = \frac{d^m x}{dt^m} + a_1\frac{d^{m-1}x}{dt^{m-1}} + \ldots + a_m x. \tag{8.2}$$

The operator (8.2) is regular if and only if its characteristic equation

$$\chi(\lambda) \equiv \lambda^m + a_1\lambda^{m-1} + \dots + a_m = 0 \tag{8.3}$$

has no zero or pure imaginary roots.

The roots of equation (8.3) fall naturally into two groups, those situated in the left half-plane and those in the right. Let α_- be the largest real part of those roots lying in the left half-plane, and α_+ the smallest real part of the roots in the right half-plane. If either group of roots is empty, we set $\alpha_- = -\infty$ or $\alpha_+ = \infty$, as the case may be.

Let α_- be a root of the polynomial $\chi(\lambda)$ if α_- is finite, and α_+ a root of the polynomial if α_+ is finite. Then we say that the polynomial $\chi(\lambda)$ is tame.

For example, the cubic polynomial

$$\chi(\lambda) = \lambda^3 + a\lambda^2 + b\lambda + c$$

is tame if the coefficients satisfy one of the inequalities

$$c \neq 0, \quad (2a^3 - 9ab + 27c)^2 + 4(3b - a^2)^3 \leqslant 0$$

or

$$ac > 0, \qquad a(2a^3 - 9ab + 27c) \leqslant 0.$$

Theorem 8.2. If the Green's function of a regular ap-operator (8.2) is of constant sign, its characteristic polynomial is tame.

8.3. Set

$$Q = \begin{Vmatrix} 0 & -1 & 0 & \dots & 0 \\ 0 & 0 & -1 & \dots & 0 \\ \cdot & \cdot & \cdot & \cdot & \cdot \\ 0 & 0 & 0 & \dots & -1 \\ a_m & a_{m-1} & a_{m-2} & \dots & a_1 \end{Vmatrix}$$

and

$$e_1 = \{1, 0, \dots, 0\}, \dots, e_m = \{0, 0, \dots, 1\}.$$

The Cauchy function $k(t-s)$ of an operator (8.2) with constant coefficients is defined as

$$k(t-s) = (e^{-Q(t-s)}e_m, e_1). \tag{8.4}$$

Clearly,

$$k(0) = k'(0) = \dots = k^{(m-2)}(0) = 0, \qquad k^{(m-1)}(0) = 1.$$

If all roots of the characteristic polynomial $\chi(\lambda)$ fall in the same half-plane, then the Green's function $G(t, s)$ of the ap-operator (8.2) with constant coefficients has a simple expression in terms of its Cauchy function. If $\chi(\lambda)$ is a Hurwitz polynomial, then

$$G(t, s) = \begin{cases} k(t-s) & \text{for} \quad t \geqslant s, \\ 0 & \text{for} \quad t < s, \end{cases}$$

but if $\chi(-\lambda)$ is a Hurwitz polynomial, then

$$G(t, s) = \begin{cases} 0 & \text{for} \quad t \geqslant s, \\ -k(t-s) & \text{for} \quad t < s. \end{cases}$$

Thus, we have the following propositions.

a) If $\chi(\lambda)$ is a Hurwitz polynomial, the Green's function $G(t, s)$ cannot be nonpositive. It is nonnegative if and only if the Cauchy function $k(t)$ is nonnegative for all $t \geqslant 0$.

b) If $\chi(-\lambda)$ is a Hurwitz polynomial and m is even, then the Green's function cannot be nonnegative. It is nonpositive if and only if the Cauchy function $k(t)$ is nonnegative for all $t \leqslant 0$.

c) If $\chi(-\lambda)$ is a Hurwitz polynomial and m is odd, then the Green's function $G(t, s)$ cannot be nonpositive. It is nonnegative if and only if the Cauchy function $k(t)$ is nonpositive for all $t \leqslant 0$.

Every regular ap-operator (8.2) with constant coefficients may be factored as a superposition $L = L_1 L_2$ of two regular ap-operators with constant coefficients such that the characteristic polynomial of one has all its roots in the right half-plane and the other in the left. The Green's function $G(t, s)$ of the ap-operator (8.2) may be represented as the superposition

$$G(t, s) = \int_{-\infty}^{\infty} G_2(t, \sigma) G_1(\sigma, s)\, d\sigma$$

of the Green's functions $G_1(t, s)$ and $G_2(t, s)$ of L_1 and L_2. Thus, $G(t, s)$ is of constant sign if $G_1(t, s)$ and $G_2(t, s)$ are of constant sign or, equivalently, if the Cauchy functions of L_1 and L_2 are of constant sign (for $t > 0$ and for $t < 0$, respectively).

If all roots of the characteristic polynomial lie in the right half-plane, then in order to check for the sign of the Green's function, we can reverse the direction of time in the differential operator. When this is done, if the order of the operator is even, the sign of the

Green's function remains unchanged; if the order is odd, the Green's function changes sign. In either case the resultant characteristic polynomial is a Hurwitz polynomial. Therefore we confine ourselves to an investigation of operators (8.2) with Hurwitz characteristic polynomials.

8.4. A solid cone $K \subset R^m$ is said to be admissible if it lies in the half-space $(u, e_1) \geqslant 0$ and $e_m \in K$.

Theorem 8.3. The Cauchy function $k(t)$ of a differential operator (8.2) with Hurwitz characteristic polynomial is nonnegative if and only if there exists an admissible cone K such that $e^{-Qt}K \subset K$ $(0 < t < \infty)$.

This theorem reduces proof of the nonnegativity of the Green's function to construction of admissible cones. In certain cases such a construction can be realized. We cite one of the relevant results.

Let $\chi(\lambda)$ be a tame Hurwitz polynomial and let λ_0 be its largest real root. We say that $\chi(\lambda)$ is completely tame if the following conditions hold:

1. λ_0 is a simple root and all other roots satisfy the inequality $\operatorname{Re} \lambda < \lambda_0$.

2. There exists a polynomial $\varkappa(\lambda)$ of degree $m-2$ with negative real roots satisfying

$$\operatorname{Re} \frac{\varkappa(i\omega)}{\chi(\lambda_0 + i\omega)} < 0 \qquad (-\infty < \omega < \infty).$$

The polynomial $\chi(\lambda)$ is said to be completely tame in the limit if it is the limit of a sequence of completely tame polynomials.

Theorem 8.5. Suppose the characteristic polynomial $\chi(\lambda)$ of the ap-operator (8.2) with constant coefficients may be represented as $\chi(\lambda) = \chi_1(\lambda)\chi_2(\lambda)$ where $\chi_1(\lambda)$ and $\chi_2(-\lambda)$ are Hurwitz polynomials which are either completely tame or completely tame in the limit.

Then the ap-operator (8.2) is regular, its Green's function is of constant sign and

$$a_m G(t, s) \geqslant 0 \qquad (-\infty < t,\ s < \infty).$$

This theorem implies that the condition of Theorem 8.2 is not only necessary but also sufficient for constancy of sign of the Green's function, provided the degrees of the polynomials are at most 4.

8.5. We now consider ap-operators

$$Lx = \frac{d^m x}{dt^m} + a_1(t)\frac{d^{m-1}x}{dt^{m-1}} + \dots + a_m(t)\,x \tag{8.5}$$

with variable coefficients. We state the general comparison principle.

Let

$$L_0 x = \frac{d^m x}{dt^m} + b_1(t)\frac{d^{m-1}x}{dt^{m-1}} + \dots + b_m(t)\,x \tag{8.6}$$

be a given regular operator with Green's function $G_0(t, s)$ of constant sign.

Theorem 8.6. Suppose there exists a function $u_0(t) \in B^m(R^1)$ *such that*

$$u_0(t) \gg 0, \quad \varepsilon L_0 u_0(t) \gg 0, \quad \varepsilon L u_0(t) \gg 0, \tag{8.7}$$

where $\varepsilon = 1$ if $G_0(t, s) \geqslant 0$ *and* $\varepsilon = -1$ *if* $G_0(t, s) \leqslant 0$. *Suppose that* $LG_0(t, s) \leqslant 0$ *for* $t \neq s$.

Then the ap-operator 8.5 is regular, its Green's function is of constant sign and it satisfies the inequality

$$\varepsilon G(t, s) \geqslant \varepsilon G_0(t, s) \qquad (-\infty < t, s < \infty).$$

Note that, under the conditions of Theorem 8.6, the trivial solution of the equation $Lx = 0$ is exponentially stable if the trivial solutions of the equation $L_0 x = 0$ is exponentially stable.

8.6. We may take various special classes of ap-operators (for example, operators with constant coefficients) to serve as comparison operators (8.6) and select suitable ap-functions $u_0(t)$ satisfying (8.7), thus obtaining various criteria for regularity of ap-operators (8.5) and for constancy of sign of their Green's functions. A few important examples follow.

a) Set

$$\chi(t; \lambda) = \lambda^m + a_1(t)\lambda^{m-1} + \dots + a_m(t) \tag{8.8}$$

and

$$\chi_0(\lambda) = (\lambda - \lambda_1)\dots(\lambda - \lambda_m),$$

where

$$\lambda_1 > \lambda_2 > \dots > \lambda_k > 0 > \lambda_{k+1} > \dots > \lambda_m.$$

Let

$$H(t, s) = \begin{cases} -\sum_{j=k+1}^{m} \frac{\chi(t;\lambda_j)}{\chi_0'(\lambda_j)} e^{\lambda_j(t-s)} & \text{for } t \geqslant s, \\ \sum_{j=1}^{k} \frac{\chi(t;\lambda_j)}{\chi_0'(\lambda_j)} e^{\lambda_j(t-s)} & \text{for } t < s. \end{cases}$$

Theorem 8.7. Suppose that

$$H(t, s) \geqslant 0 \qquad (-\infty < t,\ s < \infty)$$

and

$$(-1)^k a_m(t) \gg 0.$$

Then the ap-operator (8.5) is regular and its Green's function $G(t, s)$ *satisfies the inequality*

$$a_m(t)\, G(t, s) \geqslant 0 \qquad (-\infty < t,\ s < \infty).$$

b) Suppose the roots $\lambda_1(t), \ldots, \lambda_m(t)$ of the "characteristic polynomial" (8.8) are simple and real and that they satisfy the inequalities

$$\alpha_j \leqslant \lambda_j(t) \leqslant \beta_j \qquad (j = 1, \ldots, m),$$

where

$$\beta_1 > \alpha_1 > \ldots > \beta_k > \alpha_k > 0 > \beta_{k+1} > \alpha_{k+1} > \ldots > \beta_m > \alpha_m.$$

Then the ap-operator (8.5) is regular and its Green's function satisfies the inequality

$$(-1)^k G(t, s) \geqslant 0 \qquad (-\infty < t,\ s < \infty).$$

c) Let $\lambda_1, \ldots, \lambda_m$ be arbitrary numbers. Set

$$\Delta^0(t; \lambda_1) = \chi(t; \lambda_1),$$

$$\Delta^1(t; \lambda_1, \lambda_2) = \frac{\Delta^0(t; \lambda_1) - \Delta^0(t; \lambda_2)}{\lambda_1 - \lambda_2},$$

. . . .

$$\Delta^{m-1}(t; \lambda_1, \ldots, \lambda_m) = \frac{\Delta^{m-2}(t; \lambda_1, \ldots, \lambda_{m-1}) - \Delta^{m-2}(t; \lambda_2, \ldots, \lambda_m)}{\lambda_1 - \lambda_m},$$

where $\chi(t, \lambda)$ is the "polynomial" (8.8).

Theorem 8.9. Suppose there exist negative numbers $\lambda_1 \ldots \lambda_{m-1}$ *such that*

$$\Delta^j(t; \lambda_1, \ldots, \lambda_{j+1}) \leqslant 0 \qquad (j = 0, 1, \ldots, m-2; \ -\infty < t < \infty).$$

Let there exist an ap-function $u_0(t) \in B^m(R^1)$ *such that*

$$u_0(t) \gg 0, \quad Lu_0(t) \gg 0, \quad L_{m-1}u_0(t) \gg 0,$$

where

$$L_{m-1} = \left(\frac{d}{dt} - \lambda_1\right) \ldots \left(\frac{d}{dt} - \lambda_{m-1}\right).$$

Then the ap-operator (8.5) is regular and its Green's function is nonnegative.

Under these conditions, the trivial solution of the equation $Lx = 0$ *is exponentially stable.*

§ 9. Second-order scalar ap-operators

9.1. This section is devoted to investigation of the Green's function of an ap-operator

$$Lx = \frac{d^2x}{dt^2} + p(t)\frac{dx}{dt} + q(t)x. \tag{9.1}$$

The solutions of the homogeneous equation

$$Lx = 0 \tag{9.2}$$

are said to be nonoscillatory if each of them vanishes at most once on $(-\infty, \infty)$.

Theorem 9.1. Let the ap-operator (9.1) be regular.

Then its Green's function is of constant sign if and only if the solutions of equation (9.2) are nonoscillatory on $(-\infty, \infty)$.

As a result, tests for nonoscillation of solutions of equation (9.2) are of great interest.

In several instances they may be determined quite simply by the method of test functions. For example, each of the following conditions is sufficient for the solutions of equation (9.2) to be nonoscillatory:

a) There exists a continuous function $z_0(t)$ on $(-\infty, \infty)$ satisfying the inequality

$$D^* z_0(t) \geqslant z_0^2(t) - p(t)\, z_0(t) + q(t) \qquad (-\infty < t < \infty), \quad (9.3)$$

where $D^* z_0(t)$ is the upper right derivative of $z_0(t)$.

b) There exists a function $x_1(t)$, positive on $(-\infty, \infty)$, such that

$$\int_t^{\infty} \frac{1}{x_1^2(s)} e^{-\int_0^s p(\tau)\, d\tau} ds < \infty$$

and for any $t_1 < t_2$

$$\int_t^{t_2} x_2(t)\, Lx \ (t)\, dt \leqslant 1.$$

where

$$x_2(t) = x_1(t) \int_t^{\infty} \frac{1}{x_1^2(s)} e^{\int_s^t p(\tau)\, d\tau} ds.$$

9.2. We now state the fundamental results.

Theorem 9.8. The ap-operator (9.1) is regular and its Green's function is nonnegative if and only if the following conditions hold:

1. The solutions of equation (9.2) are nonoscillatory on $(-\infty, \infty)$.

2. There exists an ap-function $x_0(t) \in B^2(R^1)$ *satisfying the inequalities*

$$x_0(t) \gg 0, \quad Lx_0(t) \geqslant 0 \qquad (Lx_0(t) \not\equiv 0). \qquad (9.4)$$

Theorem 9.9. The ap-operator (9.1) is regular and its Green's function is nonpositive if and only if there exists an ap-function $x_0(t) \in B^2(R^1)$ *satisfying the inequalities*

$$x_0(t) \gg 0, \quad Lx_0(t) \leqslant 0 \qquad (Lx_0(t) \not\equiv 0). \qquad (9.5)$$

Note that conditions (9.5) imply that the solutions of equation (9.2) are nonoscillatory. The inequalities (9.4) do not imply this.

9.3. In investigating the operators (9.1), a transition is often made to the Riccati equation (as in the theory of second-order linear differential equations). One therefore examines properties of solutions to the Riccati equation with ap-coefficients, differential inequalities with the Riccati operator, and the like.

Note, in conclusion, that in conditions 1 and 2 of Theorem 9.8, $M(p) \neq 0$. If $M(p) > 0$, these conditions imply that the trivial solution of the homogeneous equation $Lx = 0$ is exponentially stable; if $M(p) < 0$, the trivial solution is exponentially stable with the decrease of time.

The trivial solution of an equation $Lx = 0$ *with nonoscillatory solutions is exponentially stable if and only if there exists an ap-function* $\varphi(t) \in B^1(R^1)$ *with nonnegative mean* $M(\varphi)$ *such that*

$$M(\varphi) \leqslant \frac{1}{2} M(p)$$

and

$$\frac{d\varphi(t)}{dt} \leqslant \varphi^2(t) - p(t)\varphi(t) + q(t) \qquad (-\infty < t < \infty),$$

where the last inequality is strict for certain values of t.

CHAPTER 3. GLOBAL THEOREMS ON AP-SOLUTIONS OF NONLINEAR EQUATIONS

§10. General existence theorems

10.1. We investigate the existence of ap-solutions for nonlinear equations

$$Lx = f(t, x), \tag{10.1}$$

where L is an ap-operator

$$Lx = \frac{d^m x}{dt^m} + A_1(t)\frac{d^{m-1}x}{dt^{m-1}} + \ldots + A_m(t)x, \tag{10.2}$$

and the vector-function $f(t, x)$ is uniformly almost periodic with respect to t. If the ap-operator L is regular and $G(t, s)$ is its Green's function, then the problem of ap-solutions is equivalent to the nonlinear integral equation

$$x(t) = \Pi x(t), \tag{10.3}$$

where

$$\Pi x(t) = \int_{-\infty}^{\infty} G(t, s) f[s, x(s)]\, ds. \tag{10.4}$$

The operator Π acts continuously in the space $B(R^n)$.

It is natural to try using various fixed point principles in order to investigate equation (10.3). Among the tools that come to mind, in particular, are topological methods developed for equations with completely continuous operators (Schauder principle, rotation of vector fields, etc.). It turns out that this approach is inadequate, as the operators (10.4) are not in general completely compact in $B(R^n)$.

In the investigation of equation (10.3) we may thus apply only principles which do not involve complete continuity of the operator Π.

In simple cases, it suffices to use the contracting-mapping principle or its direct generalizations.

In this book, the basic tools for obtaining global theorems on the existence of ap-solutions is construction of conic intervals in $B(R^n)$ which are invariant under Π, successive examination for monotonicity and concavity of the operator Π on these conic intervals and, finally, application of general theorems of nonlinear analysis.

The most complicated problem is to find an appropriate cone $\hat{K} \subset B(R^n)$ with respect to which the conic intervals are constructed, and to determine the monotonicity and concavity of the operators. We confine ourselves to cones K with respect to which the Green's function is of constant sign. In the case of scalar differential equations, the only $\hat{K}$ we consider is the cone of nonnegative ap-functions.

By the above technique, we obtain several efficient existence theorems when $\alpha f(t, x)$ (where $\alpha = 1$ or $\alpha = -1$ depending on whether the Green's function is, respectively, nonnegative or nonpositive) is, as the case may be, monotone, concave or convex with respect to the space variables. In applications to special equations, we must overcome various specific difficulties.

In several cases it is convenient to pass from equation (10.1) to an integral equation other than (10.3). In order to do this, we consider the auxiliary ap-operator

$$L_1 x = Lx + A(t)x$$

with Green's function $G_1(t, s)$. The ap-matrix $A(t)$ is chosen so that, first, the Green's function $G_1(t, s)$ is of constant sign with respect to some cone and, second, the nonlinear function $\alpha[A(t)x + f(t, x)]$ (where α is the "sign" of $G_1(t, s)$) is "well-behaved." The problem of ap-solutions of equation (10.1) is equivalent to the integral equation

$$x(t) = \Pi_1 x(t), \tag{10.5}$$

where

$$\Pi_1 x(t) = \int_{-\infty}^{\infty} G_1(t, s)\{A(s)x(s) + f[s, x(s)]\}\,ds. \tag{10.6}$$

The integral equation (10.5) is used, as a rule, when $f(t, x)$ is a convex function.

10.2. Suppose a semi-order $\leqslant$ is introduced in a Banach space E via a solid normal cone K. A nonlinear operator Π acting in E is *positive* if $\Pi K \subset K$. Π is *monotone* if $x \leqslant y$ implies $\Pi x \leqslant \Pi y$. A *conic interval* $\langle u, x\rangle$ $(u \leqslant v)$ is the set of $x \in E$ such that $u \leqslant x \leqslant v$.

Theorem 10.2. Suppose the operator Π is monotone on the conic interval $\langle u, v\rangle$ and leaves it invariant. Let

$$\Pi y - \Pi x \leqslant S(y - x) \qquad (u \leqslant x \leqslant y \leqslant v), \tag{10.7}$$

where S is a positive linear operator with spectral radius less than 1.

Then the equation $x = \Pi x$ has a unique solution x^* on $\langle u, v\rangle$, and the successive approximations

$$x_{n+1} = \Pi x_n \qquad (n = 0, 1, 2, \ldots) \tag{10.8}$$

converge to this solution for any initial approximation $x_0 \in \langle u, v\rangle$.

An operator Π is said to be *concave* on the conic interval $\langle u, v\rangle$ $(u, v \in K_{\text{in}})$ if it is monotone on $\langle 0, v\rangle$ and

$$\Pi(\tau x) \geqslant \tau \Pi x \qquad (0 \leqslant \tau \leqslant 1,\ x \in \langle u, v\rangle).$$

An operator Π which is concave on $\langle u, v\rangle$ $(u, v \in K_{\text{in}})$ is said to be *uniformly concave* if for any $\tau \in (0, 1)$ there exists $\eta = \eta(\tau) > 0$ such that

$$\Pi(\tau x) \geqslant \tau(1 + \eta)\Pi x \qquad (x \in \langle u, v\rangle).$$

Theorem 10.3. Suppose Π leaves the conic interval $\langle u, v\rangle$ $(u, v \in K_{\text{in}})$ invariant and is uniformly concave there.

Then the equation $x = \Pi x$ *has a unique solution* x^* *on* $\langle u, v\rangle$ *and the successive approximations (10.8) converge to this solution for any initial approximation* $x_0 \in \langle u, v\rangle$.

A nonlinear operator Π is said to be strongly positive if $\Pi x \gg 0$ for $x \in K$ and $x \neq 0$. An operator Π is said to be uniformly concave on K if it is uniformly concave on every conic interval $\langle u, v\rangle$ $(u, v \in K_{\text{in}})$.

If Π *is strongly positive and uniformly concave on* K *and there exist nonzero* $u, v \in K$ *such that* $u \leqslant \Pi u$ *and* $v \geqslant \Pi v$, *then the equation* $x = \Pi x$ *has a unique nontrivial solution in* K.

In studying uniformly concave operators, it is convenient to view the conic interval $\langle u, v\rangle$ $(u, v \in K_{\text{in}})$ as a space with metric

$$\rho(x, y) = \min\{\alpha\colon\ x \leqslant e^{\alpha} y,\ y \leqslant e^{\alpha} x\}.$$

Theorem 10.3 is then a corollary of the following assertion.

Generalized contracting-mapping principle. Suppose the operator Π *maps the metric space* $\mathfrak{R}$ *onto itself and*

$$\rho(\Pi x, \Pi y) \leqslant q(\alpha, \beta)\,\rho(x, y) \qquad (\alpha \leqslant \rho(x, y) \leqslant \beta),$$

where $q(\alpha, \beta) < 1$ *for* $0 < \alpha \leqslant \beta < \infty$.

Then the equation $x = \Pi x$ *has a unique solution* x^* *in* $\mathfrak{R}$, *which is the common limit of all successive approximations (10.8) for any initial approximation* x_0 *in* $\mathfrak{R}$.

10.3. We present a typical application of Theorem 10.2, together with the technique of test functions. We let α denote the sign of the Green's function $G(t, s)$ of the operator (10.2).

Suppose the ap-functions $u(t)$, $v(t) \in B^m(R^n)$ satisfy the inequalities

$$\alpha L u(t) \leqslant \alpha f[t, u(t)], \quad \alpha L v(t) \geqslant \alpha f[t, v(t)]. \qquad (10.9)$$

Let

$$0 \leqslant \alpha f(t, y) - \alpha f(t, x) \leqslant A(t)(y - x) \qquad (u(t) \leqslant x \leqslant y \leqslant v(t)).$$

Then equation (10.1) has a unique ap-solution on $\langle u(t), v(t)\rangle$ if there exists an ap-function $x_0(t) \in B^m(R^n)$ such that

$$x_0(t) \gg 0, \qquad \alpha L x_0(t) \gg A(t) x_0(t). \qquad (10.10)$$

10.4. We cite one application of Theorem 10.3.

Let $K(t)$ be a solid cone-valued ap-function. We say that the function $g(t, x)$ is uniformly concave on $K(t)$ if, for every t, it is monotone in x with respect to the cone $K(t)$ and to each triple of numbers $\tau \in (0, 1)$; $r, r_1 > 0$ there corresponds an $\eta = \eta(\tau; r, r_1) > 0$ such that

$$g(t, \tau x) - (1 + \eta)\tau g(t, x) \in K(t) \qquad (-\infty < t < \infty)$$

for all x ($\|x\| \leqslant r$) contained in $K(t)$ together with a spherical neighborhood of radius r_1. A uniformly concave function $g(t, x)$ is said to be *regularly concave* on $K(t)$ if, for any t belonging to some set dense on $(-\infty, \infty)$, we have $g(t, x) \neq 0$ ($x \in K(t)$, $x \neq 0$).

Suppose that the Green's function of the operator (10.2) is strongly positive or strongly negative with respect to $K(t)$ and that $\alpha f(t, x)$ is regularly concave on $K(t)$. Let there exist nonzero ap-functions $u(t)$, $v(t) \in K(t)$ belonging to $B^m(R^n)$ and satisfying (10.9). Then equation (10.1) has a unique nontrivial ap-solution $x^*(t)$ such that $u(t) \leqslant x^*(t) \leqslant v(t)$ $(-\infty < t < \infty)$.

In § 10 we presented the method of minorants and majorants; this method simplifies the search for functions satisfying (10.9). Several examples are examined.

§ 11. Stability of ap-solutions of equations with monotone and concave nonlinearities

11.1. In this section we assume that $f(t, x)$ has a derivative $f'_x(t, x)$ which is a matrix-function, jointly continuous in all its variables, uniformly with respect to t when x belongs to any fixed ball.

An ap-operator L is said to be *stable* (*unstable*) if the trivial solution of the equation $Lx = 0$ is stable (unstable).

Consider the ap-operator

$$L[x^*]\,x = Lx - f'_x[t, x^*(t)]\,x, \tag{11.1}$$

where $x^*(t)$ is an ap-solution of the equation

$$Lx = f(t, x). \tag{11.2}$$

Equation (11.2), having ap-coefficients and almost periodic (in t) right-hand side, has a pleasant property, namely when the ap-operator $L[x^*]$ is regular, the theorems on stability in the first approximation are applicable:

If $L[x^*]$ is stable, then the solution $x^*(t)$ of equation (11.2) is exponentially stable; if the ap-operator $L[x^*]$ is unstable, then the solution $x^*(t)$ is also unstable.

To all intents and purposes, the ap-solutions $x^*(t)$ of equation (11.2), while known to exist by the assertions of § 10, are unknown. Thus the ap-operators (11.2) cannot actually be constructed, and we have to deduce their regularity, stability or instability from indirect information on their structure. We base this on the reasoning outlined below.

Given a family of ap-operators

$$L_\mu x = \frac{d^m x}{dt^m} + A_1(t;\mu)\frac{d^{m-1}x}{dt^{m-1}} + \ldots + A_m(t;\mu)\,x \quad (0 \leqslant \mu \leqslant 1), \quad (11.3)$$

with matrix-coefficients $A_i(t;\ \mu)$, uniformly continuous with respect to μ. The dichotomy theorems imply that *either all the ap-operators L_μ are stable, or all of them are unstable, provided they are regular.* Thus, *in order to prove the stability or instability of the ap-operator L_1, it suffices to prove that the operator L_0 has the same property and that every equation*

$$L_\mu x = f(t), \quad (11.4)$$

with arbitrary $f(t) \in B(R^n)$ has at least one solution bounded on $(-\infty, \infty)$. The problem of stability is thus reduced to proving existence theorems.

Let us return to the operator (11.1). From the above principle of extension of stability with respect to a parameter it follows that the operator $L[x^*]$ is regular and that the operators L and $L[x^*]$ are both stable or both unstable, provided the spectral radius of the linear integral operator

$$Sx(t) = \int_{-\infty}^{\infty} G(t,s)\, f'_x[s, x^*(s)]\, x(s)\, ds \quad (11.5)$$

is less than 1.

Suppose that the Green's function $G(t, s)$ of the operator L is of constant sign with respect to a solid cone-valued ap-function $K(t)$ and that the function $\alpha f(t, x)$ (α is the "sign" of $G(t, s)$) is monotone with respect to $K(t)$. Then the operator $L[x^*]$ is regular and it is stable or unstable according as the operator L is stable or unstable, provided there exists a function $x_0(t) \in B^m(R^n)$ such that

$$x_0(t) \gg 0, \quad \alpha L x_0(t) - \alpha f'_x[t, x^*(t)]\, x_0(t) \gg 0. \quad (11.6)$$

The examination of stability is thus reduced to verification of rather crude inequalities such as (11.6) (or to estimation of the spectral radius of the operator (11.5), hence substantially simplifying the problem and allowing a wide use of different majorants and minorants for the function $f(t, x)$.

11.2. The principle presented in subsection 11.1 permits us to determine the stability of the ap-solutions whose existence was established in § 10. For example:

Under the hypotheses of the assertions in subsections 10.3 and 10.4 of this summary, the ap-solution $x^*(t) \in \langle u(t), v(t)\rangle$ *is exponentially stable if the operator* L *is stable, unstable if* L *is unstable.*

11.3. In stability theory it is important to estimate domains in the phase space for which the following holds: Trajectories of the differential equation which start from points in these domains converge asymptotically to the solution $x^*(t)$ under investigation.

Consider the system

$$\frac{dx}{dt} + A(t)x = f(t, x). \tag{11.7}$$

A stable solution $x^*(t)$ of this system is said to be s t a b l e i n t h e c o n e $K(t)$ if all solutions with initial values in K_{in} converge asymptotically to $x^*(t)$ as $t \to \infty$.

In this section we introduce a special "plug method" for investigating the stability of solutions of equation (11.7) in a cone when the equation has concave nonlinearities. This method is based on examination of the properties of the translation operator $V(t, s)x$ of system (11.7). We check the translation operator for monotonicity, positivity and concavity (with respect to the space variables x). Then we examine the solutions $x_1(t)$ and $x_2(t)$ with initial values $x_1(0) = \alpha x^*(0)$ and $x_2(0) = \beta x^*(0)$, where $0 < \alpha < 1 < \beta$ and $x^*(t)$ is an ap-solution with initial value in K_{in}. A simple check shows that

$$\lim_{t\to\infty} |x_1(t) - x^*(t)| = \lim_{t\to\infty} |x_2(t) - x^*(t)| = 0.$$

Hence the stability of $x^*(t)$ in the cone follows trivially from the monotonicity of the translation operator.

Application of the plug method is simple, though tedious. In particular, one sees that under the conditions of the assertions in subsection 10.4, applied to (11.7), the ap-solution $x^*(t)$ is stable in the cone $K(0)$.

An essentially more difficult problem is the stability in a cone of a solution $x^*(t)$ of a higher order system

$$\frac{d^m x}{dt^m} + A_1(t)\frac{d^{m-1}x}{dt^{m-1}} + \dots + A_m(t)x = f(t, x). \tag{11.8}$$

To solve it, one usually passes to the first-order system

$$\frac{du}{dt} + Q(t)u = g(t, u). \tag{11.9}$$

It turns out that, under the conditions of the assertion in subsection 10.4, the solution $u^*(t)$ of (11.9) corresponding to the ap-solution $x^*(t)$ of (11.8) is stable in a special cone $\tilde{K}_{\max}(0)$.

The cone $\tilde{K}_{\max}(0)$ in the space R^{mn} is defined via a solid cone-valued ap-function $K(t) \subset R^n$ (with respect to which the corresponding Green's function is of constant sign, and the corresponding nonlinearity is monotone and concave) by the equality

$$\tilde{K}_{\max}(0) = \{u:\ u \in R^{mn},\ [U(\sigma, 0)u]_1 \in K(\sigma) \quad \text{for} \quad \sigma \geqslant 0\}.$$

Here $[v]_1$ denotes the vector in R^n consisting of the first n components of a vector $v = \{v_1, \ldots, v_m\} \in R^{mn}$, and $U(t, s)$ is the translation operator of the linear homogeneous equation

$$\frac{du}{dt} + Q(t)u = 0.$$

11.4. *The Green's function of a regular ap-operator* L *may be nonnegative with respect to the solid cone-valued ap-function* $K(t)$ *for both stable and unstable* L. *The Green's function may be nonpositive only if* L *is unstable.*

§ 12. Almost periodic oscillations in automatic control systems

12.1. We consider the system

$$\left.\begin{aligned} \frac{d\xi}{dt} &= \lambda_0\xi + d_1\varphi(\sigma), \\ \frac{dx}{dt} &= Bx + d\varphi(\sigma), \end{aligned}\right\} \tag{12.1}$$

where

$$\sigma = c_1\xi + (c, x). \tag{12.2}$$

Here ξ is a scalar variable, $x \in R^m$: λ_0, c_1, d_1 are scalars, $\lambda_0 < 0$, c_1 and d_1 are positive; B is a Hurwitz matrix of order m whose eigenvalues satisfy the inequality $\operatorname{Re}\lambda < \lambda_0$; d and c are vectors, such that

$$(d, c) < 0, \quad (d, c) + d_1 c_1 \geqslant 0, \quad (Bd, c) > \lambda_0 (d, c)$$

and

$$\operatorname{Re}\left([(\lambda_0 + i\omega) I - B]^{-1} d, c\right) < 0 \qquad (-\infty < \omega < \infty).$$

The scalar function $\varphi(\sigma)$ is known as the characteristic function of the control system; one usually assumes that it is continuous, and that

$$\sigma\varphi(\sigma) \geqslant 0 \qquad (-\infty < \sigma < \infty). \tag{12.3}$$

We shall also assume that $\varphi(\sigma)$ is monotone and sufficiently smooth.

Set

$$\alpha_0 = -\left(B^{-1}d, c\right) - \frac{d_1 c_1}{\lambda_0}.$$

It can be shown that under our assumptions $\alpha_0 > 0$.

The origin is an equilibrium state of system (12.1). If $\alpha_0\varphi'(0) < 1$ (or $\alpha_0\varphi'(0) = 1$, $\varphi''(0) = 0$ and $\varphi'''(0) < 0$, etc.) then this equilibrium state is asymptotically stable; but if $\alpha_0\varphi'(0) > 1$ or $\alpha_0\varphi'(0) = 1$ and $\varphi''(0) \neq 0$ (or $\alpha_0\varphi'(0) = 1$, $\varphi''(0) = 0$ and $\varphi'''(0) > 0$, etc.) then the origin is an unstable equilibrium state.

The other equilibrium states of system (12.1) are given by

$$\xi_0 = -\frac{d_1}{\lambda_0}\varphi(\sigma_0), \qquad x_0 = -\varphi(\sigma_0) B^{-1}d, \tag{12.4}$$

where σ_0 are the nonzero roots of the scalar equation

$$\sigma = \alpha_0\varphi(\sigma). \tag{12.5}$$

If, for example, $\varphi(\sigma)$ is convex ($\varphi'(\sigma) > 0$ and $\varphi''(\sigma) > 0$) and

$$\lim_{\sigma\to-\infty} \sigma^{-1}\varphi(\sigma) = 0, \qquad \lim_{\sigma\to\infty} \sigma^{-1}\varphi(\sigma) = \infty, \tag{12.6}$$

then equation (12.5) has one nonzero root σ_0. The corresponding equilibrium state (12.4) is stable if the origin is an unstable equilibrium state and vice versa.

12.2. We turn to an analysis of the perturbed system

$$\left.\begin{aligned} \frac{d\xi}{dt} &= \lambda_0\xi + d_1\varphi(\sigma) + f_1(t), \\ \frac{dx}{dt} &= Bx + d\varphi(\sigma) + f(t). \end{aligned}\right\} \tag{12.7}$$

Here $f_1(t)$ and $f(t)$ are ap-functions (the first being scalar, the second taking values in R^m). We assume that $\varphi(\sigma)$ is monotone and convex and that $\alpha_0\varphi'(0) < 1$. Then the unperturbed system (12.1) has two equilibrium states the origin being stable and the other unstable.

Let σ^* be a maximum point of the function

$$\psi(\sigma) = \sigma - \alpha_0\varphi(\sigma) \qquad (-\infty < \sigma < \infty),$$

and let $\beta^* = \psi(\sigma^*)$.

Set

$$N\{f_1, f\} = \sup_{-\infty < t < \infty} \int_{-\infty}^{t} \{c_1 e^{\lambda_0 (t-s)} f_1(s) + (e^{B(t-s)} f(s), c)\}\, ds.$$

Theorem. If the perturbations $f_1(t)$ and $f(t)$ satisfy the estimate

$$N\{f_1, f\} < \beta^*,$$

then the perturbed system (12.7) has exactly two ap-solutions; one is exponentially stable and the other unstable.

Under these conditions, the stable ap-solution is the only ap-solution $\{\xi^*(t), x^*(t)\}$ satisfying the inequality

$$\sup_{-\infty < t < \infty} \{c_1\xi^*(t) + (x^*(t), c)\} < \sigma^*.$$

12.3. An automatic control system (12.1) is said to be *completely controllable* if the vectors $\{d_1, d\}, \{\lambda_0 d_1, Bd\}, \ldots, \{\lambda_0^m d_1, B^m d\}$ form a basis in R^{m+1}. Every completely controllable system can be described by a scalar equation of the form

$$\frac{d^{m+1}x}{dt^{m+1}} + a_1 \frac{d^m x}{dt^m} + \ldots + a_{m+1}x = \varphi(\sigma), \qquad (12.8)$$

where

$$\sigma = b_{m+1}x + b_m x' + \ldots + b_1 x^{(m)}.$$

The characteristic polynomial

$$\chi(\lambda) = \lambda^{m+1} + a_1\lambda^m + \ldots + a_{m+1}$$

of the operator on the left of equation (12.8) is a Hurwitz polynomial with negative leading root λ_0 (all its other roots λ are such that $\operatorname{Re}\lambda < \lambda_0$). The polynomial

$$\varkappa(\lambda) = b_1\lambda^m + b_2\lambda^{m-1} + \ldots + b_{m+1}$$

is also important in the investigation of equation (12.8).

In subsection 12.1 we made several assumptions concerning the coefficients of system (12.1). In the case of equation (12.8), these assumptions become inequalities:

$$b_1 \geqslant 0, \qquad b_2 > (a_1 + \lambda_0)\, b_1, \tag{12.9}$$

$$\varkappa(\lambda_0) > 0, \qquad 2\varkappa'(\lambda_0)\,\chi'(\lambda_0) < \varkappa(\lambda_0)\,\chi''(\lambda_0) \tag{12.10}$$

and for all nonzero ω

$$\operatorname{Re}\frac{\varkappa(\lambda_0 + i\omega)}{\chi(\lambda_0 + i\omega)} < 0. \tag{12.11}$$

These conditions imply that a_{m+1}, $b_{m+1} > 0$.

Let $\varphi(\sigma)$ satisfy the conditions of subsection 12.2 ($\varphi(\sigma)$ is monotone and convex, and the limit equalities (12.6) hold) and let $b_{m+1}\varphi'(0) < a_{m+1}$. Let σ^* be a nonzero root of the equation

$$b_{m+1}\varphi'(\sigma) = a_{m+1}.$$

Theorem 12.4. Let $h(t)$ be an almost periodic perturbation satisfying the inequality

$$\sup_{-\infty < t < \infty} h(t) < \frac{a_{m+1}}{b_{m+1}}\sigma^* - \varphi(\sigma^*).$$

Then the perturbed equation

$$\frac{d^{m+1}x}{dt^{m+1}} + a_1\frac{d^m x}{dt^m} + \ldots + a_{m+1}x = \varphi(\sigma) + h(t)$$

has exactly two ap-solutions; one is exponentially stable and the other unstable.

12.4. The above method for examining ap-solutions of automatic control systems is quite general. It is applicable even in the case of nonlinear characteristic functions. For example, we quote an assertion concerning the ap-solutions of Duffing's equation

$$\frac{d^2x}{dt^2} + c\frac{dx}{dt} + x = \beta x^3 + h(t), \tag{12.12}$$

where $\beta > 0$, $h(t)$ is almost periodic:

If $c \geqslant 2$ *and*

$$|h(t)| < \frac{2}{3\sqrt{3\beta}} \qquad (-\infty < t < \infty),$$

then Duffing's equation (12.12) has exactly three ap-solutions, $x_1(t)$, $x_2(t)$ *and* $x_3(t)$, *satisfying the inequalities*

$$x_1(t) < -\frac{1}{\sqrt{3\beta}} < x_2(t) < \frac{1}{\sqrt{3\beta}} < x_3(t) \qquad (-\infty < t < \infty).$$

$x_1(t)$ *and* $x_3(t)$ *are unstable, while* $x_2(t)$ *is exponentially stable.*

Equation (12.12) has no ap-solutions other than $x_1(t)$, $x_2(t)$ *and* $x_3(t)$.

§13. Positive almost-periodic solutions of second-order equations

13.1. In this section we consider scalar differential equations

$$\frac{d^2x}{dt^2} + p(t)\frac{dx}{dt} + q(t)x = f(t, x) \tag{13.1}$$

and

$$\frac{d^2x}{dt^2} + p(t)\frac{dx}{dt} + q(t)x = f(t, \sigma), \tag{13.2}$$

where

$$\sigma = \alpha(t)x + \frac{dx}{dt}, \tag{13.3}$$

and $\alpha(t)$ is an ap-function in $B^1(R^1)$. In equations (13.1) and (13.2), the coefficients $p(t)$ and $q(t)$ are almost periodic; the function $f(t, u)$ is jointly continuous in $u \geqslant 0$, $-\infty < t < \infty$ and uniformly almost periodic in t; for positive u, there exists a positive continuous derivative $f_u(t, u)$ and it is almost periodic with respect to t, uniformly for u in every interval $[a, b] \subset (0, \infty)$. The mean value $M(p)$ of the function $p(t)$ is assumed to be nonnegative.

The equations under consideration are assumed to have a trivial solution ($f(t, 0) \equiv 0$). The existence and stability of nontrivial ap-solutions are examined by the techniques already described.

13.2. In order to formulate the fundamental results, it is convenient to single out two special classes of functions.

We say that $f(t, u) \in \mathfrak{F}_1$, if i) the function is concave in u uniformly in t, i. e., for every $\tau \in (0, 1)$ and every $u > 0$

$$f(t, \tau u) \geqslant (1+\eta)\tau f(t, u) \qquad (-\infty < t < \infty),$$

where $\eta = \eta(\tau, u) > 0$; ii) there exists $t_0 \in (-\infty, \infty)$ such that for all positive u,

$$uf'_u(t_0, u) - f(t_0, u) < 0,$$

and, iii):

$$\lim_{u \to +0} \inf_{-\infty < t < \infty} u^{-1} f(t, u) = \infty$$

and

$$\lim_{u \to \infty} \sup_{-\infty < t < \infty} u^{-1} f(t, u) = 0.$$

A good example of a function in $\mathfrak{F}_1$ is $u^{\nu}, 0 < \nu < 1$.

We say that $f(t, u) \in \mathfrak{F}_2$ if i) the function is convex in u uniformly in t, i. e., for any $\tau \in (0, 1)$ and every $u > 0$

$$f(t, \tau u) \leqslant (1-\eta)\tau f(t, u) \qquad (-\infty < t < \infty),$$

where $\eta = \eta(\tau, u) > 0$; ii) there exists $t_0 \in (-\infty, \infty)$ such that for positive u

$$uf'_u(t_0, u) - f(t_0, u) > 0;$$

iii) for every positive u

$$\inf_{-\infty < t < \infty} f(t, u) > 0$$

and, iv):

$$\lim_{u \to +0} \sup_{-\infty < t < \infty} u^{-1} f(t, u) = 0$$

and

$$\lim_{u \to \infty} \inf_{-\infty < t < \infty} u^{-1} f(t, u) = \infty.$$

Typical functions in the class $\mathfrak{F}_2$ are u^ν for $\nu > 1$, $e^u - u - 1$, $u \ln(1+u)$, and so on.

13.3. We first consider equations with concave nonlinearities $f(t, u) \in \mathfrak{F}_1$. We assume that the ap-operator

$$Lx = \frac{d^2x}{dt^2} + p(t)\frac{dx}{dt} + q(t)x$$

is regular and its Green's function nonnegative. Then:

Equation (13.1) has exactly one nontrivial nonnegative ap-solution. This ap-solution is strictly positive and exponentially stable in Lyapunov's sense.

Let us turn to equation (13.2). Suppose, in addition, that $\alpha(t)$ in (13.3) satisfies the differential inequality

$$\frac{d\alpha(t)}{dt} \leqslant \alpha^2(t) - p(t)\alpha(t) + q(t) \qquad (-\infty < t < \infty), \tag{13.4}$$

which is not an identity, and that the mean value $M(\alpha)$ satisfies the inequality

$$M(\alpha) \geqslant \frac{1}{2} M(p). \tag{13.5}$$

Under these assumptions, equation (13.2) has a unique ap-solution in the class of nonzero functions belonging to $B^1(R^1)$ and satisfying the inequalities

$$x(t) \geqslant 0, \ \alpha(t)x(t) + \frac{dx(t)}{dt} \geqslant 0 \qquad (-\infty < t < \infty). \tag{13.6}$$

This ap-solution is strictly positive and exponentially stable.

13.4. Now let us consider equations with convex nonlinearities $f(t, u) \in \mathfrak{F}_2$. We assume that the coefficients $p(t)$ and $q(t)$ satisfy the inequality

$$M^2(p) - M(p^2) + 4M(q) > 0.$$

Under these assumptions, equation (13.1) has a unique nontrivial nonnegative ap-solution. This ap-solution is strictly positive and unstable for both increasing and decreasing time.

Let us turn to equation (13.2). Suppose, in addition, that $\alpha(t)$ in (13.3) satisfies (13.4) and that $M(\alpha) > 0$. Then:

Equation (13.2) has a unique ap-solution in the class of nonzero functions belonging to $B^1(R^1)$ and satisfying (13.6). This nontrivial ap-solution is strictly positive and unstable for both increasing and decreasing time.

CHAPTER 4. EQUATIONS WITH A SMALL PARAMETER

§ 14. Linear equations

14.1. We first examine the linear equation

$$Lx = f(t;\ \varepsilon) \qquad (\varepsilon > 0) \tag{14.1}$$

with a regular ap-operator

$$Lx = \frac{dx}{dt} + A(t)\,x \tag{14.2}$$

and almost periodic right-hand side.

We investigate conditions for the validity of the equality

$$\lim_{\varepsilon \to 0} \left\| \int_{-\infty}^{\infty} G(t, s) f(s;\ \varepsilon)\, ds \right\|_{B(R^n)} = 0, \tag{14.3}$$

where $G(t, s)$ is the Green's function of the ap-operator (14.2). Clearly, (14.3) holds if the functions $f(t, \varepsilon)$ converge to zero as $\varepsilon \to 0$ in the norm of the space $B(R^n)$. This assumption, however, is not valid even in the simplest cases (for example, even when the averaging method of Bogolyubov and Krylov is applicable). Hence less restrictive conditions for the validity of (14.30) are of great importance.

We say that $f(t; \varepsilon)$ *converges integrally to zero* as $\varepsilon \to 0$ if for every positive T

$$\lim_{\varepsilon \to 0} \sup_{|t-s| \leqslant T} \left| \int_{s}^{t} f(\tau;\ \varepsilon)\, d\tau \right| = 0. \tag{14.4}$$

If, moreover,

$$\| f(t;\ \varepsilon \|_{B(R^n)} \leqslant m < \infty \qquad (0 < \varepsilon \leqslant \varepsilon_0),$$

we say that $f(t; \varepsilon)$ *converges regularly to zero* as $\varepsilon \to 0$.

A sufficient condition for (14.3) to hold is that $f(t;\ \varepsilon)$ converge regularly to zero as $\varepsilon \to 0$; a necessary condition for (14.3) to hold is that the function converge integrally to zero.

In particular, if $f(t;\,\varepsilon)=f_0(t/\varepsilon)$, where $f_0(t)$ has zero mean, then (14.3) holds.

14.2. Consider the family of ap-operators

$$L(\varepsilon)\,x=\frac{dx}{dt}+A(t;\;\varepsilon)\,x \qquad (0<\varepsilon<\varepsilon_0). \tag{14.5}$$

We say that the ap-matrices $A(t;\;\varepsilon)$ converge regularly to an ap-matrix $A(t)$ as $\varepsilon\to 0$ if, for every positive T,

$$\lim_{\varepsilon\to 0}\;\sup_{|t-s|\leqslant T}\left|\int_s^t [A(\tau;\;\varepsilon)-A(\tau)]\,d\tau\right|=0$$

and if

$$|A(t;\;\varepsilon)|\leqslant m<\infty \qquad (-\infty<t<\infty,\;0<\varepsilon<\varepsilon_0).$$

Theorem 14.1. Suppose the operator (14.2) is regular. Let the ap-matrices $A(t;\varepsilon)$ converge regularly to the ap-matrix as $\varepsilon\to 0$.

Then for sufficiently small ε the ap-operators (14.5) are regular and

$$\lim_{\varepsilon\to 0}\;\sup_{-\infty<s,\,t<\infty} e^{\gamma_0|t-s|}\,|G(t,s;\,\varepsilon)-G(t,s)|=0,$$

where $G(t,s;\varepsilon)$, $G(t,s)$ are the Green's functions of (14.5) and (14.2), respectively, and γ_0 is a positive number.

Under these conditions, we have that for small ε the operators (14.5) are stable (unstable) if (14.2) is stable (unstable).

Now consider the equation

$$\frac{dx}{dt}=\varepsilon A_1(t)\,x+\varepsilon B(t;\;\varepsilon)\,x \qquad (0<\varepsilon<\varepsilon_0). \tag{14.6}$$

Here $A_1(t)$ and $B(t;\;\varepsilon)$ are ap-matrices and

$$\lim_{\varepsilon\to 0}\;\sup_{-\infty<t<\infty}|B(t;\;\varepsilon)|=0. \tag{14.7}$$

Let A_0 denote the mean value of the matrix $A_1(t)$:

$$A_0=\lim_{T\to\infty}\frac{1}{2T}\int_{-T}^{T}A_1(t)\,dt.$$

Theorem 14.3. If A_0 is a Hurwitz matrix, then for sufficiently small $\varepsilon > 0$ the trivial solution of equation (14.6) is exponentially stable. But if A_0 has at least one eigenvalue in the right half-plane, then for sufficiently small ε the trivial solution of equation (14.6) is unstable.

Investigating the trivial solution of equations (14.6) for stability is much more complicated if its matrix has no eigenvalues in the right half-plane, but has eigenvalues on the imaginary axis.

14.3. Suppose equation (14.6) is of the form

$$\frac{dx}{dt} = \varepsilon A_1(t)\, x + \varepsilon^2 A_2(t)\, x + \dots + \varepsilon^k A_k(t)\, x + \varepsilon^{k+1} C(t;\ \varepsilon)\, x, \quad (14.8)$$

where $A_1(t), \dots, A_k(t)$ are matrices whose elements are trigonometric polynomials and $C(t; \varepsilon)$ are ap-matrices, uniformly continuous in t with respect to $\varepsilon \in [0, \varepsilon_0]$. We introduce the method of Bogolyubov and Shtokalo for the construction of a substitution

$$x = [I + \varepsilon Y_1(t) + \dots + \varepsilon^k Y_k(t)]\, y \quad (14.9)$$

(where $Y_1(t), \dots, Y_k(t)$ are ap-matrices), by which system (14.8) takes the form

$$\frac{dy}{dt} = \varepsilon B_1 y + \varepsilon^2 B_2 y + \dots + \varepsilon^k B_k y + \varepsilon^{k+1} D(t;\ \varepsilon)\, y, \quad (14.10)$$

an equation with constant matrices $B_1, \dots, B_k$ and an ap-matrix $D(t;\ \varepsilon)$ having the same properties as $C(t;\ \varepsilon)$.

Suppose that B_1 has zero as a simple eigenvalue and that all its other eigenvalues lie in the left-half-plane. Then for small ε the matrix

$$B(\varepsilon) = B_1 + \varepsilon B_2 + \dots + \varepsilon^{k-1} B_k \quad (14.11)$$

has a simple eigenvalue $\lambda(\varepsilon)$, which is an analytic function of ε and vanishes for $\varepsilon = 0$. For small ε, all other eigenvalues of $B(\varepsilon)$ lie in the left half-plane. The eigenvalue $\lambda(\varepsilon)$ may be expressed as a series

$$\lambda(\varepsilon) = a_1 \varepsilon + a_1 \varepsilon^2 + \dots + a_j \varepsilon^j + \dots, \quad (14.12)$$

whose coefficients are easily determinable by perturbation theory.

Let a_{j_0} be the first nonvanishing coefficient of (14.12) and let $j_0 \leqslant k - 1$.

Theorem 14.4. If $a_{j_0}<0$, the ap-operators

$$L(\varepsilon)x=\frac{dx}{dt}-[\varepsilon A_1(t)+\ldots+\varepsilon^k A_k(t)+\varepsilon^{k+1}C(t;\ \varepsilon)]x \quad (14.13)$$

are regular and stable for small positive ε. If $a_{j_0}>0$, the ap-operators (14.3) are regular and unstable for small positive ε.

§ 15. Bifurcation of almost-periodic solutions

15.1. Consider the system

$$\frac{dx}{dt}=f(t,\ x;\ \mu), \quad (15.1)$$

where the right-hand side is almost periodic in t and dependent on a scalar parameter μ. We assume that the right-hand side is sufficiently smooth with respect to all variables. Suppose that for $\mu=\mu_0$, the system has an ap-solution $x^*(t)$, and that the ap-operator

$$L_0x=\frac{dx}{dt}-f'_x[t,\ x^*(t);\ \mu_0]x \quad (15.2)$$

is regular. Then system (15.1) has a unique family of ap-solutions $x(t;\ \mu)$, continuous in μ, defined for μ sufficiently close to μ_0, and equal to $x^*(t)$ for $\mu=\mu_0$. To construct $x(t;\mu)$, we go from (15.1) to the nonlinear integral equation

$$x(t)=\int_{-\infty}^{\infty}G_0(t,\ s)\left\{f[s,\ x(s);\ \mu]-f'_x[s,\ x^*(s);\ \mu_0]x(s)\right\}ds,$$

where $G_0(t,\ s)$ is the Green's function of the ap-operator (15.2), noting that its solution is the uniform limit of the successive approximations

$$x_{n+1}(t;\ \mu)=\int_{-\infty}^{\infty}G_0(t,s)\left\{f[s,x_n(s;\ \mu);\ \mu]-f'_x[s,x^*(s);\ \mu_0]x_n(s;\ \mu)\right\}ds$$

with the initial approximation $x_0(t;\ \mu)=x^*(t)$.

The problem of extending ap-solutions becomes substantially more difficult when the ap-operator (15.2) is not regular. Here we are faced with the problem of bifurcation of ap-solutions, which has not been sufficiently studied.

We consider a special case of bifurcation, the problem of ap-solutions originating from equilibrium states.

Suppose that system (15.1) has a trivial solution for all μ. A number μ_0 is said to be a **bifurcation point** if for any $\eta > 0$ there exists $\mu \in (\mu_0 - \eta, \mu_0 + \eta)$ such that system (15.1) has a nontrivial solution $x(t; \mu)$ satisfying

$$\| x(t; \mu) \|_{B(R^n)} < \eta.$$

A number μ *can be a bifurcation point only if the ap-operator*

$$L(\mu) x = \frac{dx}{dt} - f'_x(t, 0; \mu) x \tag{15.3}$$

is not regular. Let μ_0 be such a number. Let $\mu = \mu_0 + \varepsilon$ and write system (15.1) as

$$\frac{dx}{dt} = g(t, x; \varepsilon). \tag{15.4}$$

The problem is to determine conditions for the existence of small nontrivial ap-solutions of system (15.1) for small ε.

15.2. Let equation (15.4) be of the form

$$\frac{dx}{dt} = \varepsilon A_1(t) x + \ldots + \varepsilon^k A_k(t) x + \varepsilon^{k+1} C(t; \varepsilon) x + \\ + \varepsilon F(t, x; \varepsilon) + \varepsilon \omega(t, x; \varepsilon). \tag{15.5}$$

Here the matrices $A_1(t), \ldots, A_k(t)$ and $C(t; \varepsilon)$ have the same properties as the corresponding coefficients of equation (14.8); $F(t, x; \varepsilon)$ is a form of degree $m \geqslant 2$ with respect to the space variables; its coefficients are trigonometric polynomials, continuous in ε uniformly with respect to t; the remainder term $\omega(t, x; \varepsilon)$ contains the terms which vanish to order greater than m with the space variables.

Suppose that the matrix (14.11) was found via the substitution (14.9). Let zero be a simple eigenvalue of the matrix B_1, all other eigenvalues lying in the left half-plane. Let e_0 and g_0 denote eigenvectors of the matrices B_1 and B_1^* corresponding to the zero eigenvalue, normalized so that $(e_0, g_0) = 1$.

Let a_{j_0} be the first nonzero coefficient in the expansion (14.12) for the largest eigenvalue $\lambda(\varepsilon)$ of the matrix (14.11), and let $j_0 \leqslant k - 1$. Set

$$\bar{F}(x; \varepsilon) = \lim_{T \to \infty} \frac{1}{T} \int_0^T F(t, x; \varepsilon)\, dt \qquad (x \in R^n)$$

and

$$\bar{\omega}(x;\ \varepsilon) = \lim_{T\to\infty} \frac{1}{T} \int_0^T \omega(t, x;\ \varepsilon)\, dt \qquad (x \in R^n).$$

If we substitute a "slower" time variable $\tau = \varepsilon t$ into equation (15.5), then the new equation is defined for $\varepsilon \neq 0$. For $\varepsilon = 0$, it is natural to assume that this equation becomes the "averaged" equation

$$\frac{dx}{d\tau} = B_1 x + \bar{F}(x;\ 0) + \bar{\omega}(x;\ 0). \tag{15.6}$$

Theorem 15.1. Let

$$\alpha_0 = (\bar{F}(e_0;\ 0),\ g_0) \neq 0.$$

Then there exist a ball $V(r_0) = \{x(t):\ x(t) \in B(R^n),\ \|x\| < r_0\}$ *and* $\varepsilon_0 > 0$ *such that the following assertions hold:*

1. Equation (15.6) has no nontrivial solutions in the ball $V(r_0)$.

2. If m *is even, then for* $0 < \varepsilon < \varepsilon_0$ *equation (15.5) has a unique nontrivial ap-solution in* $V(r_0)$*; this solution is exponentially stable if* $a_{j_0} > 0$ *and unstable if* $a_{j_0} < 0$.

3. If m *is odd and* $\alpha_0 a_{j_0} < 0$*, then for* $0 < \varepsilon < \varepsilon_0$ *equation (15.5) has exactly two nontrivial solutions in* $V(r_0)$*, which are exponentially stable if* $a_{j_0} > 0$*, and unstable if* $a_{j_0} < 0$.

4. If m *is odd and* $\alpha_0 a_{j_0} > 0$*, then for* $0 < \varepsilon < \varepsilon_0$ *equation (15.5) has no nontrivial solutions in* $V(r_0)$.

Though the formulation of this theorem is simple, its proof requires several special techniques, apart from the application of integral equations and the construction of various invariant cones.

15.3. We consider, as an example, an equation

$$\ddot{\theta} + 2\alpha\varepsilon\dot{\theta} + [\varepsilon^2 k^2 - \varepsilon p''(t)] \sin\theta = 0 \tag{15.7}$$

describing oscillations of a pendulum with a vibrating point of suspension, and investigate the problem of ap-solutions originating from the upper equilibrium position ($\theta = \pi$).

The upper equilibrium position is exponentially stable if $M[(p')^2] > k^2$. If $M[(p')^2] < k^2$, it is unstable.

Thus, for small ε, ap-solutions near the upper equilibrium position (but distinct from it) may appear only when $M[(p')^2] = k^2$. The general situation is as follows:

If the upper equilibrium position is unstable, then for small $\varepsilon > 0$*, there are no ap-oscillations in its neighborhood; if the upper equilibrium position is stable, then in each of its neighborhoods there are exactly two ap-solutions for small* ε*, both solutions being unstable.*

To examine the stability of the upper equilibrium position of the pendulum when

$$M\,[(p')^2] = k^2, \tag{15.8}$$

it is convenient to use the substitutions (14.9). For example, from (15.8) and $M\,[p(p')^2] < 0$ it follows that, for small positive ε, the upper equilibrium position is stable, and from (15.8) and $M\,[p(p')^2] > 0$ that it is unstable.

§ 16. Bifurcation of ap-solutions of singularly perturbed second-order equations

16.1. Consider the problem of small nontrivial ap-solutions of the scalar equation

$$\varepsilon \frac{d^2x}{dt^2} + \frac{dx}{dt} + q\,(t)\,x = a\,(t)\,x^n + \omega\,(t, x) \tag{16.1}$$

describing the oscillations of a small weight suspended on an elastic element with nonlinear restoring force. Here ε is a small positive parameter, $q(t) \in B^2(R^1)$, $a(t) \in B^1(R^1)$, $n \geqslant 2$, $\omega(t, x)$ is almost periodic in t and contains terms which vanish to order greater than n with x. We assume that $M(q) = 0$ *(this is a necessary condition for the existence of small nontrivial ap-solutions of equation (16.1) for small* ε*)*, and that the indefinite integrals of certain ap-functions involving $a(t)$ and $q(t)$, with zero mean, are ap-functions.

Two ap-functions $x_1(t)$ and $x_2(t)$ are said to be *comparable* if either $x_1(t) \geqslant x_2(t)$ or $x_1(t) \leqslant x_2(t)$ for all t.

Theorem 16.1. Let $q(t) \not\equiv 0$ *and*

$$M\,(b) = \lim_{T \to \infty} \frac{1}{T} \int_0^T a\,(t)\, e^{-(n-1)\int_0^t q\,(s)\,ds}\, dt \neq 0.$$

Then there exist a ball $V(r_0)$ *and* $\varepsilon_0 > 0$ *such that the following assertions hold.*

1. For $0 < \varepsilon < \varepsilon_0$*, the trivial solution of (16.1) is exponentially stable.*

2. If n *is even, then for* $0 < \varepsilon < \varepsilon_0$ *equation (16.1) has at least one nontrivial unstable ap-solution in* $V(r_0)$*. All ap-solutions* $x(t;\,\varepsilon)$ *in* $V(r_0)$ *are pairwise comparable, and* $M\,(b)\,x(t;\,\varepsilon) > 0$ $(-\infty < t < \infty)$.

3. If n is odd and $M(b) > 0$, then for $0 < \varepsilon < \varepsilon_0$ equation (16.1) has at least one positive and at least one negative unstable ap-solution in $V(r_0)$. All ap-solutions in $V(r_0)$ are of constant sign and pairwise comparable.

4. If n is odd and $M(b) < 0$, then for $0 < \varepsilon < \varepsilon_0$ equation (16.1) has no nontrivial ap-solutions in $V(r_0)$.

16.2. It is natural to suppose that, under the conditions of part 2 of Theorem 16.1, equation (16.1) has only one nontrivial ap-solution in the ball $V(r_0)$, and, under the conditions of part 3, exactly two such solutions. We have proved this only under the additional assumption that either $a(t) \gg 0$ or $a(t) \ll 0$.

16.3. Theorem 16.1 can be extended to the more general equations

$$\varepsilon\,[m_0(t) + m(t; \varepsilon)]\,\frac{d^2x}{dt^2} + \frac{dx}{dt} + [q_0(t) + \varepsilon q_1(t) + \varepsilon q(t; \varepsilon)]\,x = \\ = [a_0(t) + a(t; \varepsilon)]\,x^n + \omega(t, x; \varepsilon).$$

In this case, in all the assumptions above, the functions $q(t)$ and $a(t)$ are to be replaced by $q_0(t)$ and $a_0(t)$; $m_0(t)$ is now assumed to be strictly positive; and the condition $q(t) \not\equiv 0$ is replaced by the nonvanishing of the mean value of the function

$$\lambda_0(t) = q_1(t) - m_0(t)\,\frac{dq_0(t)}{dt} + m_0(t)\,q_0^2(t).$$

As an example, we consider the small ap-oscillations described in the horizontal plane by a pendulum of slowly varying length $l(\varepsilon t)$ when the point of suspension vibrates according to the law $\varepsilon^{-1}p(\varepsilon t)$.

BIBLIOGRAPHY*

Aizerman, M. A. and F. R. Gantmakher

1. Absolute Stability of Control Systems. — Moscow, Izdatel'stvo Akad. Nauk SSSR, 1963. (Russian)

Amerio, L.

1. Soluzioni quasiperiodiche, o limitati, di sistemi differenziali non lineari quasi-periodici, o limitati. — Ann. Mat. Pura Appl. 39 (1955), 97–119.

Arnol'd, V. I.

1. Small denominators and problems of stability of motion in classical and celestial mechanics. — Uspekhi Mat. Nauk 18, No. 6 (114) (1963). (Russian)

Bakhtin, I. A.

1. On a class of equations with positive operators. — Dokl. Akad. Nauk SSSR **117**, No. 1 (1957), 13–16. (Russian)

Bakhtin, I. A., M. A. Krasnosel'skii and V. Ya. Stetsenko

1. On the continuity of linear positive operators. — Sibirsk. Mat. Zh. 3, No. 1 (1962), 156–160. (Russian)

Barbashin, E. A.

1. Introduction to Stability Theory. — Moscow, "Nauka," 1967. (Russian)

Berezhnoi, V. N. and Yu. S. Kolesov

1. On a criterion for stability of solutions of n-th order linear differential equations. — Prikl. Mat. Mekh. 33, No. 3 (1969). (Russian)

Bochner, S.

1. Beiträge zur Theorie der fastperiodischen Funktionen, I. Funktionen einer Variabeln. — Math. Ann. 96 (1927), 119–147.

Bogolyubov, N. N.

1. Sur l'approximation des fonctions par les sommes trigonométriques. — Dokl. Akad. Nauk SSSR (A) (1930), 147–152.
2. On Some Statistical Methods in Mathematical Physics. — Kiev, Izdatel'stvo Akad. Nauk UkrSSR, 1945. (Russian)
3. The theory of perturbations in nonlinear mechanics. — Sbornik Trudov Instituta Stroitel'noi Mekhaniki Akad. Nauk UkrSSR, No. 14 (1950), 9–34. (Russian)

Bogolyubov, N. N. and Yu. A. Mitropol'skii

1. Asymptotic Methods in Nonlinear Oscillation Theory. — Moscow, Fizmatgiz, 1963. (Russian)

Bohl, P.

1. Über eine Differentialgleichung der Störungstheorie. — J. Reine Angew. Math. **131** (1906), 268–321.

Bohr, H.

1. Zur Theorie der fastperiodischen Funktionen, I. — Acta Math. 45 (1925), 29–127.
2. Zur Theorie der fastperiodischen Funktionen, II. — Acta Math. 46 (1925). 101–214.

* [Names of periodicals are abbreviated as in Mathematical Reviews.]

3. Fastperiodische Funktionen. — Berlin, Springer, 1932. [English translation: Almost Periodic Functions. — New York, Chelsea, 1947.]

Brayton, R.K. and J. Moser

1. A theory of nonlinear networks. I. — Quart. Appl. Math. 22 (1964).

Burd, V.Sh.

1. On the problem of branching of almost periodic solutions of nonlinear ordinary differential equations. — Dokl. Akad. Nauk SSSR **159**, No. 2 (1964), 239—242. (Russian)

Burd, V.Sh., Yu.S. Kolesov and M.A. Krasnosel'skii

1. Nonlocal problems of the theory of almost periodic solutions of differential equations. — In: Third All-Union Conference on Theoretical and Applied Mechanics (Abstracts of Papers), p. 53. Moscow, 1968. (Russian)
2. Investigation of the Green's function of differential operators with almost periodic coefficients. — Izv. Akad. Nauk SSSR Ser. Mat. **33**, No. 5 (1969). (Russian)
3. On the existence and constancy of sign of Green's functions of higher-order scalar equations with almost periodic coefficients. — Izv. Akad. Nauk SSSR **33**, No. 6 (1969). (Russian)
4. Bifurcation of almost periodic solutions of singularly perturbed differential equations. — Uspekhi Mat. Nauk **25**, No. 1 (151) (1970). (Russian)

Burd, V.Sh. and T. Sabirov

1. On the stability of almost periodic branching solutions of certain systems of differential equations. — Dokl. Akad. Nauk SSSR **176**, No. 5 (1967), 991—993. (Russian)

Burd, V.Sh., P.P. Zabreiko, Yu.S. Kolesov and M.A. Krasnosel'skii

1. The averaging principle and bifurcation of almost periodic solutions. — Dokl. Akad. Nauk SSSR **187**, No. 6 (1969), 1219—1221. (Russian)
2. Small combination oscillations and the averaging principle. — In: Fifth International Conference on Nonlinear Oscillations, p. 40. Kiev, 1969. (Russian)

Coddington, E.A. and N. Levinson

1. Theory of Ordinary Differential Equations. — New York, McGraw-Hill, 1955.

Collatz, L.

1. Funktionanalysis und numerische Mathematik. — Berlin, Springer, 1964.

Coppel, W.A.

1. Dichotomies and reducibility. — J. Differential Equations **3**, No. 3 (1967), 500—521.

Corduneanu, C.

1. Functii aproape-periodice. — Bucharest, Editur Acad. RPR, 1961.

Demidovich, B.P.

1. On bounded solutions of a certain nonlinear system of ordinary differential equations. — Mat. Sb. **40** (82), No. 1 (1956), 73—94. (Russian)
2. Lectures on the Mathematical Theory of Stability. — Moscow, "Nauka," 1967. (Russian)

Dunford, N. and J. Schwartz

1. Linear Operators, Part I — General Theory. — New York, Interscience Publishers, 1958.

Erugin, N.P.

1. Reducible systems. — Trudy Mat. Inst. Akad. Nauk SSSR **13** (1946). (Russian)
2. Linear Systems of Ordinary Differential Equations. — Minsk, Izdatel'stvo Akad. Nauk BSSR, 1963. (Russian)

Esclangon, E.

1. Sur les intégrales bornées d'une équation différentielle linéaire. — C.R. Acad. Sci. Paris 160 (1915), 475—478.

Favard, J.

1. Sur les équations différentielles à coëfficients presque-périodiques. — Acta Math. **51** (1927), 31—81.

Flatto, L. and N. Levinson
1. Periodic solutions of singularly perturbed equations. — J. Math. Mech. 4 (1955), 943—950.
Friedrichs, K.O., P. le Corbeiller, N. Levinson and J.J. Stoker
1. Non-linear Mechanics. — Brown University Lecture Notes, 1942—1943.
Gantmakher, F.R.
1. Theory of Matrices. — Moscow, "Nauka," 1967. (Russian)
Gantmakher, F.R. and V.A. Yakubovich
1. Absolute stability of nonlinear control systems. — In: Proceedings of Second All-Union Conference on Theoretical and Applied Mechanics, Vol. 1, 30—63, 1963. (Russian)
Gel'fand, I.M.
1. Lectures on Linear Algebra. — Moscow, "Nauka," 1966. (Russian)
Halanay, A.
1. Soluţii aproape-periodice ale ecuatiei Riccati. — Studii şi cercetări Mat. Acad. RPR IV, Nos. 3—4 (1953), 345—354.
2. Almost periodic solutions of nonlinear systems with a small parameter. — Zh. Chistoi i Prikl. Mat. Acad. RPR 1, No. 2 (1956), 49—60. (Russian)
3. Teoria calitativă a ecuaţiilor differenţiale, Bucharest, Acad. RPR, 1963.
Hale, J.K.
1. Oscillations in Nonlinear Systems. — New York, McGraw-Hill, 1963.
Hale, J.K. and G. Seifert
1. Bounded and almost periodic solutions of singularly perturbed equations. — J. Math. Anal. Appl. 3 (1961), 18—24.
Hardy, G., J. Littlewood and G. Pólya
1. Inequalities. — Cambridge, Cambridge University Press, 1952.
Hausdorff, F.
1. Mengenlehre (3. Aufl.). — New York, Dover, 1944.
Kalman, R.E.
1. Liapunov functions for the problem of Lurie in automatic controls. — Proc. Nat. Acad. Sci. U.S.A. 49 (1963), 201—205.
Kapitsa, P.L.
1. Dynamic stability of a pendulum with vibrating point of suspension. — Zh. Eksper. Teoret. Fiz. 21, No. 5 (1951), 588—597. (Russian)
Kolesov, Yu.S.
1. Positive Periodic Solutions of Systems of Ordinary Differential Equations. — Dissertation, Voronezh, 1966. (Russian)
2. On conditions for the existence of an invariant cone for a certain class of n-th order differential equations. — Trudy Sem. Funktsional. Analiz. (Voronezh), No. 9 (1967) (Russian)
Kolesov, Yu.S. and A.Yu. Levin
1. On the sign of the Green's function of certain periodic boundary-value problems. Problemy matematicheskogo analiza slozhnykh sistem (Voronezh), No. 1 (1967), 40—43. (Russian)
Kolmogorov, A.N.
1. On the conservation of quasi-periodic motions under a small variation of the Hamiltonian. — Dokl. Akad. Nauk SSSR 98, No. 4 (1954). (Russian)
Krasnosel'skii, M.A.
1. Positive Solutions of Operator Equations. — Moscow, Fizmatgiz, 1962. (Russian)
2. The Translation Operator along Trajectories of Differential Equations. — Moscow, "Nauka," 1966. (Russian)

Krasnosel'skii, M.A. and L.A. Ladyzhenskii
1. Structure of the spectrum of positive inhomogeneous operators. — Trudy Moscow. Mat. Obshch. 3 (1954), 321–346. (Russian)
Krasnosel'skii, M.A. and V.Ya. Stetsenko
1. On the theory of equations with concave operators. — Sibirsk. Mat. Zh. 10, No. 3. (1969), 565–572. (Russian)
Krasnosel'skii, M.A., G.I. Vainikko, P.P. Zabreiko, Ya.B. Rutitskii and V.Ya. Stetsenko
1. Approximate Solution of Operator Equations. — Moscow, "Nauka," 1969. (Russian)
Krasovskii, N.N.
1. Some Problems in the Theory of Stability of Motion. — Moscow, Fizmatgiz, 1959. (Russian)
Krein, M.G.
1. Lectures on the Theory of Stability of Solutions of Differential Equations in a Banach Space. — Kiev, Izdatel'stvo Akad. Nauk UkrSSR, Institut Matematiki, 1964. (Russian)
Krein, M.G. and M.A. Rutman
1. Linear operators leaving invariant a cone in a Banach space. — Uspekhi Mat. Nauk **3**, No. 1 (23) (1948), 3–95. (Russian)
Landau, E.
1. Über einen Satz von Herrn Esclangon. — Math. Ann. **102** (1930).
Landau, L.D. and E.M. Lifshits
1. Mechanics. — Moscow, Fizmatgiz, 1958. (Russian)
La Vallée-Poussin, C. de
1. Sur les fonctions presques-périodiques de H. Bohr. — Ann. Soc. Sci. Bruxelles Sér. I **47** (1927), 141–158.
Lefschetz, S.
1. Stability of Nonlinear Control Systems. — New York, Academic Press, 1965.
Levin, A.Yu.
1. An integral nonoscillation criterion for the equation $\ddot{x} + q(t)x = 0$. — Uspekhi Mat. Nauk **20**, No. 2 (122) (1965), 244–266. (Russian)
2. Nonoscillation of solutions of the equation $x^{(n)} + p_1(t)x^{(n-1)} + \ldots + p_n(t)x = 0$. — Uspekhi Mat. Nauk **24**, No. 2 (146) (1969), 43–96. (Russian)
Levitan, B.M.
1. Almost Periodic Functions. — Moscow, Gostekhizdat, 1953. (Russian)
Lur'e, A.I.
1. Some Nonlinear Problems in the Theory of Automatic Control. — Moscow, Gostekhizdat, 1951. (Russian)
Lyapunov, A.M.
1. The General Problem of Stability of Motion. — Moscow, Gostekhizdat, 1950. (Russian)
Malkin, I.G.
1. Theory of Stability of Motion. — Moscow, Fizmatgiz, 1966. (Russian)
2. Some Problems of Nonlinear Oscillation Theory. — Moscow, Gostekhizdat, 1956. (Russian)
Markus, L. and R.A. Moore
1. Oscillation and disconjugacy for linear differential equations with almost periodic coefficients. — Acta Math. **96** (1956), 99–123.
Massera, J.L.
1. Un criterio di existencia de soluciones casi-periodicas de ciertos sistemas de ecuaciones differenciales casi-periodicas. — Bol. Fac. Ingen. Agrimens. Montevideo 6, No. 11 (1958), 345–349.

Massera, J.L. and J.J. Schäffer
1. Linear differential equations and functional analysis, I. — Ann. Math. **67**, No. 3 (1958), 517—573.
2. Linear Differential Equations and Function Spaces. — New York, Academic Press, 1966.

Meyers, P.R.
1. A converse to Banach's contraction theorem. — J. Res. Nat. Bur. Standards Sect. B **71**, Nos. 2—3 (1967), 73.

Mitropol'skii, Yu. A.
1. Nonstationary Processes in Nonlinear Oscillatory Systems. — Kiev, Izdatel'stvo Akad. Nauk UkrSSR, 1955. (Russian)

Moser, J.
1. A rapidly convergent iteration method and nonlinear differential equations. — Uspekhi Mat. Nauk **23**, No. 4 (142) (1968). 179—238. [Russian translation of: Ann. Scuola Norm. Sup. Pisa (3) 20 (1966), 265—315, 499—535.]
2. On the expansion of quasi-periodic motions in convergent power series. — Uspekhi Mat. Nauk **24**, No. 2 (146) (1969), 165—211. [Augmented Russian translation of: Convergent series expansions for quasi-periodic motions. — Math. Ann. **169** (1967), 136—176.]
3. On nonoscillating networks. — Quart. Appl. Math. **25** (1967), 1—9.
4. Combination tones for Duffing's equation. — Comm. Pure Appl. Math. **18**, Nos. 1—2 (1965), 167—181.

Naumov, B.N. and Ya.Z. Tsypkin
1. A frequency criterion for absolute stability of processes in nonlinear automatic control systems. — Avtomat. i Telemekh. 25, No. 6 (1964). (Russian)

Perov, A.I.
1. Periodic, almost periodic and bounded solutions of differential equations. — Dokl. Akad. Nauk SSSR **132**, No. 3 (1960). (Russian)

Perron, O.
1. Über Stabilität und asymptotisches Verhalten der Integrale von Differentialgleichungssysteme. — Math. Z. **29**, No. 1 (1928).

Petrovskii, I.G.
1. Lectures on the Theory of Ordinary Differential Equations. — Moscow, "Nauka," 1964. (Russian)

Pliss, V.A.
1. Some Problems in the Theory of Stability of Motion in the Large. — Izdatel'stvo Leningradskogo Gosudarstvennogo Universiteta, 1958. (Russian)

Pokrovskii, A.V.
1. Almost-periodic functions with values in a variable cone in a Banach Space. — Trudy Sem. Funktsional. Anal. (Voronezh), No. 11 (1968) (Problemy Matematicheskogo Analiza Slozhnykh Sistem, No. 3), 199—204. (Russian)

Pyatnitskii, E.S.
1. New research on the absolute stability of automatic control systems. — Avtomat. i Telemekh. **29**, No. 6 (1968), 5—36. (Russian)

Shtokalo, I.Z.
1. Linear Differential Equations with Variable Coefficients. — Kiev, Izdatel'stvo Akad. Nauk UkrSSR, 1960. (Russian)

Sobol', I.M.
1. Extremal solution of the Riccati equation and its application to investigation of the solutions of second-order linear differential equations. — Uchen. Zap. Moskov. Gos. Univ. Mat. **5**, No. 155 (1952). (Russian)

Stoker, J.J.
1. Nonlinear Vibrations in Mechanical and Electrical Systems. — New York, Interscience Publishers, 1950.

Weyl, H.

1. Integralgleichungen und fastperiodische Funktionen. — Math. Ann. **97** (1927), 338—356.

Wiener, N.

1. Generalised harmonic analysis. — Acta Math. **55** (1930), 117—258.

Yakubovich, V. A.

1. Solution of certain matrix inequalities arising in automatic control theory. — Dokl. Akad. Nauk SSSR **143**, No. 6 (1962), 1304—1307. (Russian)
2. The method of matrix inequalities in stability theory of nonlinear control systems. Absolute stability of forced oscillations. — Avtomat. i Telemekh. 25, No. 7 (1964), 1017—1029. (Russian)
3. Frequency conditions for absolute stability of steady-state regimes and forced oscillations. — In: Proceedings of International Conference on Multidimensional and Discrete Systems of Automatic Control, Prague, 1965. (Russian)
4. Periodic and almost-periodic limiting regimes of control systems with several (generally discontinuous) nonlinearities. — Dokl. Akad. Nauk SSSR **171**, No. 3 (1967), 533—536. (Russian)

Zubov, V.I.

1. Oscillations in Nonlinear and Controllable Systems. — Sudpromgiz. 1962.

SUBJECT INDEX